AGRICULTURAL WASTE

*Threats and Technologies for
Sustainable Management*

AGRICULTURAL WASTE

*Threats and Technologies for
Sustainable Management*

Edited by
Rouf Ahmad Bhat, PhD
Khalid Rehman Hakeem, PhD
Humaira Qadri, PhD
Moonisa Aslam Dervash, PhD

APPLE
ACADEMIC
PRESS

First edition published 2022

Apple Academic Press Inc.
1265 Goldenrod Circle, NE,
Palm Bay, FL 32905 USA
4164 Lakeshore Road, Burlington,
ON, L7L 1A4 Canada

CRC Press
6000 Broken Sound Parkway NW,
Suite 300, Boca Raton, FL 33487-2742 USA
2 Park Square, Milton Park,
Abingdon, Oxon, OX14 4RN UK

© 2022 Apple Academic Press, Inc.

Apple Academic Press exclusively co-publishes with CRC Press, an imprint of Taylor & Francis Group, LLC

Library and Archives Canada Cataloguing in Publication

Title: Agricultural waste : threats and technologies for sustainable management / edited by Rouf Ahmad Bhat, PhD, Khalid Rehman Hakeem, PhD, Humaira Qadri, PhD, Moonisa Aslam Dervash, PhD.

Other titles: Agricultural waste (Boca Raton, Fla.)

Names: Bhat, Rouf Ahmad, 1981- editor. | Hakeem, Khalid Rehman, editor. | Qadri, Humaira, editor. | Dervash, Moonisa Aslam, editor.

Series: Innovations in physical chemistry.

Description: First edition. | Includes bibliographical references and index.

Identifiers: Canadiana (print) 20210102055 | Canadiana (ebook) 20210102233 | ISBN 9781771889636 (hardcover) | ISBN 9781003105046 (ebook)

Subjects: LCSH: Agricultural wastes. | LCSH: Sustainable agriculture. | LCSH: Agricultural innovations.

Classification: LCC TD930 .A37 2021 | DDC 628/.74—dc23

Library of Congress Cataloging-in-Publication Data

CIP data on file with US Library of Congress

ISBN: 978-1-77188-963-6 (hbk)
ISBN: 978-1-77463-785-2 (pbk)
ISBN: 978-1-00310-504-6 (ebk)

Dedication

Dedicated to our beloved parents…

About the Editors

Rouf Ahmad Bhat, PhD
Assistant Professor, Cluster University Srinagar, Jammu and Kashmir, India

Rouf Ahmad Bhat, PhD, is an Assistant Professor at Cluster University Srinagar, Jammu and Kashmir, India, where he specializes in limnology, toxicology, phytochemistry, and phytoremediation. Dr. Bhat has been teaching graduate and postgraduate students of environmental sciences for the past three years. He is an author of more than 50 research papers and 15 book chapters and has published more than 10 books with international publishers. He has presented and participated in numerous state, national and international conferences, seminars, workshops, and symposia. Dr. Bhat has worked as an Associate Environmental Expert for the World Bank-funded Flood Recovery Project and also as Environmental Support Staff for Asian Development Bank (ADB)-funded development projects. He has received awards, appreciation, and recognition for his services to the science of water testing and air and noise analysis. He has served as an editorial board member and reviewer for several international journals. Dr. Bhat is still writing and experimenting with diverse capacities of plants for use in aquatic pollution.

Khalid Rehman Hakeem, PhD
Professor, King Abdulaziz University,
Jeddah, Saudi Arabia

Khalid Rehman Hakeem, PhD, is a Professor at King Abdulaziz University, Jeddah, Saudi Arabia. After completing his doctorate (botany, with a specialization in plant eco-physiology and molecular biology) from Jamia Hamdard, New Delhi, India, he worked as a lecturer at the University of Kashmir, Srinagar, India. At Universiti Putra Malaysia, Selangor, Malaysia, he was a Postdoctorate Fellow and Fellow Researcher (Associate Professor) for several years. Dr. Hakeem has more than 10 years of teaching and research experience in plant eco-physiology, biotechnology and molecular biology, medicinal plant research, plant-microbe-soil interactions, as well as in environmental studies. Currently, he is engaged in studying the plant processes at eco-physiological as well as molecular levels. Dr. Hakeem is the recipient of several fellowships at both national and international levels. He has served as a visiting scientist at Jinan University, Guangzhou, China. Currently, he is involved with a number of international research projects with different government organizations. To date, Dr. Hakeem has authored and edited more than 55 books with international publishers. He also has to his credit more than 120 research publications in peer-reviewed international journals and 60 book chapters in edited volumes with international publishers. At present, Dr. Hakeem serves as an editorial board member and reviewer for several high-impact international scientific journals. He is included in the advisory board of Cambridge Scholars Publishing, UK. He is also a fellow of the Plantae group of the American Society of Plant Biologists, member of the World Academy of Sciences, member of the International Society for Development and Sustainability, Japan, and member of the Asian Federation of Biotechnology, Korea.

Humaira Qadri, PhD
Head, Department of Environment and Water Management, Cluster University Srinagar, Sri Pratap College Campus, Jammu and Kashmir, India

Humaira Qadri, PhD, has been actively involved in teaching postgraduate students of environmental science for the past ten years at the Sri Pratap College Campus of Cluster University Srinagar, Jammu and Kashmir, India, where she also heads the Department of Environment and Water Management. She has published scores of papers in international journals and has more than ten books with national and international publishers. She is also the reviewer for various international journals and is the principal investigator of some major projects on phytoremediation. She is guiding a number of research students for PhD programs and has supervised more than 60 master's dissertations. She also has been on the scientific board of various international conferences and holds life memberships of various international organizations. With a number of national scientific events to her credit, she is an active participant in national and international scientific events and has organized a number of national conferences on science. A gold medalist at her master's level, she earned a number of awards and certificates of merit. Her specialization is in limnology, nutrient dynamics, and phytoremediation.

Moonisa Aslam Dervash, PhD
Cluster University Srinagar,
Sri Pratap College Campus,
Jammu and Kashmir, India

Moonisa Aslam Dervash, PhD, has been actively involved in teaching graduate and postgraduate students of environmental science at Sri Pratap College Campus, Cluster University Srinagar, Jammu and Kashmir, India. She has published many papers in international journals and has more than three books with national and international publishers. She is also a reviewer for various international journals. During her education, she has received a number of awards and certificates of merit. Her specialization is in measofuana and carbon sequestration.

Contents

Contributors

Rukhsana Akhtar
Department of Biochemistry, University of Kashmir, Hazratbal Srinagar – 190006, Jammu, and Kashmir, India

Rohaya Ali
Department of Biochemistry, University of Kashmir, Srinagar – 190006, Jammu and Kashmir, India, E-mail: rohayaali01@gmail.com

Shafat Ali
Centre of Research for Development, University of Kashmir, Srinagar, Jammu and Kashmir, India

Syed Rouhullah Ali
College of Agricultural Engineering and Technology, SKUAST K – 190025, Srinagar, Jammu and Kashmir, India

Ifra Ashraf
College of Agricultural Engineering and Technology, Sher-e-Kashmir University of Agricultural Sciences and Technology of Kashmir Shalimar Campus, Srinagar, Jammu and Kashmir, India

Rezwana Assad
Department of Botany, University of Kashmir, Srinagar – 190006, Jammu and Kashmir, India, Phone: +91-7889481075, E-mail: rezumir@gmail.com

Muhammad Ashar Ayub
Institute of Soil and Environmental Sciences, University of Agriculture, Faisalabad – 38040, Pakistan

Iqra Bashir
Department of Botany, University of Kashmir, Srinagar – 190006, Jammu and Kashmir, India

Rouf Ahmad Bhat
Division of Environmental Sciences, Sher-e-Kashmir University of Agricultural Sciences and Technology, Shalimar, Srinagar – 190025, Jammu and Kashmir, India, Phone: +91-7006655833, E-mail: rufi.bhat@gmail.com

Shakeel Ahmad Bhat
College of Agricultural Engineering and Technology, SKUAST K – 190025, Srinagar, Jammu and Kashmir, India

Showkat Ahmad Bhat
Department of Biochemistry, Government Medical College (GMC), Karan Nagar, Srinagar – 190010, Jammu, and Kashmir, India

Monica Butnariu
Banat's University of Agricultural Sciences and Veterinary Medicine, "King Michael I of Romania" from Timisoara – 300645, CaleaAradului 119, Timis, Romania, E-mail: monicabutnariu@yahoo.com

Alina Butu
National Institute of Research and Development for Biological Sciences, SplaiulIndependentei, 296, Bucharest – 060031, Romania

Amir Hussain Dar
Department of Food Technology, Islamic University of Science and Technology, Awantipora, Pulwama, Jammu and Kashmir, India

Mehraj U. Din Dar
Department of Soil and Water Engineering, Punjab Agricultural University, Ludhiana – 141004, Punjab, India, E-mail: mehrajudindar24@gmail.com

Moonisa Aslam Dervash
Division of Environmental Sciences, Sher-e-Kashmir University of Agricultural Sciences and Technology, Kashmir Shalimar, Jammu and Kashmir – 190025, India

Zia Ur Rahman Farooqi
Institute of Soil and Environmental Sciences, University of Agriculture, Faisalabad – 38040, Pakistan, Phone: +923336809609, E-mail: ziaa2600@gmail.com

Bashir Ahmad Ganai
Center of Research for Development, University of Kashmir, Hazratbal Srinagar – 190006, India

Hilal Ahmad Ganaie
Department of Zoology, Government Degree College (Boys), Pulwama – 192301, Jammu and Kashmir, India, E-mail: hilalganie@hotmail.com

Syed Maqbool Geelani
Division of Environmental Science, Sher-e-Kashmir University of Agricultural Sciences and Technology Kashmir, Shalimar, Jammu and Kashmir – 190025, India

Barkat Hussain
Division of Entomology, Sher-e-Kashmir University of Agricultural Sciences and Technology, Kashmir Shalimar, Jammu and Kashmir – 190025, India

Ijaz Hussain
Bahauddin Zakariya University, Bahadur Sub-Campus, College of Agriculture, Layyah, Pakistan

Muhammad Mahroz Hussain
Institute of Soil and Environmental Sciences, University of Agriculture, Faisalabad – 38040, Pakistan

Muhammad Ijaz
Bahauddin Zakariya University, Bahadur Sub-Campus, College of Agriculture, Layyah, Pakistan

Rehan Khan
Department of Nano-Therapeutics, Institute of Nano-Science and Technology (DST-INST), Mohali, Punjab, India

Jasbir Kour
Cytogenetics and Molecular Biology Research Laboratory, Center of Research for Development (CORD), University of Kashmir, Srinagar – 190006, Jammu and Kashmir, India

Ajaz Ahmad Kundoo
Division of Entomology, Sher-e-Kashmir University of Agricultural Sciences and Technology, Kashmir Shalimar, Jammu and Kashmir – 190025, India

Sabhiya Majid
Department of Biochemistry, Government Medical College (GMC), Karan Nagar, Srinagar – 190010, Jammu, and Kashmir, India

Showkat Hamid Mir
Department of Botany, University of Kashmir, Srinagar – 190006, Jammu and Kashmir, India

Umair Mubarak
Institute of Soil and Environmental Sciences, University of Agriculture, Faisalabad – 38040, Pakistan

Falak Mushtaq
Cytogenetics and Molecular Biology Research Laboratory, Center of Research for Development (CORD), University of Kashmir, Srinagar – 190006, Jammu and Kashmir, India

Muntazir Mushtaq
Division of Biotechnology, Sher-e-Kashmir University of Agricultural Sciences and Technology, Jammu and Kashmir – 190025, India

Ahmad Nawaz
Bahauddin Zakariya University, Bahadur Sub-Campus, College of Agriculture, Layyah, Pakistan

Rumisa Nazir
Department of Environmental Science, Government College for Women Nawakadal, Srinagar, Jammu and Kashmir, India

Mansha Nisar
Department of Environmental Sciences, Sri-Pratap College, Srinagar – 190001, Jammu, and Kashmir, India, E-mail: manshanisar@gmail.com

Shauket Ahmed Pala
Section of Mycology and Plant Pathology, Department of Botany, University of Kashmir, Hazratbal Srinagar – 190006, Jammu and Kashmir, India

Iflah Rafiq
Department of Botany, University of Kashmir, Srinagar – 190006, Jammu and Kashmir, India

Iqra Rasheed
Bahauddin Zakariya University, Bahadur Sub-Campus, College of Agriculture, Layyah, Pakistan

Irfan Rashid
Department of Botany, University of Kashmir, Srinagar – 190006, Jammu and Kashmir, India

Nowsheeba Rashid
Amity Institute of Food Technology, Amity University Noida, Uttar Pradesh, India

Saiema Rasool
Forest Biotech Laboratory, Department of Forest Management, Faculty of Forestry, University of Putra Malaysia, Serdang, Selangor – 43400, Malaysia

Shabhat Rasool
Department of Biochemistry, Government Medical College (GMC), Karan Nagar, Srinagar – 190010, Jammu, and Kashmir, India

Muneeb U. Rehman
Department of Biochemistry, Government Medical College (GMC), Karan Nagar, Srinagar – 190010, Jammu, and Kashmir, India; Department of Clinical Pharmacy, College of Pharmacy, King Saud University, Riyadh – 11451, Saudi Arabia, E-mail: muneebjh@gmail.com

Zafar Ahmad Reshi
Department of Botany, University of Kashmir, Srinagar – 190006, Jammu and Kashmir, India

Ioan Sarac
Banat's University of Agricultural Sciences and Veterinary Medicine, "King Michael I of Romania" from Timisoara – 300645, CaleaAradului 119, Timis, Romania

Aamir Ishaq Shah
Department of Hydrology, Indian Institute of Technology, Roorkee – 247667, Uttarakhand, India

Naseer Ue Din Shah
Cytogenetics and Molecular Biology Research Laboratory, Center of Research for Development
(CORD), University of Kashmir, Srinagar – 190006, Jammu and Kashmir, India

Muhammad Shahid
Department of Bioinformatics and Biotechnology, Government College University,
Faisalabad – 38000, Pakistan

Dig Vijay Singh
Department of Environmental Science, Babasaheb Bhimrao Ambedkar Central University,
Lucknow – 226025, Uttar Pradesh, India

Irshad Ahmad Sofi
Department of Botany, University of Kashmir, Srinagar – 190006, Jammu and Kashmir, India

Ramona Stef
Banat's University of Agricultural Sciences and Veterinary Medicine, "King Michael I of Romania"
from Timisoara – 300645, CaleaAradului 119, Timis, Romania

Muhammad Tahir
Department of Environmental Sciences, COMSATS University Islamabad, Vehari Campus, Pakistan,
E-mail: muhammad_tahir@ciitvehari.edu.pk

Sami Ul-Allah
Bahauddin Zakariya University, Bahadur Sub-Campus, College of Agriculture, Layyah, Pakistan

Adil Farooq Wali
RAK College of Pharmaceutical Sciences, RAK Medical and Health Sciences University,
Ras Al Khaimah, P.O. Box – 11172, United Arab Emirates

Abdul Hamid Wani
Section of Mycology and Plant Pathology, Department of Botany, University of Kashmir,
Hazratbal Srinagar – 190006, Jammu and Kashmir, India

Hilal Ahmad Wani
Department of Biochemistry, Government Medical College (GMC), Karan Nagar,
Srinagar – 190010, Jammu, and Kashmir, India

Mohsin Zafar
Department of Soil and Environmental Sciences, University of the Poonch Rawalakot, Azad – 12350,
Jammu and Kashmir, Pakistan

Nukshab Zeeshan
Institute of Soil and Environmental Sciences, University of Agriculture, Faisalabad – 38040, Pakistan

Abbreviations

AA	amino acids
AD	anaerobic digestion
AFOs	animals feeding operations
AHP	analytic hierarchy process
AM	arbuscular mycorrhizal
AMF	arbuscular mycorrhizal fungi
AWs	agricultural wastes
BES	bio-electrochemical systems
BGA	blue-green algae
Bt	*Bacillus thuringensis*
BTEX	benzene-toluene-ethylbenzene-xylenes
CH_4	methane
CHP	combination of heat and power
CO_2	carbon dioxide
CR	crop residues
DTPA	diethylenetriaminepentaacetate
E	energy
EB	enhanced bioremediation
EC	electro-coagulation
EDDS	ethylenediamine disuccinate
EDTA	ethylenediaminetetraacetate
EET	extracellular electron transfer
EM	effective microorganisms
EMP	Embden-Meyerhoff-Parnas
EPA	Environmental Protection Agency
EQIP	environmental quality incentives program
FAO	Food and Agriculture Organization
FBM	fish bone meal
Fe^{3+}	iron
FRTR	Federal Remediation Technologies Roundtable
GMO	genetically modified organism
GMP	genetically modified plants
HMP	hexosomonophosphate
IAA	indole-3-acetic acid
IBA	indole butyric acid
IPM	integrated pest management
ISB	*in situ* bioremediation
ISR	induced systemic resistance
LA	lactic acid
MBM	meal bone meal

MCA	multi-criteria assessment
MGDA	methylglycine diacetate
MMO	methane monooxygenase
Mn^{4+}	manganese
MO	microorganisms
MSW	municipal solid waste
N	nitrogen
NH_3	ammonia
NO_3-	nitrate
O_2	oxygen
OCPs	organochlorine pesticides
ORPs	oxidation decrease possibilities
OWA	ordered weighted average
OWs	organic wastes
P	phosphorus
PAHs	polycyclic aromatic hydrocarbons
PCBs	polychlorinated biphenyls
PCE	perchloroethylene
PGPR	plant growth-promoting rhizobacteria
PGPS	plant growth-promoting streptomycetes
PM	poultry manure
POPs	persistent organic pollutants
PRBs	permeable reactive barrier
PROMETHEE	Preference Ranking Organization Method for Enrichment Evaluation
PSM	phosphate solubilizing microorganisms
RAS	recirculating aquaculture systems
SAW	simple additive weighting
SCB	single-cell biomass
SCP	single-cell proteins
SERB	surfactant-enhanced-bioremediation
SHF	separate hydrolysis and fermentation
SIRA	sequential and integrated remediation approach
SITE	superfund innovative technology evaluation
SMF	submerged fermentation
SO_4^{2-}	sulfate
SSB	silicate solubilizing bacteria
SSF	simultaneous saccharification and fermentation
SVE	soil vapor aspiration
TCE	trichloroethylene
TOC	total organic content
TOPSIS	Technique for Order Preference by Similarity to Ideal Solution
UCIL	Union Carbide India Ltd.
USGS	US geological survey
WHO	World Health Organization
XOS	xylooligosaccharides

Preface

As the human population mushrooms, the world advances towards more agricultural development to fulfill the food requirements of this huge populace. However, the agricultural development is usually accompanied by waste from the irrational application of intensive farming methods and the abuse of chemicals used in cultivation, remarkably affecting rural environments in particular and the global environment. Agricultural wastes (AWs) are non-product outputs of production and processing of agricultural products that may contain material that can benefit man but whose economic values are less than the cost of collection, transportation, and processing for beneficial use. Agricultural waste, otherwise called agro-waste, is comprised of animal waste, food processing waste, crop waste, and hazardous and toxic agricultural waste. Estimates of agricultural waste arising are rare, but they are generally thought of as contributing a significant proportion of the total waste matter in the developed world. In recent years, the quantity of agricultural waste has been rising rapidly all over the world. As a result, agricultural waste's environmental problems and negative impacts are drawing more and more attention. The problems posed by the agri-waste are threatening ecological stability as well as environmental sustainability, which makes it imperative to analyze and understand the problems associated with agricultural waste and the steps to be taken to tackle it. There is also a need to adopt proper approaches to reduce and reuse agricultural waste.

The present book is an attempt to highlight the issues of agricultural waste as well as the technologies, techniques, and strategies that can be used to manage agricultural waste. The book contains 14 chapters. Chapter 1 deals with the source and impacts of agri-waste on the environment with a deeper insight into these waste management.

Agricultural wastes find diverse use in enhancing the quality of the environment. Chapter 2 deals with the various uses of agri-waste with special emphasis on their applications in plant-soil systems. Agricultural practices like mechanical tillage, mono-cropping, application of agro-chemicals, irrigation with waste, and industrial waters affect soil health and productivity. All these disturb vital soil attributes, ultimately leading to reduced soil fertility and decreased crop yields. Chapter 3 details these issues highlighting the impact of various agricultural practices on soil health.

Pesticides are being extensively used in modern agriculture. However, their severe impacts on the environment are inflexible. There is a growing need for sustainable techniques to battle the effects of pesticides. Chapter 4 elaborates the global scenario of remediation techniques to combat the pollution effects of pesticides.

Innovative engineering based technologies are getting a major boost for improving water quality in regions affected by waterlogging and poor drainage problems. Agricultural drainage bioreactors are a major enhancement in this direction. Chapter 5 examines the plan and establishment of bioreactors, gives an overall view towards improved denitrification treatment of agricultural drainage, and explains various factors impacting the nitrate removal.

Biocontrol agents have generated considerable interest in terms of organic agriculture and environmental safety. Chapter 6 deals with the common biocontrol agents as well as the important approaches in biocontrol. Ecofriendly practices as if vermicomposting is a viable means of trans-forming various organic wastes (OWs) into products that can be used safely and beneficially as bio-fertilizers and soil conditioners. Chapter 7 details the science of vermicomposting for achieving sustainability of agriculture as well as the environment.

Chapter 8 gives a detailed insight of green chemistry into biorefineries, which in the present times serves as a path for achieving the objective of green and sustainable products.

As the quantum of agricultural waste increases, there is a growing need for more ecologically sound methods for the degradation of this waste. Microbes are playing a remarkable role in this area. Chapter 9 elaborates on the microbial interventions and biochemistry pathways for the degradation of agricultural waste.

The role and value of biofertilizers in improving and enhancing the quality of the soil and giving "eco-accommodating" natural agro-input are discussed in Chapter 10.

While synthetic fertilizers are posing harmful impacts to the environment and human health, the need to find organic alternatives increases each day. Chapter 11 details the importance of organic fertilizers like manure, biochar, and composts.

Agricultural waste utilization technology must either use the residues rapidly, or store the residues under conditions that do not cause spoilage or render the residues unsuitable for processing to the desired end product. In this direction, Chapter 12 discusses the mushroom cultivation technology used for the conversion of agro-industrial wastes into useful products.

Some biotechnological interventions in agricultural waste management are highlighted and discussed in Chapter 13, while Chapter 14 aims to present the methods and techniques of restoration and reconstruction by means of bioremediation in order to establish specific technologies for depollution and bioremediation.

The present book is an important reference source, highlighting the issues of agricultural waste and the technologies, techniques, and strategies that can be used to manage the agricultural waste. This book is a valuable reference for academicians, researchers, students, professionals, and policymakers who can benefit from the book's innovative content. Suggestions for the improvement of the book are always welcome.

—Editors

Rouf Ahmad Bhat, PhD
Khalid Rehman Hakeem, PhD
Humaira Qadri, PhD
Moonisa Aslam Dervash, PhD

CHAPTER 1

Agricultural Waste: Sources, Implications, and Sustainable Management

DIG VIJAY SINGH,[1] ROUF AHMAD BHAT,[2] and SYED MAQBOOL GEELANI[2]

[1] Department of Environmental Science, Babasahib Bhimrao Ambedkar University, Lucknow, Uttar Pradesh, India

[2] Division of Environmental Science, Sher-e-Kashmir University of Agricultural Sciences and Technology, Shalimar, Jammu and Kashmir – 190025, India

ABSTRACT

Agriculture involves diverse fields that produce an enormous quantity of waste at a rapid rate. Improper disposal of waste can further degrade the health of the ecosystem by causing eutrophication and releasing greenhouse gases into the environment. Increasing reliability on the fertilizers not only elevates the production cost but can also have serious impacts on the quality of the environment. Waste from agriculture is mostly organic in nature, and with the help of a suitable technique, can be easily converted into a rich nutrient source for soil health. Agriculture waste like crop residue, weeds, food processing, livestock, and poultry waste has immense potential in enriching soil fertility. To lower nutrient deficiency of soils, it is important to rely on organic manure, which improves the crop yield and has no detrimental impact on the environment.

1.1 INTRODUCTION

Agricultural wastes (AWs) are produced during various agricultural activities that can be used for different purposes (Skoulou and Zabaniotou, 2007). The generation of the waste from agricultural activities is enormous in quantity as most of India's population is dependent upon agriculture for livelihood (Obi et al., 2016). Cultivation of different crops results in waste production, which can be processed into a useful resource by suitable techniques (Smit and Nasr, 1992). The increasing population followed by the shrinking of agricultural land has put more pressure on the limited resource to increase production, ultimately leading to the degradation of soil structure (Bhatt et al., 2016). Organic wastes (OWs) generated from the agricultural industry can be used as a soil amendment that can improve the soil water retention, nutrient content in the soil, and organic matter in the soil (Cooperband, 2002). Continuous application of OW in compost form has proved very beneficial to the soil in the long run as it can increase organic matter content up to 15 cm in the soil (Diacono and Montemurro, 2011). AWs are produced during crop cultivation, harvesting, and processing of grain/fruits. The composition of waste produced from the agriculture depends upon the types of crop, agricultural industry and can be in solid or liquid form (Sarkar et al., 2012). AW consist of animal waste (manure, animal carcasses), food processing waste (only 20% of maize is canned and 80% is waste), crop waste, and hazardous and toxic agricultural waste (pesticides) (Fontenot and Jurubescu, 1980; Ezejiofor et al., 2014; Obi et al., 2016). The estimate of the quantity of waste generated from agricultural sources is very rare due to the diversity of crops and different sectors like horticulture, apiculture, and sericulture. Increasing production in the agriculture sector has also resulted in simultaneous waste generation at an increasing rate. As per one estimate, about 998 million tonnes are produced annually, out of which 798.4 million tonnes is organic waste. Waste is generated along with crops, and the generated waste can be used for different purposes in the agriculture field (Sadh et al., 2018). Less focus is being provided on the conversion of waste into useful products in developing countries as the waste produced is burned that become a serious pollution source in India (Agarwal et al., 2015). Waste produced due to agriculture activities, if properly managed, can become a significant source of nutrients and organic matter for the soil.

1.2   WASTES PRODUCED DURING CULTIVATION ACTIVITIES

From seed sowing to the harvesting of crops, various types of weeds and disease-causing organisms also grow, which can significantly impact crop yield (Oerke, 2006). In order to protect crops from pests and weeds, several kinds of pesticides are used to control or arrest their growth. The application of pesticides has a negative effect on the other beneficial microbes, plants, and animals (Subhani et al., 2000). The use of pesticides can enhance crop growth, thus can result in the increased waste generation (Obi et al., 2016).

Pesticides are also persistent and can have a long-term impact on the environment (Gill and Garg, 2014). Another issue is the plastic bottles in which pesticides are stored can also become a serious pollutant as it is estimated that about 1.8% remain in these bottles and can become a serious pollutant in other ecosystems. Residues produced during crop cultivation are mostly burned in developing countries that can degrade the quality of the atmosphere (Satyendra et al., 2013). Fertilizers are used to maintain the availability of nutrients for crops, but excessive uses of such chemicals degrade the quality of other ecosystems (Geng et al., 2019). The application of fertilizers and pesticide has a positive effect on crop yield, but with simultaneous production of crop waste, that can become a serious problem in the future. Waste produced can be converted into a useful product by composting, and the compost produced can be easily used in the soil and have a positive effect on soil and crop health (Bernal et al., 2009).

1.2.1 LIVESTOCK WASTE

Livestock waste can be a solid or liquid form that, if not properly managed, can lead to numerous problems. Solid waste from livestock is manure that being rich in the nutrient can improve the health of the soil and reduce the use of inorganic fertilizers as a nutrient source (Ahmad et al., 2016). All essential macronutrients and micronutrients are present in poultry wastes and can become efficient organic manure for agriculture crops (Dalolio et al., 2017). Due to the presence of good nutrient content in poultry manure (PM), improper management can cause leaching of nutrients, which results in groundwater pollution and surface water eutrophication (Gerber et al., 2007). PM is rich in phosphorus that has a positive effect on the growth and productivity of crops (Farhad et al., 2009). During livestock production, many freshwaters are wasted for washing of cages and livestock, thus resulting in the production of wastewater. Greenhouse gas like methane is also released during livestock production that can pose a threat to the survival of living organisms in the biosphere (Hatchett, 2004). Apart from greenhouse gas, hydrogen sulfide and ammonia are also produced that can significantly impact the environment (Brouček and Čermák, 2015). Bad odor is also produced from the livestock rearing house; it can create problems for the people residing in the nearby area (Ritter, 1989). The intensity of the smell depends on animal density, ventilation, temperature, and humidity. OW can create a serious problem by causing disease in humans as well as animals as germs and parasites easily grow on this kind of waste (Gutberlet and Uddin, 2018).

1.2.2   AVAILABILITY OF VEGETABLE AND FRUIT WASTE

Vegetables and fruits are produced in large quantities in India as most of the produced vegetables and fruits are consumed at the local level. Due to bumper generation of fruits and vegetables, a lot of waste is also generated that can be used as manure or substrate for biofuel production (Jahid et al., 2018). In Maharashtra, on an annual basis, the highest amount of fruit and vegetable waste is produced. A huge quantity of vegetable and fruit waste is available for recycling, but such decomposable material is not recycled in such a way so that dependence on inorganic nutrients can be minimized (Chatterjee et al., 2017). India being an agriculture-rich country, generates enormous waste from the agriculture sector, but the proper management of the generated waste is lacking that adds more problems in the environment conservation (Agarwal et al., 2015).

1.2.3   WASTE FROM AQUACULTURE

Growing aquaculture has resulted in an increase in feed use to increase production for achieving the increasing demands (Merino et al., 2012). The amount of feed used in aquaculture is directly related to the waste generation in this system (van Rijn, 2013). In aquaculture, the major waste produced is metabolic waste, which can be dissolved or suspended from water. Properly managed aquaculture farms can results in the production of $3/10^{th}$ of feces as solid waste (Dauda et al., 2018). Quantity of the solid waste generated is dependent upon the feeding rate (Eding et al., 2006) that is directly dependent upon temperature that is higher the temperature higher is the feeding rate with simultaneous solid waste generation and vice versa. In an aquaculture farm, proper flow of water is necessary to minimize the fragmentation of fish feces, and feces collected can reduce the OW content in the farm (Dauda et al., 2018). The problem related to animal feed is the limited availability of protein content, and efforts are made to find an alternative protein source that can meet the growing demand (Kaushik and Seiliez, 2010). OW from crops is rich in fibers but has low protein, starch, and fat content.

1.2.4   AGRO-INDUSTRIAL WASTE

Demand for different types of food products is increasing with the increasing population and changing living standards (Kearney, 2010). In order to fulfill the growing demands, several industries are opened to increase the availability of various food products. During the production of various food products, waste is also generated in enormous quantities, which is degrading the quality of the environment (Sadh et al., 2018). It is estimated that about 1/5th of the fruits and vegetables produced annually in India are wasted. An increase in production also leads to an increase in waste production that can have detrimental effects on the environment (Sagar et al., 2018). Apart from organic waste, wastewater is also produced from these industries, which are high in nutrient content; suspended solids and organic matter thereby can lead to the degradation of surface water quality (Noukeu et al., 2016). Waste produced from agro-industries can provide a better opportunity in recycling as a huge quantity of fruits and vegetables are used, which leads to the production of waste simultaneously (Sadh et al., 2018). Huge waste generated can be converted into a valuable and nutrient-rich product by vermicomposting (Rupani et al., 2010). The valuable product can acts as a sustainable nutrient source by elevating soil fertility in an eco-friendly manner. The different sources of AWs are presented in Figure 1.1.

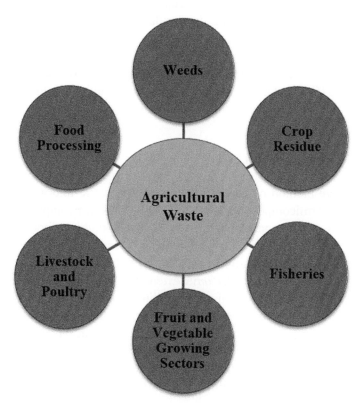

FIGURE 1.1 Various sources of agricultural wastes.

1.3   OW MANAGEMENT

Management of waste from different agricultural sources is a demand of the time, as without treatment, disposal of waste creates a lot of problems to the environment (Brunner and Rechberger, 2016). Waste from agriculture is disposed of in an unscientific manner in most of the countries of the world (Taiwo, 2011). It is important to consider the produced waste as a rich nutrient resource than only the impact of fertilizer can be minimized. OW by composting can be converted into manure, and using as manure in the field has positive effects on the soil as well as crop health (Jack and Thies, 2006). Improper disposal of agricultural waste can cause air (odor formation), water (eutrophication), and soil (decrease in microbial poll) pollution. Improper disposal also increases the incidence of diseases as waste becomes the nutrient source for various pathogens, which can cause several problems in living organisms. The demand for food is increasing with the rising population, which leads to a huge gap between production and demand as agricultural land is shrinking day-by-day (Nath et al., 2015). Thus it is important to recover nutrients from the waste material in order to sustain the growing nutrient demand. Recycling of the OW and using processed waste as a nutrient source is a viable and economical option to overcome the rising nutrient gap (Chew et al., 2019). It is important to devise a holistic approach in which all OW types can be managed in the proper way.

1.3.1   VERMI-COMPOSTING

The effective method of converting agricultural waste into a nutrient-rich product is vermicomposting (Suthar, 2009). In vermicomposting, earthworms are used to decompose waste material from kitchens, food-processing units, agricultural fields into a form that can be easily used in the field as a potent nutrient source (Sharma et al., 2005). Earthworm species like *Eisenia foetida, Eudrilus eugeniae, Perionyx excavates,* and *Decogester bolaui* are mostly used in India because of easy handling, short life span, the capability to thrive in extreme weather conditions, high multiplication rate, as well as cost-effective (Domínguez, 2018). Vermicomposting is one of the easy methods available for the conversion of waste into a useful product (Shukla, 2018). Agriculture waste rich in organic material can be used efficiently in order to avoid wastage of useful nutrients from the crop residue (Cerda et al., 2018).

1.3.2 ADSORBENTS IN THE ELIMINATION OF HEAVY METALS

Industrialization and urbanization have resulted in an increase in the release of several heavy metals into the environment (Singh et al., 2011). Heavy metals are toxic as cannot be degraded into non-toxic forms, thus have a significant impact on flora and fauna living in the heavy metal contaminated environment (Duruibe et al., 2007). Currently, adsorption is used to remove heavy metals from the contaminated environment (Lakherwal, 2014). By using adsorption, different waste of agricultural origin can be used as an effective adsorbent for heavy metal removal from the contaminated ecosystem (Demirbas, 2008). Agriculture waste like sugarcane bagasse, horticulture waste, rice husk, sawdust has proven to be low-cost adsorbent of heavy metals and can remove heavy metals in an effective way (Saka et al., 2012).

1.3.3 MANURE APPLICATION

Animal manure can be used as a nutrient source in the field that significantly helps to improve soil health (Timsina, 2018). Manure is not only the source of macronutrients (nitrogen, phosphorus, potassium) but also micronutrients, thereby help in maintaining the fertility of the soil in a cost-effective manner. Manure application to the soil increases nutrient and moisture content of the soil as well as improves the health of the soil (Wang et al., 2016). Processed waste can be used as a soil amendment (composts, biosolids) and is known to increase the organic matter as well as the nutrient status of the soil (Cooperband, 2002). The amount of nutrients contributed to the soil by amendment is totally dependent upon the composition of amendments. Waste material as soil amendments has tremendous advantages as waste material has good nutrient content and also can be used for reclamation projects (Sydnor and Redente, 2002). Using waste as soil amendments will reduce the burden of OW accumulation and also improves the soil health by retaining moisture and elevation of nutrient content (Fidelis and Rao, 2017). AW acts as a soil amendment (Parr and Hornick, 1992).

1.3.4 METHANE PRODUCTION

Manure can be easily used for the production of methane gas in an economical way (Møller et al., 2004). Methane gas produced can be used as fuel for heating purposes. Methane production from manure occurs in anaerobic conditions in which waste is first converted into organic acids, and then methanogen uses organic acids and leads to the formation of methane gas (Li et al., 2011). Anaerobic digestion (AD) of waste is essential for the large-scale disposal of poultry, swine, and dairy waste (Sakar et al., 2009). AD results in odor reduction, waste stabilization, and also retain the nutrient value of processed waste (Wilkie, 2005a, b). The different technologies for the degradation of organic waste for sustainable development are depicted in Figure 1.2.

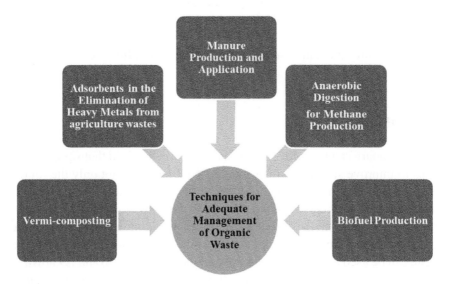

FIGURE 1.2   Environmental sound technology for sustainable management of agricultural wastes.

1.3.5   BIOFUEL PRODUCTION

Agro-industrial waste can be used to produce biofuel, thus providing alternative fuel sources compared to fossil fuel (Leiva-Candia et al., 2014). With the increasing burden on fossil fuels, biofuel production as an alternative energy source is also increasing to overcome dependence on fossil fuels. The utilization of agricultural waste as a substrate for biofuel can be economical as dependence on forest for biomass can be easily reduced (Mohammed et al., 2018). Crops are cultivated in less time and waste are also produced in large amount, thus increasing OW's availability as a substrate for biofuel production (Lee et al., 2019). Vegetable waste is also an effective OW for biofuel production by fermentation technique using *Saccharomyces cerevisiae* (Hossain et al., 2017). In order to reduce the enormous quantity of OW generation, biofuel is produced from agricultural waste and can help to conserve the environment by increasing the production of eco-friendly fuel (Chandra et al., 2012). Weeds are also produced in large quantities and the crops and have shown biofuel producing potential, thus becoming the potent substrate due to fast growth and short life cycle (Chandel and Singh, 2011). The soil's fertility and health are deteriorating at an alarming rate, but OW can help overcome these issues and enhance soil properties in an eco-friendly way.

1.4 CONCLUSION

Waste from agriculture industries is disposed in an unscientific manner in most of the countries of the world. Inadequate management of this waste put tremendous pressure on the quality of the environment (air, water, and soil). Furthermore, OW increases the incidence of diseases as waste becomes the nutrient source for various disease-spreading microbes. Thus, it is important to manage and recover essential constituents from AWs in a sustainable manner. Therefore, it is important to devise holistic approaches in which all the types of OW can be managed in an adequate manner. Recycling the OW and using processed waste as a nutrient source is a viable and economical option to overcome the rising nutrient gap.

KEYWORDS

- **agricultural waste**
- **biofuel**
- **heavy metals**
- **organic wastes**
- **recycling**
- **sustainable management**

REFERENCES

Agarwal, R., Chaudhary, M., & Singh, J., (2015). Waste management initiatives in India for human well-being. *European Scientific Journal.*

Ahmad, A. A., Radovich, T. J. K., Nguyen, H. V., Uyeda, J., Arakaki, A., Cadby, J., Paull, R., Sugano, T., & Teves, G., (2016). Use of organic fertilizers to enhance soil fertility, plant growth, and yield in a tropical environment. *Organic Fertilizers – From Basic Concepts to Applied Outcomes.*

Bernal, M. P., Alburquerque, J. A., & Moral, R., (2009). Composting of animal manures and chemical criteria for compost maturity assessment: A review. *Bioresource Technology, 100*(22), 5444–5453.

Bhatt, R., Kukal, S. S., Busari, M. A., Arora, S., & Yadav, M., (2016). Sustainability issues on rice-wheat cropping system. *International Soil and Water Conservation Research, 4*(1), 64–74.

Brouček, J., & Čermák, B., (2015). Emission of harmful gases from poultry farms and possibilities of their reduction. *Ekologia, 34*(1), 89–100.

Brunner, P. H., & Rechberger, H., (2016). *Practical Handbook of Material Flow Analysis: For Environmental, Resource, and Waste Engineers.* CRC Press.

Cerda, A., Artola, A., Font, X., Barrena, R., Gea, T., & Sánchez, A., (2018). Composting of food wastes: Status and challenges. *Bioresource Technology, 248*, 57–67.

Chandel, A. K., & Singh, O. V., (2011). Weedy lignocellulosic feedstock and microbial metabolic engineering: Advancing the generation of 'biofuel'. *Applied Microbiology and Biotechnology, 89*(5), 1289–1303.

Chandra, R., Takeuchi, H., & Hasegawa, T., (2012). Methane production from lignocellulosic agricultural crop wastes: A review in context to second generation of biofuel production. *Renewable and Sustainable Energy Reviews, 16*(3), 1462–1476.

Chatterjee, R., Gajjela, S., & Thirumdasu, R. K., (2017). Recycling of organic wastes for sustainable soil health and crop growth. *Int. J. Waste Resour., 7*(296), 2.

Chew, K. W., Chia, S. R., Yen, H. W., Nomanbhay, S., Ho, Y. C., & Show, P. L., (2019). Transformation of biomass waste into sustainable organic fertilizers. *Sustainability, 11*(8), 2266.

Cooperband, L., (2002). *Building Soil Organic Matter with Organic Amendments.* Centre for Integrated Agricultural Systems, College of Agricultural and Life Sciences, University of Wisconsin, Madison.

Dalólio, F. S., Da Silva, J. N., De Oliveira, A. C. C., Tinôco, I. D. F. F., Barbosa, R. C., De Oliveira, R. M., Albino, L. F. T., & Coelho, S. T., (2017). Poultry litter as biomass energy: A review and future perspectives. *Renewable and Sustainable Energy Reviews, 76*, 941–949.

Dauda, A. B., Ajadi, A., Tola-Fabunmi, A. S., & Akinwole, A. O., (2018). Waste production in aquaculture: Sources, components and managements in different culture systems. *Aquaculture and Fisheries.*

Demirbas, A., (2008). Heavy metal adsorption onto agro-based waste materials: A review. *Journal of Hazardous Materials, 157*(2, 3), 220–229.

Diacono, M., & Montemurro, F., (2011). Long-term effects of organic amendments on soil fertility. In: *Sustainable Agriculture* (Vol. 2, pp. 761–786). Springer, Dordrecht.

Domínguez, J., (2018). Earthworms and vermicomposting. *Earthworms the Ecological Engineers of Soil.*

Duruibe, J. O., Ogwuegbu, M. O. C., & Egwurugwu, J. N., (2007). Heavy metal pollution and human biotoxic effects. *International Journal of Physical Sciences*, *2*(5), 112–118.

Eding, E. H., Kamstra, A., Verreth, J. A. J., Huisman, E. A., & Klapwijk, A., (2006). Design and operation of nitrifying trickling filters in recirculating aquaculture: A review. *Aquacultural Engineering*, *34*(3), 234–260.

Ezejiofor, T. I. N., Enebaku, U. E., & Ogueke, C., (2014). Waste to wealth-value recovery from agro-food processing wastes using biotechnology: A review. *British Biotechnology Journal*, *4*(4), 418–481.

Farhad, W., Saleem, M. F., Cheema, M. A., & Hammad, H. M., (2009). Effect of poultry manure levels on the productivity of spring maize (*Zea mays* L.). *J. Anim. Plant Sci.*, *19*(3), 122–125.

Ferronato, N., & Torretta, V., (2019). Waste mismanagement in developing countries: A review of global issues. *International Journal of Environmental Research and Public Health*, *16*(6), 1060.

Fidelis, C., & Rao, B. R., (2017). Enriched cocoa pod composts and their fertilizing effects on hybrid cocoa seedlings. *International Journal of Recycling of Owin Agriculture*, *6*(2), 99–106.

Fontenot, J. P., & Jurubescu, V., (1980). Processing of animal waste by feeding to ruminants. In: *Digestive Physiology and Metabolism in Ruminants* (pp. 641–662). Springer, Dordrecht.

Geng, Y., Cao, G., Wang, L., & Wang, S., (2019). Effects of equal chemical fertilizer substitutions with organic manure on yield, dry matter, and nitrogen uptake of spring maize and soil nitrogen distribution. *Plos One*, *14*(7), e0219512.

Gerber, P., Opio, C., & Steinfeld, H., (2007). *Poultry Production and the Environment: A Review* (p. 153). Animal production and health division, Food and Agriculture Organization of the United Nations, Vialedelle Terme di Caracalla.

Gill, H. K., & Garg, H., (2014). Pesticides: Environmental impacts and management strategies. In: *Pesticides-Toxic Aspects*. Intech Open.

Gutberlet, J., & Uddin, S. M. N., (2018). Household waste and health risks affecting waste pickers and the environment in low-and middle-income countries. *International Journal of Occupational and Environmental Health*, 1–12.

Hatchett, A. N., (2004). Bovines and global warming: How the cows are heating things up and what can be done to cool them down. *Wm. a Mary Envtl. L. & Poly. Rev.*, *29*, 767.

Hossain, N., Zaini, J. H., & Mahlia, T. M. I., (2017). A review of bioethanol production from plant-based waste biomass by yeast fermentation. *Int. J. Technol.*, *8*(1), 5–18.

Jack, A. L., & Thies, J. E., (2006). Compost and vermicompost as amendments promoting soil health. *Biological Approaches to Sustainable Soil Systems*, 453–466.

Jahid, M., Gupta, A., & Kumar, D., (2018). Production of bioethanol from fruit wastes (banana, papaya, pineapple and mango peels) under milder conditions. *Journal of Bioprocessing and Biotechniques*, *8*(3), 1–11.

Kaushik, S. J., & Seiliez, I., (2010). Protein and amino acid nutrition and metabolism in fish: Current knowledge and future needs. *Aquaculture Research*, *41*(3), 322–332.

Kearney, J., (2010). Food consumption trends and drivers. *Philosophical Transactions of the Royal Society B: Biological Sciences*, *365*(1554), 2793–2807.

Lakherwal, D., (2014). Adsorption of heavy metals: A review. *International Journal of Environmental Research and Development*, *4*(1), 41–48.

Lee, S. Y., Sankaran, R., Chew, K. W., Tan, C. H., Krishnamoorthy, R., Chu, D. T., & Show, P. L., (2019). Waste to bioenergy: A review on the recent conversion technologies. *BMC Energy*, *1*(1), 4.

Leiva-Candia, D. E., Pinzi, S., Redel-Macías, M. D., Koutinas, A., Webb, C., & Dorado, M. P., (2014). The potential for agro-industrial waste utilization using oleaginous yeast for the production of biodiesel. *Fuel*, *123*, 33–42.

Li, Y., Park, S. Y., & Zhu, J., (2011). Solid-state anaerobic digestion for methane production from organic waste. *Renewable and Sustainable Energy Reviews*, *15*(1), 821–826.

Marshall, R. E., & Farahbakhsh, K., (2014). Systems approaches to integrated solid waste management in developing countries. *Waste Management*, *33*(4), 988–1003.

Merino, G., Barange, M., Blanchard, J. L., Harle, J., Holmes, R., Allen, I., Allison, E. H., et al., (2012). Can marine fisheries and aquaculture meet fish demand from a growing human population in a changing climate? *Global Environmental Change*, *22*(4), 795–806.

Mohammed, N. I., Kabbashi, N., & Alade, A., (2018). Significance of agricultural residues in sustainable biofuel development. *Agricultural Waste and Residues*.

Møller, H. B., Sommer, S. G., & Ahring, B. K., (2004). Methane productivity of manure, straw and solid fractions of manure. *Biomass and Bioenergy*, *26*(5), 485–495.

Nath, R., Luan, Y., Yang, W., Yang, C., Chen, W., Li, Q., & Cui, X., (2015). Changes in arable land demand for food in India and China: A potential threat to food security. *Sustainability*, *7*(5), 5371–5397.

Noukeu, N. A., Gouado, I., Priso, R. J., Ndongo, D., Taffouo, V. D., Dibong, S. D., & Ekodeck, G. E., (2016). Characterization of effluent from food processing industries and stillage treatment trial with *Eichhornia crassipes* (Mart.) and *Panicum maximum* (Jacq.). *Water Resources and Industry*, *16*, 1–18.

Obi, F. O., Ugwuishiwu, B. O., & Nwakaire, J. N., (2016). Agricultural waste concept, generation, utilization and management. *Nigerian Journal of Technology*, *35*(4), 957–964.

Oerke, E. C., (2006). Crop losses to pests. *The Journal of Agricultural Science*, *144*(1), 31–43.

Parr, J. F., & Hornick, S. B., (1992). Agricultural use of organic amendments: A historical perspective. *American Journal of Alternative Agriculture*, *7*(4), 181–189.

Ritter, W. F., (1989). Odor control of livestock wastes: State-of-the-art in North America. *Journal of Agricultural Engineering Research*, *42*(1), 51–62.

Rupani, P. F., Singh, R. P., Ibrahim, M. H., & Esa, N., (2010). Review of current palm oil mill effluent (POME) treatment methods: Vermicomposting as a sustainable practice. *World Applied Sciences Journal*, *11*(1), 70–81.

Sadh, P. K., Duhan, S., & Duhan, J. S., (2018). Agro-industrial wastes and their utilization using solid state fermentation: A review. *Bioresources and Bioprocessing*, *5*(1), 1.

Sagar, N. A., Pareek, S., Sharma, S., Yahia, E. M., & Lobo, M. G., (2018). Fruit and vegetable waste: Bioactive compounds, their extraction, and possible utilization. *Compr. Rev. Food Sci. Food Saf.*, *17*, 512–531.

Saka, C., Şahin, Ö., & Küçük, M. M., (2012). Applications on agricultural and forest waste adsorbents for the removal of lead (II) from contaminated waters. *International Journal of Environmental Science and Technology*, *9*(2), 379–394.

Sakar, S., Yetilmezsoy, K., & Kocak, E., (2009). Anaerobic digestion technology in poultry and livestock waste treatment: A literature review. *Waste Management and Research*, *27*(1), 3–18.

Sarkar, N., Ghosh, S. K., Bannerjee, S., & Aikat, K., (2012). Bioethanol production from agricultural wastes: An overview. *Renewable Energy*, *37*(1), 19–27.

Satyendra, T., Singh, R. N., & Shaishav, S., (2013). Emissions from crop/biomass residue burning risk to atmospheric quality. *International Research Journal of Earth Sciences, 1,* 24–30.

Sharma, S., Pradhan, K., Satya, S., & Vasudevan, P., (2005). Potentiality of earthworms for waste management and in other uses: A review. *The Journal of American Science, 1*(1), 4–16.

Shukla, N., (2018). *To Determine the Value Addition in the Compost and Vermicompost Produced from Kitchen Waste.*

Singh, R., Gautam, N., Mishra, A., & Gupta, R., (2011b). Heavy metals and living systems: An overview. *Indian Journal of Pharmacology, 43*(3), 246.

Skoulou, V., & Zabaniotou, A., (2007). Investigation of agricultural and animal wastes in Greece and their allocation to potential application for energy production. *Renewable and Sustainable Energy Reviews, 11*(8), 1698–1719.

Smit, J., & Nasr, J., (1992). Urban agriculture for sustainable cities: Using wastes and idle land and water bodies as resources. *Environment and Urbanization, 4*(2), 141–152.

Subhani, A., El-ghamry, A. M., Changyong, H., & Jianming, X., (2000). Effects of pesticides (Herbicides) on soil microbial biomass: A review. *Pakistan Journal of Biological Sciences, 3*(5), 705–709.

Suthar, S., (2009). Vermicomposting of vegetable-market solid waste using *Eisenia fetida*: Impact of bulking material on earthworm growth and decomposition rate. *Ecological Engineering, 35*(5), 914–920.

Sydnor, M. E., & Redente, E. F., (2002). Reclamation of high-elevation, acidic mine waste with organic amendments and topsoil. *Journal of Environmental Quality, 31*(5), 1528–1537.

Taiwo, A. M., (2011). Composting as a sustainable waste management technique in developing countries. *Journal of Environmental Science and Technology, 4*(2), 93–102.

Timsina, J., (2018). Can organic sources of nutrients increase crop yields to meet global food demand? *Agronomy, 8*(10), 214.

Van, R. J., (2013). Waste treatment in recirculating aquaculture systems. *Aquacultural Engineering, 53,* 49–56.

Wang, X., Jia, Z., Liang, L., Yang, B., Ding, R., Nie, J., & Wang, J., (2016). Impacts of manure application on soil environment, rainfall use efficiency and crop biomass under dryland farming. *Scientific Reports, 6*(1).

Wilkie, A. C., (2005a) Anaerobic digestion: Biology and benefits. *Dairy Manure Management: Treatment, Handling, and Community Relations,* 63–72.

Wilkie, A. C., (2005b). Anaerobic digestion of dairy manure: Design and process considerations. *Dairy Manure Management: Treatment, Handling, and Community Relations, 301*(312), 301–312.

CHAPTER 2

Agricultural Wastes and Its Applications in Plant-Soil Systems

ZIA UR RAHMAN FAROOQI, UMAIR MUBARAK, NUKSHAB ZEESHAN, MUHAMMAD MAHROZ HUSSAIN, and MUHAMMAD ASHAR AYUB

Institute of Soil and Environmental Sciences, University of Agriculture, Faisalabad – 38040, Pakistan, Phone: +923336809609, E-mail: ziaa2600@gmail.com (Z. U. R. Farooqi)

ABSTRACT

Agriculture is the largest sector in the world, which supplies raw material to almost all the industries right from the food industry to textile, construction, mechanical, and energy to even the cosmetics industry. It produces a different type of raw materials for different industries like cereal crops, sugar cane, vegetables, and oil crops for the food industry, cotton for the textile industry, wood for construction and mechanical industries, and different plant products/extracts to make cosmetic products. Besides, the agriculture industry includes crop residues (CR), animal dung, sugarcane bagasse, livestock animal dead bodies, pruning waste, cotton stalk, etc. These wastes can be used for the betterment of soil and plant health, including energy production, by converting them into useful products. These products are produced by some processes like composting, biochar, and manure production. Some other uses of agricultural wastes (AW) include as biosorbents for heavy metals, crop residue supply to soils, mulching, and energy production from household to large scales. This chapter discusses the prospects of AW uses for soil and plant health improvement along with future directions.

2.1 INTRODUCTION

Agricultural wastes (AW) are unwanted materials formed from agricultural activities and its associated operations. These activities could be the growing of crops or livestock and herd management, and raw material processing in industries produced from agriculture (Zhou et al., 2015). Agricultural waste can also be soil sediments, nutrient, and pesticides runoff from the soil as a result of flooding or rains (Sharifi et al., 2016; Quinteros et al., 2017), different animals wastes (ElMekawy et al., 2015), crop residues (CR) (Quiñones et al., 2015), poultry wastes (Abouelenien et al., 2016), dead or slaughtered animals waste (Arshad et al., 2018), vegetables processing waste (Bakatovich et al., 2018) and water containing burned tree and crop ashes resultant of the accidental fire in agricultural fields (Rajput et al., 2016). To manage these AW, a complete understanding of their composition, physical, chemical, and biological reactions is necessary, along with the factors controlling their fates in the environment. Soil application of natural or synthetic chemicals is an important practice which supplies nutrients and improves soil fertility, soil biological and physical properties (Väisänen et al., 2016). Pyrolysis of AW is also done to achieve waste management, energy production, and biochar formation for soil application (Lee et al., 2017). CR can also be applied as such in the soils and proved an important addition for soil health and crop production (Hatfield, 2017; Gul et al., 2015). Poultry manure (PM) is also used as a nitrogen (N) source and contains up to 1.5 gkg^{-1} of N, 0.8 g potassium (K), and 0.5 gkg^{-1} phosphorus (P) (Moshia et al., 2016). The production of composts from AW for soil conditioning is also a sustainable way to manage these wastes and in alleviating salt-affected soils. By using compost, soil microbial activities, microbial C and N, and soil respiration were improved, which showed that the importance of compost (Murphy et al., 2016).

Different AW used as agricultural soil amendments include municipal sewage, agro-industrial, slaughterhouse, AWs compost, AW compost, pig slurry, digestate, and paper mill wastes to improve soil physicochemical properties and control bioavailability of heavy metals and organic contaminants. Application of these AW lowers the bioavailability of contaminants in soil and plants and their use as pollutant removal agents (Alvarenga et al., 2015). Biochars of different CR also proved effective in immobilizing metal soils. Some wastes like cow dung also used as a source of energy through biogas production (El Mekawy et al., 2015; Nandi et al., 2017), biofuel production (Prasad et al., 2017), and also used as a domestic fuel in houses for cooking (Xiao et al., 2015) as well as mulching (Chen et al., 2018). In addition, this chapter also discusses comprehensively the agricultural waste advantages, their effective uses, and its effects on soil and plant health along with the negative impact on soil properties deterioration as a result of CR burning.

2.2 TYPE OF AGRICULTURAL WASTES (AWS)

There are many types of AW used in agriculture and of different origin, mostly from crops like wheat, rice, maize, sugarcane, cotton, and millet, from forestry includes tree branches, leaves, and wood, horticulture wastes include thinning and pruning wastes, vegetable leaves, and processing and livestock wastes comprising of animal dung, dead, and slaughtered animals waste and their manures.

2.2.1 FORESTRY WASTE

Agricultural forestry is a big contributor of AW in terms of tree branches, thinning, and pruning of trees for making them tall and big canopy size leaves (Figure 2.1). Tree branches are also broken and fell down as a result of heavy rains and storms. Many temporary and perennial forestry crops and forest residues are generated each year (Nones et al., 2017) from different forest trees like eucalyptus shed leaves and its bark. Some trees shed more leaves in a specific time of the season, and some of their branches to grow taller. Forest biomass and waste degrade, slowly making the forest soils more and more productive. Organic matter is increased in these types of forests, which promotes further plantations and forest growth (James et al., 2019; Santos et al., 2016).

FIGURE 2.1 Forestry waste.

2.2.2   HORTICULTURE WASTE

Horticulture waste includes different crop wastes and by-products after their processing. For example, fruits and vegetables as consumed raw, minimally processed, as well as processed, due to their nutrients and health-promoting compounds. Thus, significant losses and waste in the fresh and processing industries are produced. The United Nations Food and Agriculture Organization (FAO) has estimated that losses and waste in fruits and vegetables are the highest among all types of foods and can reach up to 60%. The processing operations of fruits and vegetables produce significant wastes or by-products, which constitute about 25% to 30% of a whole commodity group. The waste is composed mainly of seed, skin, rind, and pomace (Sagar et al., 2018). In addition to the food crops, different ornamental plants are also a big source of AW as these plants are cut and pruned to modify for making different canopy shapes or heights after pruning (Wu et al., 2016; Roslim et al., 2018).

2.2.3   AGRONOMY WASTE

Major crops used in the food are categorized under the umbrella of agricultural crops or agronomy, including wheat, rice, and maize grown individually or in combination to other staple crops in different parts of the world. Thus, the production being high, these crops produce a huge amount of AW with food grains. When crops are harvested, a huge amount of CR are left behind in the fields (Speratti et al., 2017). Apart from the CR, other AW are corn cobs produced after the use of maize grains, cotton stalks after the picking of cotton, and other miscellaneous plant parts. Agronomical wastes also include dead plants, seeds of all kinds, and a mixture of soil runoff, leaching of nutrients, and eroded soil and nutrients (Nanda et al., 2016).

2.2.4   LIVESTOCK WASTE

Livestock production is a rapidly increasing industry, especially in developing countries, due to the increased consumption demands for meat and other dairy products. Due to this, larger quantities of animal wastes are leftover (Figure 2.2); these wastes include animal dung, dead, and slaughtered animal waste, animal blood, and different manures like poultry and pig manures (Zhu and Hiltunen, 2016; Zhu et al., 2017).

FIGURE 2.2 Livestock waste.

2.3 MANAGEMENT AND USES OF AGRICULTURAL WASTE

AW's disposal is a large environmental concern and the methods used range from simpler to complex for AW application. Interest in the potential diversion of AW for useful purposes is growing with several different methods. These include the application of AW to soils to support the soil's physical properties, increase soil organic matter, and improve soil bulk density and water holding capacity (Khaleel et al., 1981).

2.3.1 CONVENTIONAL APPROACHES

Different approaches to use AW for useful purposes include compost formation for soil application, the direct burial of CR to the soil, biochar production, biosorbent production for heavy metals removal, and biofuel production, etc. In ancient times, AW was not a problem nowadays. Farmers used to bury these wastes on the fields by using plows or allow them to naturally decompose to supply nutrients to the soil. Another approach was the CR burning on the field not associated with the smog problem. Like cotton stalks as fuelwood, AW is also practiced for an economical source of fuelwood and the good reason of less deforestation for fuelwood. Incineration was also done for the last option for the wastes, which cannot be used for any profitable or useful purpose (Girotto et al., 2015).

2.3.2 COMPOSTING

Composts from AW supply different nutrients to plants and soil (Figure 2.3). It provides organic matter to soil improving the soil capability to hold more water, bulk density, detoxify the soil from different toxic metals by adsorbing them on organic matter and ultimately improving plant growth (Zhang et al., 2017). It is proved that compost is the combination of different plant nutrients (Karak et al., 2015), but influence of composts on different toxic metals binding and crop productivity enhancement. The composts of different AW are effective in alleviating heavy metals toxicity and enhance crop production (Chen et al., 2017). There are also the procedures and techniques to make composts more effective by additional bacterial inoculum or earthworm's addition (vermin-composting) (Huang et al., 2017).

FIGURE 2.3   Composting.

2.3.3 USES AS FUEL

As there are a lot of risks associated with the use of fossil fuels, the researchers are looking for readily available fuel sources. AW is a good raw material for this. Like anaerobic digestion (AD) of cow dung with new feedstock such as CR yields into biogas which can fulfill the energy requirement of a small house to a whole community. It can replace biomass-cooking fuels and also reduce indoor air pollution. After biogas production from cow dung and CR, a by-product called digestate is also produced, which acts as a fertilizer or soil conditioner and provides the nutrients to soil. The supply of digestate to soil reduces the farmers' dependency on nitrogen phosphorus and potassium-containing synthetic fertilizers only by 0.1, 1.6%, and 31%. By using CR in combination with cow dung, their burning into the fields is avoided, which contributes to air pollution in India and climate change globally (Sfez et al., 2017). CR plays an important role in the household of the rural communities, not having the proper facilities of fuel and energy sources for cooking and heating their houses. Uses of AW for household energy and heating are about 77.5% of the total rural households in India and 55.3% in Pakistan. Due to the consistent and cheap supply of the fuel sources as AW, only 2% in India and 5–7% households in Pakistan have shifted from AW fuel to other energy sources in a decade (Ravindra et al., 2019a, b).

2.3.4 CROP RESIDUES (CR) BURNING ON FIELD

It was the easiest way to burn CR on the spot by setting fire. It has some advantages, as if it does not require any labor to collect and transport it to other place for its proper disposal. It was also thought that setting fire to the CR supplies ash to the soil is a good source of organic matter in agricultural soils. However, later on, it was observed that CR burning in the fields does not supply organic matter to the soils but decomposes the existing organic matter by increasing the soil temperature and burning the soil organic matter. In addition to the soil organic matter loss, the smoke raised from the CR burning resulted in smog condition as evident in subcontinent, especially in India and Pakistan (Ni et al., 2015). This practice is also common in China, and the emissions from the CR burning were calculated in the last 2 decades in 17 districts. It was found that 2707.34 Tg of CO_2 was emitted CR burning (Sun et al., 2016).

2.3.5 MULCHING

Mulching protects plants from extreme heat, cold, and water stress. As water and nutrient availability in dryland crops and agricultural areas is influenced by the quantity of rainfall and soil water holding capacity. Thus, to improves agricultural productivity in these areas, different CR are used as mulch to protect crop plants to secure nutrients, water, and temperature of the soil. Different materials used as mulch are CR, like rice, wheat, maize, and sugarcane. Mulching proves best in terms of improving crop production, improved biomass production, and water use efficiency with high profit and the best-harvested monetary benefits. Therefore, crop residue mulching may be practiced using AW as effective means for protecting soil deterioration, improving soil water holding capacity in soils, sustaining agricultural productivity (Thu et al., 2016). Mulching is also considered a soil management technique used to enhance soil organic matter and carbon sequestration, but it varies with CR amount, crop, and soil type (Chen et al., 2018). This technique is also used as organic amendment to improve physical, chemical, and biological attributes of agricultural soils, improving soil organic matter contents, soil moisture retention, enhanced nutrient cycling, and decreased soil loss, among other environmental and soil health benefits (Turmel et al., 2015; Ranaivoson et al., 2018).

2.3.6   BURIAL OF CROP RESIDUES (CR)

Burial of AW in fields is an ancient practice which involves plowing of the CR after the harvesting and burial of these AW contributes to crop needs of nitrogen for next crop (Jahanzad et al., 2016) and called crop residue recycling intensively applied to enhance the utilization of resources in agricultural systems. The CR recycling shows that the crop residue of sugar cane and mung bean can improve the growth and the production of sugarcane and enhance the accumulation of dry matter, nitrogen, phosphorus, and potassium in the aboveground part of sugarcane and their availability in soil (He et al., 2018). CR burial is a good source of reducing weed seed dispersal and infestation (Mohler et al., 2018). It improves soil aggregation, microbial diversity, nutrient storage, and supply as well as water-holding capacity in combination with soil carbon sequestration (Singh et al., 2015).

2.3.7   USE AS BIOSORBENT

There is an emerging trend of using crop CR and other AW as biosorption of different heavy metals and other harmful contaminants of water (Sadeek et al., 2015). Heavy metals, especially carcinogens like arsenic and cadmium, pose serious human health risks by contaminating groundwater reservoirs and food crops by bioaccumulating in edible parts. Over 170 million people have been affected by these metals due to the uptake of contaminated water and food grains. Different methods are employed to remove these metals from water and food, but these are costly reverse osmosis, ion exchange, and electrodialysis. However, the cost-effective method was developed from different AW like sugarcane bagasse, peels of various fruits, and wheat straw and used as biosorbents, offering an environment-friendly solution for the toxic metals (Shakoor et al., 2016). Industrial effluents containing azo dyes are also treated with these cost-effective and environmentally friendly processes of biosorption using AW-derived biosorbents (Lee et al., 2016; Tran et al., 2017).

2.3.8 BIOCHAR PRODUCTION

Biochar emerged as the solution to all the major soil problems ranging from nutrient deficiency, organic matter loss, carbon sequestration, carbon source, slow-releasing nutrient reservoir, and remedial measure to soil contaminants like heavy metals and pesticides. These biochars are derived or made from CR like corn and rice stalks, cattle pigs, and PMs (Liu et al., 2015). The biochar is used as a nutrient source and known as the carbon source supply and sequestering source as it stores carbon in it for longer periods and releases it extremely slow. It is also used as biosorbent material for capturing heavy metals and pesticides residues in the contaminated soil and binding them to it for longer times for safe food grains production (Zhao et al., 2018; Igalavithana et al., 2017). There are other advanced biochar types like magnetic biochar derived from various types of AW exhibiting a good magnetic property and larger surface area. These magnetic biochars showed a remarkable application as an adsorbent for various wastewater treatments (Thines et al., 2017).

2.3.9 BIOFUEL PRODUCTION (WASTE TO ENERGY)

With the increase in global energy demand and decreasing the fossil fuels, there has been increasing demand for energy sources with cheap and continuous supply. Developing countries have increased their fuel consumption due to industrial development. This increased consumption of energy sources can lead to early end of fossil fuels. The bioenergy produced from the AW biomass is being a sustainable alternate energy source which received high acceptance in various sectors includes public, industries, and government policies. Hence an economic and efficient production process is essential to commercialize AW biomass-based biofuels (Gaurav et al., 2017). The bioethanol, biogas, and electricity from rice husk are used for commercial production and production costs of biofuels are 0.27–0.82 USD/kg and it is possible to take advantage of AW as energy source in biorefinery to produce biofuels as source of energy to supply to demands in this country

(Daza Serna et al., 2016). Iran produces 520,400 tons of pistachio wastes/yr from 500,000 ha total area of production under it. By optimum use of this pistachio waste, more than 400,000 tons of biofuel can be generated in which 103.5 million cubic meters of biogas and 47.6 million liters of ethanol can be produced (Taghizadeh-Alisaraei et al., 2017).

The production of biofuels from AW to blend with gasoline is another practice being done worldwide supporting the development of rural technology with knowledge-based jobs and mitigating greenhouse gas emissions. Today, engineering for plant construction is accessible and new processes using AW have reached a good degree of maturity and high conversion yields of nearly 90%. The growth of biofuel production is expected to be growing exponentially and it is necessary to move on ahead from its very early stages to a more mature consolidated technology (Valdivia et al., 2016).

2.4 IMPACTS OF AW ON THE SOIL-PLANT SYSTEM

Agricultural waste has much importance in the current deteriorating environmental scenario where air, soil, and plant health affect the yield losses and the environmental consequences (Table 2.1). The application of agriculture waste in soil has many forms. A study was conducted to check AW's suitability in soil and biochar obtained from the dry method. Results revealed that slow pyrolysis in AW lowers the alkali and alkaline earth metal along with heavy metals load from soil. It is also economically cheaper than biochar produced by dry method (Kambo and Dutta, 2015). Compost enhances the microbial activity in soil due to which enzymatic activity gets improved. These AW are also involved in the improvement of carbon sequestration in the soils (Oo et al., 2018). Soil physiochemical properties are also enhanced which yields better productivity and yield of crop along with nutritional value improvement (Abujabhah et al., 2016). Farmers have been using organic amendments since long time ago to improve the organic matter content in soil to maintain the healthy fertility status of soil (Scotti et al., 2015). Application of AW in soil significantly improves the soil organic matter and reduces the need of nitrogen and phosphorus application to the soil without affecting the yield (Melo et al., 2018). Coconut husk, woodchips, and orange bagasse have been evaluated and found that orange bagasse biochar enhances nitrogen and phosphorus nutrition. There are numerous positive and negative impacts of AWs on different environs; some are depicted in Figure 2.4.

TABLE 2.1 Agricultural Wastes and Their Impacts on Soil and Plant Health

SL. No.	Agricultural Waste	Impacts on Soil and Plants				References
		Negative		Positive		
		Soil	Plants	Soil	Plants	
1.	Municipal solid waste	Increase the heavy metals bank in soil.	High heavy metals in roots.	Improve nutrient cycling soil enzymes activity.	Growth and yield improvement. Less heavy metals in plants	Meena et al. (2019); Carbonell et al. (2011)
2.	Farmyard manure	Alleviates water stress and heavy metals toxicity in soil	—	Enhances phosphorus use efficiency and increase nitrate concentration.	Increase micronutrients in plants	Andriamananjara et al. (2018); Suthar (2012)
3.	Sugarcane bagasse	Decreased fungal activity.	Impairs nutrient uptake in plants	Decreased in heavy metals and increase in enzymatic activity.	Enhance yield and growth.	Nie et al. (2018); Dotaniya et al. (2016)
4.	Fresh chicken manure	Decrease the antibiotic resistance.	Increases pH and disturbs equilibrium of Ca, Mg.	Improve soil microbial activity and improve biomass production.	Increase the yield.	Urra et al. (2019); Yu et al. (2018)
5.	Crop residues	Increase pest attack.	CR removal causes deleterious effects on soil and nutrient leaching.	Improve the soil physical properties, maintain soil temperature, and prevents any physical injury from raindrops.	Hinder the heavy metal uptake in contaminated soil to crops.	Cherubin et al. (2018)
6.	Animal dead bodies	Continuous application increases heavy metal load in soil.	Heavy metals accumulation in edible portion of crop.	Increase the essential nutrient concentration in soil necessary for plant growth.	Growth and yield are enhanced.	Al-Wabel et al. (2018)
7.	Poultry manure	Prolonged nutrient releases.	Extensive use causes crop burning.	Enhanced nitrate and nutrient contents in soil.	Increase the nutrient concentration in radish and tomatoes.	Aylaj et al. (2018); Adekiya et al. (2019)

TABLE 2.1 (Continued)

SL. No.	Agricultural Waste	Impacts on Soil and Plants				References
		Negative		Positive		
		Soil	Plants	Soil	Plants	
8.	Carbonized poultry manure	Nitrogen in it is not bioavailable for plant.	Available nutrients are not utilized by plants efficiently due to nitrogen deficiency in tissues.	Reduce the mass, pathogens attack, and odor in the soil system.	Enhance biomass, available phosphorus potassium.	Steiner et al. (2018)
9.	Pruning waste	Copper contamination may occur.	–	Reduce N_2O emissions, enhances carbon poll in soil.	Fix carbon, which is long term available to plants	Oo et al. (2018); Duca et al. (2016)
10.	Cotton stalk/cotton stock derived bio-char	Unstable in soil with respect to bio-char	–	Increase the carbon contents of soil significantly.	Enhance the plant growth and yield.	Song et al. (2019)
11.	Crop residues burning fly ash	Suppress the growth of soil fungi.	Increase salinity, TDS, and cations/anions concentration.	Enhance the Ca and OH ions activities in soils. Buffers soil pH.	Contamination risk to plants grown on infected soil.	Shrivastava et al. (2018); Yao et al. (2015)
12.	Maize residue	–	–	Increase soil carbon, plant biomass, and nitrogen.	Increase chlorophyll contents and yield.	Mupangwa et al. (2019)
13.	Wood-ash.	No significant effect on nutrient concentrations of foliar, litter, and stem tissue	–	Surface application of wood ash contributes nutrients to soil.	Efficient nutrient supply.	Petrovský et al. (2018)
14.	Pig manure vermicompost			Increased humic and fulvic acid concentration. Plant nutrients increased.	Increased crop growth and yield.	Atiyeh et al. (2002)

Along with positive effects some constraints are also there, for example, manure application to soil reduces the activity of Mycorrhizal fungi and reduces their root growth (Elzobair et al., 2016). The addition of compost also has some negative effects on plants. Compost addition to soil increases the organic pollutants in soil and enhances the soil electrical conductivity. It also has issue of some heavy metal contamination in soil and hinders the growth of the seed (Gallardo-Lara and Nogales, 1987).

To avoid environmental issues, AW now a day is applied to soil to better plant production. Composts, biochars, and different manure are applied to achieve the ultimate benefit of getting better crop yield (Elzobair et al., 2016). These AW when processed form, are called organic amendments. These amendments supply the nutrients to plants and help them to grow well. These AW derived amendments give the plants micro and macronutrients to support biological and grain yield. These amendments also improve plants' chlorophyll contents, improve water uptake, nutrients status, transpiration rates, and reduce heavy metals uptake in plants (Álvarez-López et al., 2016).

On the other hand, there are some disadvantages that exist when we apply AW on plants in different forms. However, there are not many. The disadvantages of using AW include increased soil pH and calcium/magnesium equilibrium disturbance in the soil also disturbs the plant growth (Urra et al., 2019), increased pest attacks, decreases fungal growth, AW release nutrients very slow and their continuous application may results in the heavy metals buildup in the soil as these amendments and wastes hinder the heavy metals uptake to plant and adsorb them on their surfaces (Aylaj et al., 2018).

2.5 SUMMARY

AWs are produced in bulk worldwide, requiring their efficient management. These wastes are produced from crop production, food crops production, and forestry and livestock animals. These wastes can play an important role in improving soil and plant health. Their use can improve soil physical properties, increase organic matter in soil, water-holding capacity, improve soil aggregation, bulk density, soil carbon sequestration, crop production, and yield. These wastes can be applied to the soil as soil amendments, conditioners, and fertilizers in the form of pig and PMs, CR burial in the soil, compost, and biochars. Other useful and environmentally friendly uses of these wastes are biosorbent production for heavy metals and pesticide remediation from soil and water. Biochar production for improving carbon sequestration and slow-releasing nutrients pools. Further research is needed to explore more efficient and convenient ways for the use of AW in daily life.

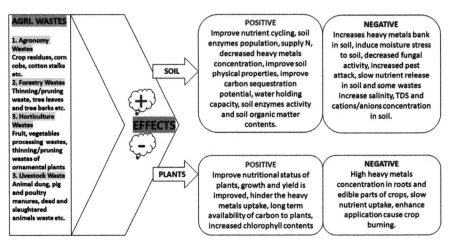

FIGURE 2.4   Types of AW and their effects on soil and plant health.

KEYWORDS

- **agriculture waste**
- **crop residues**
- **poultry manure**
- **soil properties**
- **soil/plant health**
- **waste management**

REFERENCES

Abouelenien, F., Namba, Y., Nishio, N., & Nakashimada, Y., (2016). Dry co-digestion of poultry manure with agriculture wastes. *Applied Biochemistry and Biotechnology, 178*(5), 932–946.

Adekiya, A. O., Agbede, T. M., Aboyeji, C. M., Dunsin, O., & Simeon, V. T., (2019). Biochar and poultry manure effects on soil properties and radish (*Raphanus sativus* L.) yield. *Biological Agriculture and Horticulture, 35*, 33–45.

Alisaraei, T. A., Assar, H. A., Ghobadian, B., & Motevali, A., (2017). Potential of biofuel production from pistachio waste in Iran. *Renewable and Sustainable Energy Reviews, 72*, 510–522.

Alvarenga, P., Mourinha, C., Farto, M., Santos, T., Palma, P., Sengo, J., Morais, M. C., & Cunha-Queda, C., (2015). Sewage sludge, compost and other representative organic wastes as agricultural soil amendments: Benefits versus limiting factors. *Waste Management, 40*, 44–52.

Al-Wabel, M. I., Hussain, Q., Usman, A. R. A., Ahmad, M., Abduljabbar, A., Sallam, A. S., & Ok, Y. S., (2018). Impact of biochar properties on soil conditions and agricultural sustainability: A review. *Land Degradation and Development, 29*(7), 2124–2161. doi: 10.1002/ldr.2829.

Andriamananjara, A., Rakotoson, T., Razanakoto, O. R., Razafimanantsoa, M. P., Rabeharisoa, L., & Smolders, E., (2018). Farmyard manure application in weathered upland soils of Madagascar sharply increases phosphate fertilizer use efficiency for upland rice. *Field Crops Research, 222*, 94–100.

Arshad, M., Bano, I., Khan, N., Shahzad, M. I., Younus, M., Abbas, M., & Iqbal, M., (2018). Electricity generation from biogas of poultry waste: An assessment of potential and feasibility in Pakistan. *Renewable and Sustainable Energy Reviews, 81*, 1241–1246.

Atiyeh, R. M., Lee, S., Edwards, C. A., Arancon, N. Q., & Metzger, J. D., (2002). The influence of humic acids derived from earthworm-processed organic wastes on plant growth. *Bioresource Technology, 84*, 7–14.

Aylaj, M., Lhadi, E. K., & Adani, F., (2018). Municipal waste and poultry manure compost affect biomass production, nitrate reductase activity and heavy metals in tomato plants. *Compost Science and Utilization*, 1–13.

Bakatovich, A., Davydenko, N., & Gaspar, F., (2018). Thermal insulating plates produced on the basis of vegetable agricultural waste. *Energy and Buildings, 180*, 72–82.

Carbonell, G., Imperial, R. M. D., Torrijos, M., Delgado, M., & Rodriguez, J. A., (2011). Effects of municipal solid waste compost and mineral fertilizer amendments on soil properties and heavy metals distribution in maize plants (*Zea mays* L.). *Chemosphere, 85*(10), 1614–1623.

Chen, J., Heiling, M., Resch, C., Mbaye, M., Gruber, R., & Dercon, G., (2018). Does maize and legume crop residue mulch matter in soil organic carbon sequestration? *Agriculture, Ecosystems and Environment, 265*, 123–131.

Chen, Y., Liu, Y., Li, Y., Wu, Y., Chen, Y., Zeng, G., Zhang, J., & Li, H., (2017). Influence of biochar on heavy metals and microbial community during composting of river sediment with agricultural wastes. *Bioresource Technology, 243*, 347–355.

Cherubin, M. R., Oliveira, D. M. D. S., Feigl, B. J., Pimentel, L. G., Lisboa, I. P., Gmach, M. R., Varanda, L. L., et al., (2018). Crop residue harvest for bioenergy production and its implications on soil functioning and plant growth: A review. *Scientia Agricola, 75*, 255–272.

Daza, S. L. V., Solarte, T. J. C., Serna, L., Oaiza, S., Chacón, P. Y., & Cardona, A. C. A., (2016). Agricultural waste management through energy producing biorefineries: The Colombian case. *Waste and Biomass Valorization, 7*(4), 789–798.

Dotaniya, M. L., Datta, S. C., Biswas, D. R., Dotaniya, C. K., Meena, B. L., Rajendiran, S., Regar, K. L., & Lata, M., (2016). Use of sugarcane industrial by-products for improving sugarcane productivity and soil health. *International Journal of Recycling of Organic Waste in Agriculture, 5*(3), 185–194.

Duca, D., Toscano, G., Pizzi, A., Rossini, G., Fabrizi, S., Lucesoli, G., Servili, A., et al., (2016). Evaluation of the characteristics of vineyard pruning residues for energy applications: Effect of different copper-based treatments. *Journal of Agricultural Engineering, 47*(1), 22–27.

El Mekawy, A., Srikanth, S., Bajracharya, S., Hegab, H. M., Nigam, P. S., Singh, A., Mohan, S. V., & Pant, D., (2015). Food and agricultural wastes as substrates for bioelectrochemical system (BES): The synchronized recovery of sustainable energy and waste treatment. *Food Research International, 73*, 213–225.

Gaurav, N., Sivasankari, S., Kiran, G. S., Ninawe, A., & Selvin, J., (2017). Utilization of bioresources for sustainable biofuels: A review. *Renewable and Sustainable Energy Reviews, 73*, 205–214.

Girotto, F., Alibardi, L., & Cossu, R., (2015). Food waste generation and industrial uses: A review. *Waste Management, 45*, 32–41.

Gul, S., Whalen, J. K., Thomas, B. W., Sachdeva, V., & Deng, H., (2015). Physico-chemical properties and microbial responses in biochar-amended soils: Mechanisms and future directions. *Agriculture, Ecosystems, and Environment, 206*, 46–59.

Hatfield, J. L., (2017). *Crops Residue Management*. CRC Press, Boca Raton, Florida, USA.

He, T. G., Su, L. R., Li, Y. R., Su, T. M., Qin, F., & Li, Q., (2018). Nutrient decomposition rate and sugarcane yield as influenced by mung bean intercropping and crop residue recycling. *Sugar Tech., 20*(2), 154–162.

Huang, C., Zeng, G., Huang, D., Lai, C., Xu, P., Zhang, C., Cheng, M., et al., (2017). Effect of *Phanerochaete chrysosporium* inoculation on bacterial community and metal stabilization in lead-contaminated agricultural waste composting. *Bioresource Technology, 243*, 294–303.

Igalavithana, A. D., Lee, S. E., Lee, Y. H., Tsang, D. C. W., Rinklebe, J., Kwon, E. E., & Ok, Y. S., (2017). Heavy metal immobilization and microbial community abundance by vegetable waste and pine cone biochar of agricultural soils. *Chemosphere, 174*, 593–603.

Jahanzad, E., Barker, A. V., Hashemi, M., Eaton, T., Sadeghpour, A., & Weis, S. A., (2016). Nitrogen release dynamics and decomposition of buried and surface cover crop residues. *Agronomy Journal, 108*(4), 1735–1741. doi: 10.2134/agronj2016.01.0001.

James, T. K., Ghanizadeh, H., Harrington, K. C., & Bolan, N. S., (2019). Effect on herbicide adsorption of organic forestry waste products used for soil remediation. *Journal of Environmental Science and Health, Part B.*, 1–9.

Karak, T., Sonar, I., Nath, J. R., Paul, R. K., Das, S., Boruah, R. K., Dutta, A. K., & Das, K., (2015). Struvite for composting of agricultural wastes with termite mound: Utilizing the unutilized. *Bioresource Technolog., 187*, 49–59.

Khaleel, R., Reddy, K. R., & Overcash, M. R., (1981). Changes in soil physical properties due to organic waste applications: A review. *Journal of Environmental Quality, 10*(2), 133–141.

Lee, J., Yang, X., Cho, S. H., Kim, J. K., Lee, S. S., Tsang, D. C. W., Ok, Y. S., & Kwon, E. E., (2017). Pyrolysis process of agricultural waste using CO_2 for waste management, energy recovery, and biochar fabrication. *Applied Energy, 185*, 214–222.

Lee, L. Y., Gan, S., Yin, T. M. S., Lim, S. S., Lee, X. J., & Lam, Y. F., (2016). Effective removal of acid blue 113 dye using overripe *Cucumis sativus* peel as an eco-friendly biosorbent from agricultural residue. *Journal of Cleaner Production, 113*, 194–203.

Liu, N., Charrua, A. B., Weng, C. H. L., Yuan, X., & Ding, F., (2015). Characterization of biochars derived from agriculture wastes and their adsorptive removal of atrazine from aqueous solution: A comparative study. *Bioresource Technology, 198*, 55–62.

Meena, M. D., Yadav, R. K., Narjary, B., Yadav, G., Jat, H. S., Sheoran, P., Meena, M. K., et al., (2019). Municipal solid waste (MSW): Strategies to improve salt affected soil sustainability: A review. *Waste Management, 84*, 38–53.

Mohler, C. L., Taylor, A. G., DiTommaso, A., Hahn, R. R., & Bellinder, R. R., (2018). Effects of incorporated rye and hairy vetch cover crop residue on the persistence of weed seeds in the soil. *Weed Science, 66*(3), 379–385.

Moshia, M., Khosla, R., Westfall, D., Davis, J., & Reich, R., (2016). Precision manure management on site-specific management zones: Nitrogen mineralization. *Journal of Plant Nutrition, 39*(1), 59–70.

Mupangwa, W., Thierfelder, C., Cheesman, S., Nyagumbo, I., Muoni, T., Mhlanga, B., Mwila, M., Sida, T. S., & Ngwira, A., (2019). Effects of maize residue and mineral nitrogen applications on maize yield in conservation-agriculture-based cropping systems of Southern Africa. *Renewable Agriculture and Food Systems*, 1–14.

Murphy, R. P., Montes, M. J. A., Govaerts, B., Six, J., Kessel, V. C., & Fonte, S. J., (2016). Crop residue retention enhances soil properties and nitrogen cycling in smallholder maize systems of Chiapas, Mexico. *Applied Soil Ecology, 103*, 110–116.

Nanda, S., Dalai, A. K., Berruti, F., & Kozinski, J. A., (2016). Biochar as an exceptional bioresource for energy, agronomy, carbon sequestration, activated carbon and specialty materials. *Waste and Biomass Valorization, 7*(2), 201–235.

Nandi, R., Saha, C., Huda, M., & Alam, M., (2017). Effect of mixing on biogas production from cow dung. *Eco-Friendly Agril. J., 10*(2), 7–13.

Ni, H., Han, Y., Cao, J., Chen, L. W. A., Tian, J., Wang, X., Chow, J. C., et al., (2015). Emission characteristics of carbonaceous particles and trace gases from open burning of crop residues in China. *Atmospheric Environment, 123*, 399–406.

Nie, C., Yang, X., Niazi, N. K., Xu, X., Wen, Y., Rinklebe, J., Ok, Y. S., Xu, S., & Wang, H., (2018). Impact of sugarcane bagasse-derived biochar on heavy metal availability and microbial activity: A field study. *Chemosphere, 200*, 274–282.

Nones, D. L., Brand, M. A., Ampessan, C. G. M., & Friederichs, G., (2017). Quantification of agricultural and forestry waste biomass to production of compacts for power generation. *Revista de Ciências Agroveterinária., 16*(2), 155–164.

Oo, A. Z., Sudo, S., Win, K. T., Shibata, A., & Gonai, T., (2018). Influence of pruning waste biochar and oyster shell on N_2O and CO_2 emissions from Japanese pear orchard soil. *Heliyon, 4*(3), e00568.

Petrovský, E., Remeš, J., Kapička, A., Podrázský, V., Grison, H., & Borůvka, L., (2018). Magnetic mapping of distribution of wood ash used for fertilization of forest soil. *Science of the Total Environment, 626*, 228–234.

Prasad, P., Gowda, B., Nalini, B., Ashiwini, G., Prasanna, K., & Kumar, K. R., (2017). Co-digestion of biofuel deoiled cakes with different combinations of cow dung for biogas production and nutrient rich manure. *Int. J. Curr. Microbiol. App. Sci., 6*(11), 3066–3075.

Quiñones, T. S., Retter, A., Hobbs, P. J., Budde, J., Heiermann, M., Plöchl, M., & Ravella, S. R., (2015). Production of xylooligosaccharides from renewable agricultural lignocellulose biomass. *Biofuels., 6*(3/4), 147–155.

Quinteros, E., Ribó, A., Mejía, R., López, A., Belteton, W., Comandari, A., Orantes, C. M., et al., (2017). Heavy metals and pesticide exposure from agricultural activities and former agrochemical factory in a Salvadoran rural community. *Environmental Science and Pollution Research, 24*(2), 1662–1676.

Rajput, P., Sarin, M., Sharma, D., & Singh, D., (2016). Characteristics and emission budget of carbonaceous species from post-harvest agricultural-waste burning in source region of the Indo-Gangetic plain. In: *Air Quality* (pp. 271–292). Apple Academic Press.

Ranaivoson, L., Naudin, K., Ripoche, A., Rabeharisoa, L., & Corbeels, M., (2018). Is mulching an efficient way to control weeds? Effects of type and amount of crop residue in rainfed rice-based cropping systems in Madagascar. *Field Crops Research, 217*, 20–31.

Ravindra, K., Agarwal, N., Kaur, S. M., & Mor, S., (2019a). Appraisal of thermal comfort in rural household kitchens of Punjab, India and adaptation strategies for better health. *Environment International, 124*, 431–440.

Ravindra, K., Kaur, S. M., Mor, S., & John, S., (2019b). Trend in household energy consumption pattern in India: A case study on the influence of socio-cultural factors for the choice of clean fuel use. *Journal of Cleaner Production, 213*, 1024–1034.

Roslim, R., Nordin, A. A., Mat, M. H. C., Zakaria, A., & Jaafar, M. N., (2018). Evaluation of pruning waste of *Mangifera indica* var Harumanis cultivated in greenhouse. In: *MATEC Web of Conference* (p. 6022). EDP Sciences.

Sadeek, S. A., Negm, N. A., Hefni, H. H. H., & Wahab, M. M. A., (2015). Metal adsorption by agricultural biosorbents: Adsorption isotherm, kinetic and biosorbents chemical structures. *International Journal of Biological Macromolecules, 81*, 400–409.

Sagar, N. A., Pareek, S., Sharma, S., Yahia, E. M., & Lobo, M. G., (2018). Fruit and vegetable waste: Bioactive compounds, their extraction, and possible utilization. *Comprehensive Reviews in Food Science and Food Safety, 17*(3), 512–531. doi: 10.1111/1541-4337.12330.

Santos, F. A., Alban, L., Frankenberg, C. L. C., & Pires, M., (2016). Characterization and use of biosorbents prepared from forestry waste and their washed extracts to reduce/ remove chromium. *International Journal of Environmental Science and Technology, 13*(1), 327–338.

Sfez, S., Meester, D. S., & Dewulf, J., (2017). Co-digestion of rice straw and cow dung to supply cooking fuel and fertilizers in rural India: Impact on human health, resource flows and climate change. *Science of the Total Environment, 609*, 1600–1615.

Shakoor, M. B., Niazi, N. K., Bibi, I., Murtaza, G., Kunhikrishnan, A., Seshadri, B., Shahid, M., et al., (2016). Remediation of arsenic-contaminated water using agricultural wastes as biosorbents. *Critical Reviews in Environmental Science and Technology, 46*(5), 467–499.

Sharifi, Z., Hossaini, S. M., & Renella, G., (2016). Risk assessment for sediment and stream water polluted by heavy metals released by a municipal solid waste composting plant. *Journal of Geochemical Exploration, 169*, 202–210.

Shrivastava, S., Mahish, P. K., & Ghritlahare, A., (2018). Effect of industrial fly ash on the growth of some crop field soil fungi adapted with ash content. *International Journal of Agriculture, Environment and Biotechnology, 11*(1), 203–207.

Singh, P., Heikkinen, J., Ketoja, E., Nuutinen, V., Palojärvi, A., Sheehy, J., Esala, M., et al., (2015). Tillage and crop residue management methods had minor effects on the stock and stabilization of topsoil carbon in a 30-year field experiment. *Science of the Total Environment, 518, 519*, 337–344.

Song, X., Li, Y., Yue, X., Hussain, Q., Zhang, J., Liu, Q., Jin, S., & Cui, D., (2019). Effect of cotton straw-derived materials on native soil organic carbon. *Science of the Total Environment, 663*, 38–44.

Speratti, A. B., Johnson, M. S., Martins, S. H., Nunes, T. G., & Guimarães, C. E., (2017). Impact of different agricultural waste biochars on maize biomass and soil water content in a Brazilian cerrado arenosol. *Agronomy, 7*(3), 49.

Steiner, C., Harris, K., Gaskin, J., & Das, K. C., (2018). The nitrogen contained in carbonized poultry litter is not plant available. *In Open Agriculture, 3*, 284.

Sun, J., Peng, H., Chen, J., Wang, X., Wei, M., Li, W., Yang, L., et al., (2016). An estimation of CO_2 emission via agricultural crop residue open field burning in China from 1996 to 2013. *Journal of Cleaner Production, 112,* 2625–2631.

Suthar, S., (2012). Impact of vermicompost and composted farmyard manure on growth and yield of garlic (*Allium stivum* L.) field crop. *International Journal of Plant Production, 3*(1), 27–38.

Thines, K. R., Abdullah, E. C., Mubarak, N. M., & Ruthiraan, M., (2017). Synthesis of magnetic biochar from agricultural waste biomass to enhancing route for waste water and polymer application: A review. *Renewable and Sustainable Energy Reviews, 67,* 257–276.

Thu, A. K., Thein, S. S., Myint, A. K., & Toe, K., (2016). Effectiveness of crop residues mulching on water use efficiency and productivity under different annual cropping patterns of Magway. *Journal of Agricultural Research, 3*(2), 1–6.

Tran, H. N., You, S. J., Nguyen, T. V., & Chao, H. P., (2017). Insight into the adsorption mechanism of cationic dye onto biosorbents derived from agricultural wastes. *Chemical Engineering Communications, 204*(9), 1020–1036.

Turmel, M. S., Speratti, A., Baudron, F., Verhulst, N., & Govaerts, B., (2015). Crop residue management and soil health: A systems analysis. *Agricultural Systems, 134,* 6–16.

Urra, J., Alkorta, I., Lanzén, A., Mijangos, I., & Garbisu, C., (2019). The application of fresh and composted horse and chicken manure affects soil quality, microbial composition and antibiotic resistance. *Applied Soil Ecology, 135,* 73–84.

Väisänen, T., Haapala, A., Lappalainen, R., & Tomppo, L., (2016). Utilization of agricultural and forest industry waste and residues in natural fiber-polymer composites: A review. *Waste Management, 54,* 62–73.

Valdivia, M., Galan, J. L., Laffarga, J., & Ramos, J. L., (2016). Biofuels 2020: Biorefineries based on lignocellulosic materials. *Microbial Biotechnology, 9*(5), 585–594.

Wu, Y., Wang, C., Liu, X., Ma, H., Wu, J., Zuo, J., & Wang, K., (2016). A new method of two-phase anaerobic digestion for fruit and vegetable waste treatment. *Bioresource Technology, 211,* 16–23.

Xiao, Q., Saikawa, E., Yokelson, R. J., Chen, P., Li, C., & Kang, S., (2015). Indoor air pollution from burning yak dung as a household fuel in Tibet. *Atmospheric Environment, 102,* 406–412.

Yao, Z. T., Ji, X. S., Sarker, P. K., Tang, J. H., Ge, L. Q., Xia, M. S., & Xi, Y. Q., (2015). A comprehensive review on the applications of coal fly ash. *Earth-Science Reviews, 141,* 105–121.

Yu, C. H., Wang, S. L., Tongsiri, P., Cheng, M. P., & Lai, H. Y., (2018). Effects of poultry-litter biochar on soil properties and growth of water spinach (*Ipomoea aquatica* Forsk.). *Sustainability, 10*(7), 2536.

Zhang, L., Zeng, G., Dong, H., Chen, Y., Zhang, J., Yan, M., Zhu, Y., et al., (2017). The impact of silver nanoparticles on the co-composting of sewage sludge and agricultural waste: Evolutions of organic matter and nitrogen. *Bioresource Technology, 230,* 132–139.

Zhao, C., Lv, P., Yang, L., Xing, S., Luo, W., & Wang, Z., (2018). Biodiesel synthesis over biochar-based catalyst from biomass waste pomelo peel. *Energy Conversion and Management, 160,* 477–485.

Zhou, Y., Zhang, L., & Cheng, Z., (2015). Removal of organic pollutants from aqueous solution using agricultural wastes: A review. *Journal of Molecular Liquids, 212,* 739–762.

Zhu, L. D., & Hiltunen, E., (2016). Application of livestock waste compost to cultivate microalgae for bioproducts production: A feasible framework. *Renewable and Sustainable Energy Reviews, 54,* 1285–1290.

Zhu, L. D., Li, Z. H., Guo, D. B., Huang, F., Nugroho, Y., & Xia, K., (2017). Cultivation of *Chlorella* sp. with livestock waste compost for lipid production. *Bioresource Technology, 223,* 296–300.

CHAPTER 3

Impact of Agriculture on Soil Health

MANSHA NISAR

*Department of Environmental Sciences, Sri Pratap College,
Srinagar – 190001, Jammu and Kashmir, India,
E-mail: manshanisar@gmail.com*

ABSTRACT

Agriculture is the main source of producing food by raising various crops on natural soil resources. The rising population requiring more food as a basic need has brought about intensification in the agricultural process world over – the main target of all agricultural practices being increased food production. In the process, the importance of soil resource which is much more than just food production gets neglected and soil health suffers from continuous unsustainable cultivation of crops. Farming practices of nowadays focusing more on this food production in terms of quantity, considering soil merely for producing more and more food has detrimental effects on this crop supporting resource the soil itself. Unsustainable agricultural activities have a negative effect on soil through changes in important soil health attributes, both physicochemical and biological ones. Agricultural activities affect soil health and productivity by processes of erosion, compaction, and salinization. The major practices disturbing soil health include mechanical tillage, mono-cropping, application of agrochemicals and irrigation with waste and industrial waters. All these disturb vital soil attributes like aggregate stability, pH, organic matter levels, microbial parameters, and soil fauna. Ultimately this leads to reduced soil fertility and decreased crop yields. To overcome this problem faced by agricultural soils, it is very important to understand how diverse agricultural practices impact soil health attributes. This will help to identify and develop best soil management sustainable agricultural practices. This study is a step towards it. Finally, these practices will only help us achieve the goal of increased good quality food production and proper soil management.

3.1   INTRODUCTION

Soil, the uppermost part of the lithosphere of earth, is the basic and a vital resource that is critically important for the future of the human population on this planet by functioning in the world food production. Healthy soils form an essential basis of ability of humans to feed themselves effective by being the basic source of agricultural production. However, it is the pressure to increase this agricultural production to feed the growing world population that impacts soil health. The deterioration of soil in terms of its health in an area is a serious cause of concern for human health and wellbeing. The extent of damage to soil health depends on both the area's soil characteristics and agriculture peculiarities. In fact, the extent of impact to soil health by agriculture in any part of the world depends on the type of agricultural practice followed. As per Zalidis et al. (2002), "the overall detrimental effects of agricultural practices on soil quality include erosion, desertification, salinization, compaction, and pollution." There is evidence in the form of studies for widespread agricultural support system degradation, i.e., soils by increasing intensive agricultural practices world over. Soil health is basically defined in terms of its physical, chemical, and biological characteristics-which act as indicators for soil health assessment. Poor understanding of these soil characteristics has negative ramifications related to agricultural yields (quality and quantity). Therefore, utmost importance has been given to detailed and intensive understanding of these key indicators for grasping the real extent of the impact that agriculture has on soil health (Ahamadou and Huang, 2013). The continuing degradation of agricultural soils the world over due to increasing population and consequently rising demand for food is the factor that triggers interest in studying how agriculture can impact soil. As per data by World Resource Institute (1998), "a third of Earths soil is acutely degraded due to agriculture in the form of poor soil management and unsustainable agricultural practices." Also as per the 90s data, "by 1990 poor agricultural practices had contributed to the degradation of 38% of the roughly 1.5 billion ha of cropland worldwide and since 1990 the losses have continued at a rate of 5–6 million ha annually." The loss of soil from agricultural areas is reported to be 10 to 40 times faster than the rate of soil formation imperiling food security (Pimental and Burgess, 2013). This continuing agricultural soil degradation, as revealed by data, calls for having a detailed understanding of impact of agriculture on soil health for improving soils. The chapter derives a sound concept of soil health in terms of its key indicators by reviewing the recent studies on these indicators as impacted by various agricultural practices adopted worldwide.

3.2 SOIL HEALTH AND ITS INDICATORS

Despite being a commonly used term in studies pertaining to agricultural practices and several definitions being proposed, a standard definition of soil health does not exist as such. The reason being its close association with the term "soil quality" and discussions whether the terms are synonyms or not, and they can be used interchangeably. The term soil health being introduced in the mid-1990s, with some studies using soil quality and health synonymously (Acton and Gregorich, 1995; Allen et al., 2011) and others differentiating them by including ecological attributes in soil health (Pankhurst, 1997). Many earlier studies reviewed which elaborate on the difference in the two terms consider soil quality as "being related to soil functions" (Kibblewhite et al., 2008) and soil health being an "integrative property presenting soil as a finite and dynamic living resource" (Doran and Parkin, 1996; Doran and Zeiss, 2000) with soil health focusing more on biotic components (Anderson, 2003). The estimation of soil conditions-mainly soil health through a set of specific indicators consisting of physical, chemical, and biological characteristics of soil is considered a reductionist approach to analyzing soil health (Doran and Jones, 1996; Van Camp et al., 2004; Kibblewhite et al., 2008). When studying how soil is affected by agricultural practices, considering soil health from this reductionist point of view appears to work. This so because of two reasons: it is an approachable and practical way of assessing soil conditions. Secondly, it considers soil in the background of sustainable agricultural practices. Thus, this concept gives a proper definition of soil health clearly defining soil in terms of ecological attributes, especially the biotic ones. According to Pankhurst et al. (1997), "the concept of soil health includes the soil's ecological attributes, which have implications beyond its quality on its capacity to produce particular crops and these attributes are chiefly associated with the soil biota, diversity, food web structure." Therefore, this approach of defining soil health is noteworthy in terms of consideration of agricultural impact on soil as in the current phase of increasing agricultural production through use of modern techniques, it's this vital attribute that is being ignored. The consequence being increased degradation of agricultural soils.

3.2.1 INDICATORS

Studies on the impact of agriculture on soil health gauging whether the agricultural practice maintains or degrades soil stature depend on how specific soil characteristics or properties change over time due to these agricultural practices. Thus, the understanding of this change depends on these quantifiable physical, chemical, and biological properties of soil, which are soil health indicators. In fact, studies have clearly stated that we "cannot evaluate soil health directly, it can be deduced from these quantifiable attributes-the indicators of soil health which are influenced by land use and soil management practices" (Sanchez-Maranon et al., 2002; Shukla et al., 2006). According to Allen et al. (2011) and recently Patil and Lamnganbi (2018) indicators of soil health are defined as "composite set of measurable physical, chemical, and biological attributes which relate to functional soil processes and can be used to evaluate soil health status, as affected by various management practices including agriculture." The choice of an attribute for identifying the good and key indicators of soil health is very important. To recognize and point out good and key indicator in terms of soil health assessment, it is very important to consider such an attribute or soil indicator that is easily and reproducibly measurable (Yemefack et al., 2006), very responsive to changes in management practices and provides ability of easy and practical use to agricultural specialists-as has also been suggested by several researchers (Doran and Parkin, 1996; Erkossa et al., 2007). Though for various agricultural management systems a fundamental set of soil health assessment indicators has been put forward but its problem is being beyond the expertise of the producers persists (Hamblin, 1991). Romig et al. (1995) suggested best way of identifying "key indicators of soil health by simply choosing the practical ones which have meaning to farmers and other land managers thus also proving fruitful in the assessment of sustainability of any soil management practice." According to Islam and Weil (2000), "soil properties that are neither so stable as to be insensitive to management, nor so easily changed as to give little indication of long term alterations are best for assessment of soil health."

3.2.1.1 PHYSICO-CHEMICAL INDICATORS

Various soil physical and chemical properties linked to the specific soil function of producing food in terms of affecting crop yields and growth are used to specify soil health in agroecosystems to assess the overall health of agricultural soils. These properties of soil are considered to be very important in terms of affecting crop growth and being easily affected by soil management practices (Mairura et al., 2007). Among them, key soil health physicochemical indicators in relation to agricultural systems have been identified in several studies (Gregorich et al., 1994; Herrick and Whitford, 1995; Aggelides and Londra, 2000; Arshad and Martin, 2002; Dexter, 2004; Sparling et al., 2004; Weil and Magdoff, 2004; Arias et al., 2005; Celik, 2005; McVay et al., 2006; Pattinson et al., 2008; Qi et al., 2009; Thierfelder and Wall, 2010; Oliver et al., 2013; Ghaemi et al., 2014; Zornoza et al., 2015; Takoutsing et al., 2016). Many of these studies have also reviewed the interrelations between these soil indicators in agroecosystems (Arshad and Martin, 2002; Karlen et al., 2003; Laishram et al., 2012; Zornoza et al., 2015). In understanding the impact of agriculture on soil, those studies were reviewed, which have linked soil properties to different agricultural practices and identified the potential soil health indicators accordingly. The main physicochemical indicators, which provide vital information about soil health, thus proving to be crucial in soil health assessment studies related to agricultural systems as reviewed through various studies, are given in Table 3.1.

TABLE 3.1   Important Physico Chemical Indicators Used for Soil Health Assessment in Agro Ecosystems

Soil health Indicator	Role in Agricultural System (Soil Processes Influenced)	Management Practice Impacting the Indicator	References
Aggregate stability	Determines-rooting depth, soil water distribution movement and retention, nutrient recycling, aeration and microbial activity	Tillage Crop rotation Organic farming	Pardo et al. (2000); Albiach et al. (2001); Chan et al. (2003); Six et al. (2004); Moebius et al. (2007); Riley et al. (2008); Rimal and Lal (2009); Williams and Petticrew (2009); Gomiero et al. (2011); Aziz et al. (2013)
Bulk density	Monitors-soil compaction	Tillage Organic amendment	Carter (1990); Hakansson and Lipiec (2000); Reynolds et al. (2002); Raper (2005); Tejeda et al. (2006); Pattison et al. (2008); Chaudhari et al. (2013); Ogwo and Ogu (2014)
Available water holding capacity	Soil water availability and movement, drainage rate, Water air balance	Tillage Irrigation Organic farming	Arias et al. (2005); Carter (2007); Jarvis (2007); Idowu et al. (2009); Reynolds et al. (2009); O'Farrell et al. (2010); Gupta et al. (2010); Geisseler et al. (2011); Malik et al. (2014)

TABLE 3.1 *(Continued)*

Soil health Indicator	Role in Agricultural System (Soil Processes Influenced)	Management Practice Impacting the Indicator	References
pH	Nutrient availability, Microbial activity and crop performance	Organic farming Crop rotation Monoculture	Pattison et al. (2008); Rahman et al. (2008); Gil et al. (2009); Ge et al. (2011); Swiatkiewicz and Gastol (2012); Bai et al. (2018)
Electrical conductivity	Plant and microbial activity thresholds	Tillage	Smith et al. (2002); Arnold et al. (2005); Corwin and Lesch (2005); Gil et al. (2009); Patel (2015)
Available nutrients	Capacity for growth and yield, productivity	Tillage Organic farming Fertilizer use	Compton and Boone (2000); Zalidis et al. (2002); Weil and Magdolf (2004); Zhang et al. (2006); Cantarella (2007); Uzoma et al. (2011); Sousa and Figueiredo (2016)

3.2.1.2 BIOLOGICAL INDICATORS

Biological indicators are most important in assessing soil health under changing agricultural practices as, "soil health is defined only by focusing more on the biotic components" (Anderson, 2003). In addition, many processes in soil are driven by the biota present in the soil (Ritz et al., 2009). Furthermost definitions differentiating the terms soil quality and soil health have laid emphasis on the biological component being the one primarily determining soil health. While defining soil health, scientists and researchers have laid emphasis on "ecological attributes of the soil-including mainly soil biota, biodiversity, and food web structure of soil, microbial activities with their involvement beyond just in the function of food production by soil" (Pankhurst et al., 1997). An assortment of physical, chemical, and biological indicators has already been suggested, which can be used to evaluate soil health (Doran and Parkin, 1994). However, using biological indicators for soil health assessment has the edge over using physicochemical indicators of providing consolidated information related to many environmental factors. Furthermore, assessment through biological indicators is especially relevant to studying agricultural impacts as it helps to evaluate "the key elements of soil health effected by various agricultural practices" (Pandolfini et al., 1997). Reviewing literature reveals that many biological indicators for soil health assessment have been proposed each with their own virtues and limitations – "biological indicators typically include microbial biomass and enzyme activities, soil carbon, abundance of soil fauna" (Jordan et al., 1995; Karlen et al., 1997; Staben et al., 1997; Bandick and Dick, 1999; Eivazi et al., 2003). As per studies, "physicochemical indicators of soil health have a slow response as compared to the biological ones" (Cardoso et al., 2013). Also, on the other side, evaluation of biological-microbial level indicators like biomass, decomposition rates, soil enzymes, and soil organic matter helps in rapid soil health assessment (Visser and Parkinson, 1992). Considering the role of these biological indicators, especially in food production for agriculture, the key biological indicators mostly studied to understand the impact of agriculture on soil health include soil organic matter and carbon, microbial biomass and respiration, enzyme activity in soil fauna.

Among the biological indicators of soil health, "agreeably, the single most significant one is considered to be soil organic matter" (Robinson et al., 1994). In fact, soil organic matter has been recognized as a "central indicator of soil health" by the "Soil and Water Conservation Society" in 1995. Thus organic matter can be considered as an important indicator and key determinant for soil health assessment, particularly in the case of agroecosystems because of its significant role in vital functions of soil like nutrient cycling (sink and source of carbon, nitrogen, and regulation of sulfur and phosphorus cycling), biological activity (providing substrate for microbes) and crop productivity (Karlen et al., 1997; Saggar et al., 2001; Haynes, 2008; Murphy, 2015). It has been revealed that agricultural management practices strongly affect the pool of this principal soil health indicator (Doran et al., 1998; Farquharson et al., 2003). As a driver of main soil functions of microbial activity and crop productivity in agroecosystems, decreases in soil organic matter can lead to decrease in microbial diversity and soil fertility thus deteriorating the overall health of soil in agroecosystems. It has also been reported that decline in soil organic matter deteriorates soil physical characteristics causing loss of soil structure and compaction ultimately decreasing crop yields (Weil and Magdoff, 2004). Taking into account these impacts of decreased organic matter in soil on crop productivity and health, researchers have suggested "proper management of this biological indicator to be at the heart of sustainable agriculture" (Weil, 1992). Soil organic matter has been assessed in various studies related to decreasing crop yields in agroecosystems and the impact of agriculture on overall health of soil (Haynes and Naidu, 1998; Carter, 2002; Ding et al., 2002; Jarecki and Lal, 2003; Celik, 2005; Fliebach et al., 2007; Abiven et al., 2009; Bai et al., 2018; Sarker et al., 2018).

Considering soil health assessment, organic carbon is among the important and key biological indicators in agroecosystems as agricultural practices are known to impact soil organic carbon reserves (Bruce et al., 1999). Soil organic carbon is a liable and vital indicator of soil health. Other important microbial biological indicators are dependent on it being the energy source for them. The main reasons why soil carbon content is important in agricultural soils are that carbon is transferred from soil to atmosphere and regulates nutrient supply and microbial activity (West and Marland, 2002; McLauchlan, 2007). Many studies have documented change in soil organic carbon in agroecosystems revealing that "intensive agriculture has an impact on this biological indicator with reduction in its pool over the years following cultivation" (Compton and Boone, 2000; Murty et al., 2002; Entry et al., 2004; Al-Kaisi et al., 2005; Ogle, 2005; McLauchlan, 2006; Luo et al., 2010; Cusack et al., 2013).

Microbial indicators basically an assortment of various microbial parameters are an "excellent indicator of change in soil health" (Kennedy and Papendick, 1995). This is so because compared to physico-chemical indicators of soil health microbial parameters are more susceptible and quick respondents to change inflicted upon the soil environment, thus forecasting any change in soil environment and giving an integrative measure of soil health (Masto et al., 2009). It has been established and widely accepted that "changing agricultural management practices change soil biota and thus have a significant effect on the soil microbial parameters" (Stark et al., 2007) and so a diverse set of microbial parameters are commonly used as biological indicators for soil health assessment in agricultural systems. The most commonly and widely used among them include "microbial biomass and soil respiration" (Bastida et al., 2006). Microbial biomass apart from being the most widely used biological indicator of soil health in agricultural systems also serves as a "promising indicator of soil health due to its rapid response to changes in soil use and management" (Nogueira et al., 2006). It has also been considered as a "sensitive indicator of changes in soil processes" (Saha and Mandal, 2009). Researchers (Smith and Paul, 1990; Sparling, 1997) consider "microbial biomass as a very useful tool for understanding changes in soil properties and consequently in determining the extent of soil health deterioration." In spite of being a promising tool and commonly used indicator in soil health assessment studies, there is "unavailability of its benchmark values" (Dalal, 1998), which proves to be a limitation in the use of this biological indicator. Looking from agricultural perspective in terms of food production, it's this microbial biomass, the living component of soil associated with soil organic matter, that provides nutrients for crop growth through cycling, so holding more importance in agricultural soils. Several studies have laid emphasis on assessing soil microbial biomass and have evaluated the change in this indicator under different agricultural practices (Balota et al., 2003; Livia et al., 2005; Brem Pong et al., 2008; Yusuf et al., 2009; Gosai et al., 2010; Chakraborty et al., 2011; Bowles, 2014; Amaral and Abelho, 2016; Maharjan, 2017; Sheoran et al., 2018). These studies have found that the main agricultural management practices to which this biological indicator responds include tillage practices, crop rotation, addition of manures, and fertilizers. Another commonly used biological indicator of soil health is soil respiration rate. This rate is important in case of soil health assessment as it "specifies the complete range of biological activity of soil microbes" (Doran et al., 1996). Like other biological parameters, this rate too is influenced by changing agricultural practices (Vakali et al., 2011; Yazdanpanah et al., 2016).

Another crucial biological indicator associated with microbial genesis, which plays a major role in nurturing soils, is soil enzymes. Soil enzymes are a class of enzymes occupying soil and the appraisal of activity of these resident enzymes provides a simple assessment of status of soils. They have been reported to be "useful soil quality indicators due to their vital role in soil ecology and fertility, being a practical, sensitive, and integrative tool for soil assessment and management due to their quick response to changes in soil characteristics" (Ruiz et al., 2009; Utobo and Tewari, 2015). Studies do provide a detailed understanding of the changes in the activity of various soil enzymes like amylases, dehydrogenases, phosphatases, and urease brought about by agricultural practices like cropping ones and excessive use of agrochemicals (Aon et al., 2001; Klose and Ajwa, 2004; Gianfreda et al., 2005; Alotaiba and Schoenau, 2011; Utobo and Tewari, 2015; Rao et al., 2017).

3.3 HOW AGRICULTURAL PRACTICES EFFECT SOIL HEALTH?

Present-day agricultural practices with their sole aim of maximizing yields pose a serious threat to soil health. With this aim, the dramatic agricultural changes that have been taking place in recent years constitute the predominant change of the century, intensive production in terms of "increased use of agrochemicals and commercial seeds being the major trends" (Tomich et al., 2011). This major change of the century has manifested itself in the form of massive expansion and intensification of agriculture over just a span of few years. This is a predominant change and a predominating cause of environmental degradation through its impact on soil. All these production maximizing agricultural practices alter soil mainly negatively by changing the essence of soil health—the key indicators. Among these practices that lead to mismanagement and ultimately degradation of soil are tillage, short rotation, irrigation (leading to soil salinization) and a tendency to adopt monoculture rather than crop diversity (Montgomery, 2007; Ponting, 2007). Today, the abandonment of traditional techniques with modern hallmark techniques of agriculture – the intensive-based production management ones has led to major soil degradation. These techniques have the capability to change soil profiles from inherent to anthropogenic (Zalidis et al., 2002; Borselli et al., 2006). Modern agriculture especially the current agricultural practices followed intensively to maximize yields has certain main features like selection of specific crop varieties (involving genetic modifications and hybrids), maintenance of nutrients and fertility of soils to desired levels through use of chemical fertilizers, effective control of pests by use of chemicals. These features increase agricultural production today which from an environmental perspective are main polluters of soil disturbing overall soil health. On reviewing various studies, the main practices currently in use that significantly impact the health of agricultural soils were found to be tillage involving use of machinery, mono-cropping, excessive use of agrochemicals and various irrigation practices. These agricultural practices impact soil through processes like compaction, erosion, and salinization.

3.3.1   PROCESSES INVOLVED IN THE IMPACT

3.3.1.1   TILLAGE EROSION

The main agricultural practices that degrade soil health operating through acceleration of soil erosion include tillage and certain cropping practices. These practices disturb soil structure cause soil redistribution and soil compaction contributing to increased soil erodibility. Studies have identified tillage erosion "as a major process of soil redistribution on agricultural land, as a result increasing soil susceptibility to agents of erosion like water leading to net soil loss" (Lindstrom et al., 1992; Lobb et al., 1995). Therefore, agricultural tillage is capable of causing soil erosion by action of water. Further evidence is available in literature about the escalation of this tillage induced erosion in soils through mechanized agriculture which makes use of heavy and large sized tilling tools increasing tillage depth and speed so causing more losses to soil (Van Muysen et al., 2002; Van Oost et al., 2005). On one hand, it becomes clear that modern mechanized agriculture with heavy machinery tillage involvement accelerates soil erosion. On the other hand, it has also been established that minimum till or no till practices reduces soil erosion effectively. Tillage erosion has gained importance in recent years (Czubaszek and Czubaszek, 2014) in terms of being the major operating system of soil erosion by water as it concentrates surface water runoff in certain convergent areas or patches of the field due to redistribution of soil. Through all this tillage erosion ultimately impacts agricultural soils and hence crop yield and development (Heckrath et al., 2005). Related to tillage erosion in agriculture studies have concluded that "shifting cultivation to more permanent highland cropping system leads to an increase in tillage erosion and intensity" (Turkelboom et al., 1999; Rybicki et al., 2016; Nguyen and Pham, 2018).

3.3.1.2   SOIL SALINIZATION

The process of salinization leading to agricultural soil losses and limiting productivity of crops is a serious problem arising from improper irrigational practices in agricultural fields. Arid and semi-arid soils of world are more susceptible to degradation through salinization. Salinization associated with increasing soil conductivity is a negative effect which irrigated agriculture has on soil health. Salinization process in agriculture is related to "build-up of excessive salts in the soil due to its inherent saline nature or because of improper or insufficient drainage condition of soil" (Singh, 2015). Recently, Machado, and Serralheiro (2017) reported that "process of soil salinization has accelerated because of large scale intensive farming associated with extensive irrigation of fields." However, present times necessitate the extensive development of irrigated agriculture, due to pollution and scarcity of water sources, along with population pressure. However, the use of improper high salt concentration wastewater for this purpose impacts soil negatively by risking it to salinization. This irrigation-induced soil salinization has negatively affected crop yields by deteriorating soil health ultimately decreasing overall agricultural productivity (Houk et al., 2006; Endo et al., 2011; Singh, 2015; Machado and Serralheiro, 2017). Estimates regarding area of irrigated agricultural lands salinized range from 30% to 50% in various parts of world (Flowers, 1999; Hillel, 2000; Ngigi, 2002; Gulzar et al., 2005).

3.3.2   IMPACT OF DIFFERENT TYPES OF TILLAGE ON SOIL HEALTH

Among tillage practices followed on agricultural lands, mechanical tillage, a regular one involving use of machinery for tilling, can negatively impact soil. Looking at the history of agricultural intensification, the adoption of mechanical tillage has increased efficiency of agricultural operations. However, at the same time looking at the recent record of reduced or no-tillage practices gaining importance and at comparison between the two many negative impacts of this tillage have come to light. The benefits associated with reduced-tillage in terms of soil protection from erosion with involvement of low costs are the reason no or zero till practice in agriculture fields is gaining importance with its wider adoption world over. "Many parts of the world, particularly North and South America, are reverting to reduced tillage practices" (Landers et al., 2001). In fact, "reduced tillage practices are increasing worldwide" (El Titi, 2003). Literature highlights the main benefits of this tillage being the reason as "soil and water conservation with reduced requirement of equipment, fuel, and labor" (Holland, 2004; Vogeler et al., 2006). Many studies evaluating the effect of intensive soil tilling on overall soil structure clearly reveal that "tillage practices disrupt soil struc-ture by destroying stable aggregates and promoting soil erosion, thereby resulting in overall deterioration of soil health" (Volk et al., 2004; Riley et al., 2008; Williams and Petticrew, 2009; Babujia et al., 2010; Vakali, 2011). Also higher rates of soil carbon loss have been reported in these studies under intensive tillage in comparison to no-tillage. Considering the impact of various tillage practices on soil structure and health, almost all types of practices impact soil's physical properties like density, aggregation, and soil pores negatively. However, certain tilling practices like conventional tillage are more hostile or aggressive in terms of disturbing soil structure consid-erably and also negatively affecting soil biological properties "decreases organic matter, loss of microbial communities" (Grace et al., 1994; Gupta et al., 1994; Bayer et al., 2001; Bailey and Lazarovits, 2003). While hostile practices like the conventional tillage "adversely affect long term soil productivity due to erosion and loss of organic matter in soils" (Mathew et al., 2012), less hostile ones the " minimum tillage" (Sun et al., 2010) and "no-tillage" (Balota et al., 2003; Babujia et al., 2010) have also been identi-fied for agricultural soil management. Table 3.2 highlights studies related to the effect of various tillage practices on soil health's physical and microbial indicators.

TABLE 3.2 Studies Revealing Comparative Impact of Tillage Practices on Soil Health

Tillage Practice	Soil Health Indicator	Effect	References
Conventional	Aggregate stability	Negatively	Kushwaha et al. (2001); Mikha and Rice (2004); Fontaine and Barot (2005); Fontaine et al. (2007); Jacobs et al. (2010); Salem et al. (2015)
	Water holding capacity	Negatively (lower)	Rasmussen (1999); Osunbitan et al. (2005); Bhattacharyya et al. (2006); Jacobs et al. (2010); Sharma et al. (2011)
	Soil organic carbon	Negatively (significant decrease)	Chan et al. (2002); Roscoa and Burman (2003); Wang et al. (2004); Bronick and Lal (2005); Bunemann et al. (2008); Poirier et al. (2009)
	Microbial biomass	Negatively	Alvear et al. (2005); Laudicina et al. (2011); Zuber and Villamil (2016)
Zero/ Minimum	Aggregate stability	Positively (greater soil stability rates and bulk density values)	Kushwaha et al. (2001); Liebig et al. (2004); Bhattacharyya et al. (2006); Jacobs et al. (2009); Moraru and Rusu (2010); Slawinski et al. (2012); Salem et al. (2015); Basir et al. (2017)
	Water holding capacity	Positively (faster infiltration rates)	Tebrugge and During (1999); Pagliai et al. (2000); Liebig et al. (2004); Fabrizzi et al. (2005); Bhattacharyya et al. (2006); Moraru and Rusu (2010); Sharma et al. (2011); Basir et al. (2017)
	Soil organic carbon	Positively (significant increase)	Pankhurst et al. (2000); Zibilske et al. (2002); Al-Kaisi et al. (2005); Rahman et al. (2007); Jacobs et al. (2009); Moraru and Rusu (2010); McLeod et al. (2013); Basir et al. (2016)
	Microbial biomass	Positively (greater amounts)	Alvarez and Alvarez (2000); Balota et al. (2003); Liebig et al. (2004); Alvear et al. (2005); Laudicina et al. (2011); Zuber and Villamil (2016)

So on reviewing the widespread and detailed studies pertaining to impact of tillage practices on soil heath one significant point that emerges with regard to the impact of tillage is that plowing in tillage negatively impacts soil health by modification of physical indicators (soil structure and aggregate stability) and loss of microbial characteristics (soil biota and microbial biomass) and soil carbon. Also, it becomes clear that though practices of zero or reduced tillage are considered to be the less aggressive positive impact ones-improving attributes associated with soil health like the physical ones water holding capacity through reducing runoff and increasing infiltration and storage and the microbial ones too by enhancing soil biological activities, but negatively these practices increase weed development by retention of residues and risks crops to various diseases and pests. This way, the zero or no-tillage practice gaining importance in agriculture is giving rise to another problem, the increased use of agrochemicals to fight weeds and pests. As per studies, "the greatest disadvantage of farming involving reduced or no-tillage at all is that it delays nitrogen mineralization in soils thereby leading to reduced crop development while benefiting growth of weeds" (Pekrun et al., 2003; Peigne et al., 2007; Berner et al., 2008). Experimental field studies over a time span have also clearly demonstrated that "reduced tillage has positive effects on soil microbial parameters through increased amounts and quality of soil organic matter" (Kandeler et al., 1999; De Souza Andrade et al., 2003), but its negative effects are visible in the form of "increased need for more herbicide application due to an increased weed appearance" (Dieke et al., 2008).

3.3.3   CROPPING PRACTICES AND THEIR IMPACT ON SOIL HEALTH

Increasing specialization in production process is an important aspect of current phenomena of agricultural intensification. This causes a reduction in the number of crop species being maintained ultimately resulting in monoculture or continuous cropping. According to scientists (Tilman, 1999)- "monocultures lower soil productivity and crop yields with higher vulnerability to attacks by weeds and pests." Studies have shown that all of this does occur under mono-cropping practice in agricultural fields (Acosta-Martinez et al., 2004; Gajda and Martyniuk, 2005; Nayyar et al., 2009). Crop type is an important consideration in monoculture practice that greatly influences its impact on soil function and overall health. Studies have revealed that "in comparison to continuous cropping of corn and soya bean which reduces soil organic carbon and nitrogen along with decline in soil biological activity" (Liu et al., 2005), "continuous cropping of wheat has lesser impact" (Mikhailova et al., 2000). Thus clearly the extent to which mono-cropping impacts soil health is influenced by type of crop grown continuously. In general monoculture has well known negative impacts on soil and this practice of cropping is not sustainable in the long run. The negative and most serious long term impacts of monoculture in respect of soil health include "reduction in organic matter levels and microbial biomass along with decline in activities of soil enzymes like dehydrogenases and phosphatase and increase in number of crop pests" (Liu et al., 2006).

Looking at the other side if rotation complexity of crops in agricultural system is increased, changes are made from monoculture to continuous rotation cropping prominent positive impacts on soil health become visible. The overall impact on soil health of intensification of rotation complexity in agriculture and following crop rotation is positive "positively enhancing soil organic matter, carbon, and fauna consequently increasing crop productivity" (West and Post, 2002; Jarecki and Lal, 2003; Bremer et al., 2008; Bai et al., 2018). As with mono-cropping in crop rotation to the extent of positive impact is influenced by the intercrop involved and its rotation frequency. Several studies have been carried out on this aspect to see to how percent soil carbon and nitrogen, soil enzymes and microbial communities, soil fauna are influenced by combination of different crops in crop rotation cycle or by intercropping (Witt et al., 2000; Yang and Kay, 2001; Liu et al., 2003; Zhou et al., 2011). It has been found that "diversified crop rotations involving use of crops like oilseeds and pulses make better use of available water and prevent build-up of diseases and other pests" (Bailey et al., 2001). The pronounced effect of crop rotation on agricultural soils is how this cropping practice affects biological indicators like soil carbon and microbial properties compared to physical indicators. Studies have shown that "crop rotation barely affects aggregate stability and pH – the physicochemical indicators of soil health" (Miglierina et al., 2000; Filho et al., 2002). Therefore, overall practice of crop rotation is beneficial for soil health with positive impacts majorly on soil biology (soil carbon and organic matter levels plus microbial communities) along with effective disease and weed control mechanism.

Comparing the two agricultural or cropping practices monoculture and crop rotation it has been found that, "crop rotation systems improve soil productivity as compared to monoculture systems by being more efficient at reducing long-term yield variability and increasing total soil carbon and nitrogen concentrations over time" (Varvel, 2000; Kelley et al., 2003). Researchers 'Dick (1992)' reveal that, "crop rotation promotes crop productivity by suppressing deleterious microorganisms that flourish under monoculture." The adoption of crop rotation practices has been advocated by agricultural scientists to achieve sustainable agricultural development considering its beneficial effect and positive impact on soils compared to monocropping.

3.3.4   SOIL HEALTH AND ORGANIC AMENDMENTS

Utilization of waste products with their inclusion in agriculture in the form of organic amendments to improve crop yields directly influences soil productivity and health. "Addition of organic matter is known to significantly improve soil physical properties, nutrient availability, and microbial activity" (Abiven et al., 2009; Tsiafouli et al., 2014). Various kinds of organic amendments like nitrogen amendments, animal manures, and different kinds of compost, bio-solids, and humic substances are used in farming practices with the main aim to improve food production. All forms of organic substances added increase soil carbon levels and enhance nutrient availability in the form of "soluble phosphorus and exchangeable potassium" (Bunemann et al., 2006). Studies have proved that this practice is "beneficial for agriculture in terms of improving overall soil health and promoting disease free healthy plant growth with higher productivity" (Akhtar and Malik, 2000; Lazarovits et al., 2001; Whipps, 2004). The main benefits attributed to the use of organic amendments are "an improvement in soil physicochemical characteristics including aggregation, bulk density, water holding capacity, pH, and cation exchange capacity along with enhancement of soil microbial activity, leading to increased mineral exchange between plants and soil" (Fliebach et al., 2007; Tejada et al., 2008). Studies consider organic farming as one of the best management practices due to its various advantages in improving soil health and preventing soil pollution due to agricultural activities.

3.4   CASE STUDY

A study was carried out to assess the impact of various land use/land cover classes on soils of Sindh area, in the Kashmir region of Jammu and Kashmir. Soil samples of different land use/land cover classes of the region were collected and analyzed for various physicochemical parameters. Table 3.3 presents the results of this study showing the various physicochemical characteristics of the Sindh region soils under different land use/land cover classes. This data presented clearly reveals the deteriorating conditions of agricultural soils in the region with higher pH values, lower water holding capacity, and lower contents of organic matter and carbon. The higher pH values of cultivated soils in the region are due to the region's Karewa geology with high calcium carbonate. Continuous cultivation has increased the decomposition rate of organic matter in the soils, hence decreasing soil organic matter. Since soil organic matter is a critical component impacting other important attributes of soil like water holding capacity so decreased organic matter means lower water holding capacity. In fact, it has been reported that, "each 1% of organic matter adds 1.5% of available water capacity" (Gol, 2009). Considering the available nutrients in these soils (N, P, and K), they are closely associated with organic matter should show reduced levels. But here in this case study, the agricultural soils of the Sindh region show higher levels of these nutrients the reason being increased use of agrochemical fertilizers to compensate the loss of soil productivity via organic matter and increase food production. In addition, improper irrigation of agricultural fields is done with wastewater domestic effluent discharges the result being soil health deterioration. The case study simply supports the negative impact of improper agricultural practices on overall soil productivity and health. Clearly, continuous agricultural activities in this mountainous fragile region that are too not proper and unplanned have negatively impacted the soils of the region by disturbing its physicochemical characteristics and lowering the critical organic matter levels.

TABLE 3.3 Physico-Chemical Characteristics of Soils Under Different Land Use/Land Cover Classes in Sindh Region

Land use/ Land Cover Class	pH	Electrical Conductivity (µS/cm)	Water Holding Capacity (%)	Organic Carbon (%)	Organic Matter (%)	Av. N (Kg/ ha)	Av. P (Kg/ ha)	Av.K (Kg/ ha)
Agriculture	7.60	158	49.5	1.55	2.67	376.32	38.75	383.04
Wetland	5.25	396	87.2	3.75	6.46	427.20	40.10	424.50
Bare Land	6.79	140	44.2	1.17	2.01	228.12	21.82	118.12
Built up	6.91	310	43.2	1.21	2.08	337.04	33.20	305.20
Forests	6.70	164	58.7	2.37	4.08	627.20	46.60	394.24
Pastures	6.28	145	47.8	3.03	5.22	613.36	40.80	464.80

3.5 CONCLUSION

The concluding remarks of the study are based on the substantial knowledge available on soil degradation in agricultural systems and the fact that soil degradation has taken place all over the world and is continuing as a result of unsustainable agricultural practices the sole aim being quantitative in terms of higher yields. Clearly improper agricultural management practices influence soil indicators and its overall health in various ways. Several of these management practices in the form of intensification of agriculture interventions have a negative effect on soil health like high tillage through erosion, improper irrigation through land salinization, and mono-cropping practice through impact on soil biological properties and increasing risk of diseases and pests. On the other hand, literature review also reveals that certain agricultural operations including zero or minimum tillage and farming practices such as use of organic amendments in form of locally generated organic wastes (OWs), crop rotation, proper, and adequate irrigation mechanisms, timely pruning all are positive favorable ones in terms of restoring soil organic matter, resistance to pathogens and diseases, maintenance of proper soil moisture and reducing erosion. Therefore, emphasis has been laid on adoption of these management practices in modern agriculture to improve soil health which has deteriorated to critical levels in many regions of the world.

An important general observation perceived while reviewing literature on soil health in agricultural systems in the study is that in most agricultural practices followed world over the importance of soil health is not clear and as a result is underestimated. This is attributed to lack of knowledge on the concept of soil health and its indicators. This knowledge deficiency leads to poor understanding especially among agricultural dependent economies and agricultural practicing classes thus having a negative impact on the overall agricultural system with improper management practices being in vogue without concern for vital soil resource. This ultimately results in poor quality and decreased quantity of yields in agriculture along with degradation of soil. Therefore, understanding soil health, especially in terms of its vital indicators, is important to suggest best soil management practices and future course of action regarding improving soils, which this study tries to understand. Today's world is at such a critical stage in terms of population and global food production and its likely impacts on resources like soil that it becomes imperative to realize the importance of soil health and modify agricultural practices accordingly to minimize negative impact on the main food production center the soil.

The study observing the negative impacts of agricultural practices on soil health suggests focusing on roads that lead to sustainable agriculture to improve soil health. Considering positive impacts of certain other agricultural practices on soil and suggestions revealed in various studies in terms of sustainable best agriculture and soil management practices-these include tillage conservation, intercropping, and mixed crop rotation, integrated nutrient, and organic matter management along with pest management, better coordination in application of agrochemicals and organic amendments like crop residues (CR), manures, and organic household wastes.

KEYWORDS

- **agriculture**
- **crop residues**
- **crop rotation**
- **manure**
- **organic waste**
- **urbanization**

REFERENCES

Abiven, S., Menasseri, S., & Chenu, C., (2009). The effects of organic inputs over time on soil aggregate stability: A literature analysis. *Soil Biology and Biochemistry, 41*, 1–12.

Acosta-Martinez, V., Zobeck, T. M., & Allen, V., (2004). Soil microbial, chemical and physical properties in continuous cotton and integrated crop-livestock systems. *Soil Sci. Soc. Am. J., 68*, 1875–1884.

Acton, D. F., & Gregorich, L. J., (1995). Understanding soil health. In: Acton, D. F., & Gregorich, L. J., (eds.), *The Health of Our Soils: Towards Sustainable Agriculture* (pp. 5–10). Centre for Land and Biological Resources Research, Research Branch, Agriculture and Agri-Food Canada: Ottawa.

Aggelides, S. M., & Londra, P. A., (2000). Effect of compost produced from town wastes and sewage sludge on physical properties of a loamy and a clay soil. *Bioresour. Technol., 71*, 253–259.

Ahamadou, B., & Huang, Q., (2013). Impacts of agricultural management practices on soil quality. In: Xu, J., & Sparks, D. L., (eds.), *Molecular Environmental Soil Science* (pp. 429–480). Springer-Science Dordrecht.

Akhtar, M., & Malik, A., (2000). Role of organic soil amendments and soil organisms in the biological control of plant-parasitic nematodes: A review. *Bioresour. Technol., 74*, 35–47.

Albiach, R., Canet, R., Pomares, F., & Ingelmo, F., (2001). Organic matter components and aggregate stability after the application of different amendment to a horticultural soil. *Bioresource Technology, 76*, 125–129.

Al-Kaisi, M. M., Yin, X. H., & Licht, M. A., (2005). Soil carbon and nitrogen changes as influenced by tillage and cropping systems in some Iowa soils. *Agriculture, Ecosystems and Environment, 105*, 635–647.

Allen, D. E., Singh, B. P., & Dalal, R. C., (2011). Soil health indicators under climate change: A review of current knowledge. In: Singh, B. P., Cowie, A. L.,& Chan, K. Y., (eds.), *Soil Health and Climate Change* (pp. 25–41). Springer-Verlag Berlin: Heidelberg.

Alotaiba, K. D., & Schoenau, J. J., (2011). Enzymatic activity and microbial biomass in soil amended with biofuel production by products. *Applied Soil Ecology, 48*(2), 227–235.

Alvarez, C. R., & Alvarez, R., (2000). Short-term effects of tillage systems on active soil microbial biomass. *Biol. Fert. Soils, 31,* 157–161.

Alvear, M., Rosas, A., Rouanet, J. L., & Borie, F., (2005). Effects of three soil tillage systems on some biological activities in an ultisol from southern Chile. *Soil and Tillage Research, 82*(2), 195–202.

Amaral, F., & Abelho, M., (2016). Effects of agricultural practices on soil and microbial biomass carbon, nitrogen and phosphorus content: A preliminary case study. *Web Ecol., 16,* 3–5.

Anderson, T., (2003). Microbial eco-physiological indicators to assess soil quality. *Agriculture Ecosystems andEnvironment, 98,* 285–293.

Aon, M. A., Cabello, M. N., Sarena, D. E., Colaneri, A. C., Franco, M. G., Burgos, J. L., & Cortassa, S., (2001). Spatio-temporal patterns of soil microbial and enzymatic activities in an agricultural soil. *Applied Soil Ecology, 18*(3), 239–254.

Arias, M. E., Gonzalez-Pervez, J. A., Gonzalez-Villa, F. J., & Ball, A. S., (2005). Soil health: A new challenge for microbiologists and chemists. *International Microbiology, 8,* 13–21.

Arnold, S. L., Doran, J. W., Schepers, J., & Wienhold, B., (2005). Portable probes to measure electrical conductivity and soil quality in the field. *Commun. Soil Sci. Plant Anal., 36,* 2271–2287.

Arshad, M. A., & Martin, S., (2002). Identifying critical limits for soil quality indicators in agroecosystems. *Agriculture, Ecosystems and Environment, 88,* 153–160.

Aziz, I., Mahmood, T., & Islam, K. R., (2013). Effect of long term no-till and conventional tillage practices on soil quality. *Soil and Tillage Research, 131,* 28–35.

Babujia, L. C., Hungria, M., Franchini, J. C., & Brookes, P. C., (2010). Microbial biomass and activity at various soil depths in a Brazilian oxisol after two decades of no tillage and conventional tillage. *Soil Biology andBiochemistry, 42,* 2174–2181.

Bai, Z., Caspari, T., Gonzalez, M. R., Batjes, N. H., Mader, P., Bunemann, E. K., Ron, D. G., et al., (2018). Effects of agricultural management practices on soil quality: A review of long-term experiments for Europe and China. *Agriculture, Ecosystems and Environment, 265,* 1–7.

Bailey, K. L., & Lazarovits, G., (2003). Suppressing soil-borne diseases with residue management and organic amendments. *Soil and Tillage Research, 72,* 169–180.

Bailey, K. L., Gossen, B. D., Lafond, G. P., Watson, P. R., & Derksen, D. A., (2001). Effect of tillage and crop rotation on root and foliar diseases of wheat and pea in Saskatchewan from 1991 to 1998: Univariate and multivariate analyses. *Can. J. Plant Sci., 81,* 789–803.

Balota, E. L., Colozzi, A., Andrade, D. S., & Dick, R. P., (2003). Microbial biomass in soils under different tillage and crop rotation systems. *Biol. Fert. Soils, 38,* 15–20.

Bandick, A. K., & Dick, R. P., (1999). Field management effects on soil enzyme activities. *Soil Biology andBiochemistry, 31*(11), 1471–1479.

Basir, A., Jan, M. T., Alam, M., Shah, A. S., Afridi, K., Adnan, M., Ali, K., & Mian, I. A., (2017). Impacts of tillage, subtle management and nitrogen on wheat production and soil properties. *Canadian Journal of SoilScience, 97,* 133–140.

Basir, A., Jan, M. T., Arif, M., & Khan, J. M., (2016). Response of tillage, nitrogen and stubble management on phenology and crop establishment of wheat. *Int. J. Agric. Biol., 18,* 1–8.

Bastida, F., Moreno, J. L., Hernandez, T., & Garcia, C., (2006). Microbiological degradation index of soils in a semiarid climate. *Soil Biology and Biochemistry, 38,* 3463–3473.

Bayer, C., Martin-Neto, L., Mielniczuk, J., Pillon, C. N., & Sangoi, L., (2001). Changes in soil organic matter fractions under subtropical no till cropping systems. *Soil Science Society of America Journal, 65*(5), 1473–1478.

Berner, A., Hildermann, I., Fliessbach, A., & Mader, P., (2008). Crop yield and soil quality response to reduced tillage under organic management. *Soil and Tillage Research, 101,* 89–96.

Bhattacharyya, R., Prakash, V., Kundu, S., & Gupta, H. S., (2006). Effect of tillage and crop rotation on pore size distribution and soil hydraulic conductivity in sandy clay loam soil of the Indian Himalayas. *Soil andTillage Research, 86*(2), 129–140.

Borselli, L., Torri, D., Oygarden, L., De Alba, S., Casasnovas, M. J. A., Bazzoffi, P., & Jakab, G., (2006). Soil erosion by land leveling. In: Boardman, J., & Poesen, J., (eds.), *Soil Erosion in Europe* (pp. 643–658). John Wiley and Sons, Chichester: UK.

Bowles, T. M., Acosta-Martinez, V., Calderon, F., & Jackson, L. E., (2014). Soil enzyme activities, microbial communities, and carbon and nitrogen availability in organic agroecosystems across an intensively-managed agricultural landscape. *Soil Biology and Biochemistry, 68,* 252–262.

Bremer, E., Janzen, H. H., Eliert, B. H., & McKenzie, R. H., (2008). Soil organic carbon after twelve years of various crop rotations in an aridic boroll. *Soil Science Society of America Journal, 72,* 970–974.

Brempong, A. S., Gantner, S., Adiku, S. G. K., Archer, G., Edusei, V., & Tiedje, J. M., (2008). Changes in the biodiversity of microbial populations in tropical soils under different fallow treatments. *Soil Biology andBiochemistry, 40,* 2811–2818.

Bronick, C. J., & Lal, R., (2005). Soil structure and management: A review. *Geoderma, 124,* 3–22.

Bruce, J. P., Frome, M., Haites, E., Janzen, H., Lal, R., & Paustian, K., (1999). Carbon sequestration in soils. *Journal of Soil and Water Conservation, 54,* 382–389.

Bunemann, E. K., Marschner, P., Smernik, R. J., Conyers, M., & McNeill, A. M., (2008). Soil organic phosphorus and microbial community composition as affected by 26 years of different management strategies. *Biology and Fertility of Soils, 44,* 717–726.

Bunemann, E. K., Schwenke, G. D., & Van, Z. L., (2006). Impact of agricultural inputs on soil organisms- A review. *Australian Journal of Soil Research, 44,* 379–406.

Cantarella, H., (2007). Nitrogen. In: Novais, R. F., (ed.), *Soil Fertility* (pp. 375–470). Vicosa: Brazil.

Cardoso, E. J. B. N., Vasconcellos, R. L. F., Bini, D., Miyauchi, M. Y. H., Santos, C. A., Alves, P. R. L., Paula, A. S., et al., (2013). Soil health: Looking for suitable indicators. What should be considered to assess the effects of use and management on soil health? *Scientia Agricola, 4,* 274–289.

Carter, M. R., (1990). Relative measures of soil bulk density to characterize compaction in tillage studies on fine sandy loams. *Canadian Journal of Soil Science, 70,* 425–433.

Carter, M. R., (2002). Soil quality for sustainable land management: Organic matter and aggregation interactions that maintain soil functions. *Agron. J., 94,* 38–47.

Carter, M. R., (2007). Long-term influence of compost on available water capacity of a fine sandy loam in a potato rotation. *Canadian Journal of Soil Science, 87,* 535–539.

Celik, I., (2005). Land-use effects on organic matter and physical properties of soil in a southern Mediterranean highland of Turkey. *Soil and Tillage Research, 83,* 270–277.

Chakraborty, A., Chakrabarti, K., Chakraborty, A., & Ghosh, S., (2011). Effect of long-term fertilizers and manure application on microbial biomass and microbial activity of a tropical agricultural soil. *Biology andFertility of Soils, 47,* 227–233.

Chan, K. Y., Heenan, D. P., & Oates, A., (2002). Soil carbon fractions and relationship to soil quality under different tillage and stubble management. *Soil and Tillage Research, 63,* 133–139.

Chan, K. Y., Heenan, D. P., & So, H. B., (2003). Sequestration of carbon and changes in soil quality under conservation tillage on light-textured soils in Australia: A review. *Australian Journal of ExperimentalAgriculture, 43,* 325–334.

Chaudhari, P. R., Ahire, D. V., Ahire, V. D., Chkravarty, M., & Maity, S., (2013). Soil bulk density as related to soil texture, organic matter content and available total nutrients of Coimbatore soil. *International Journal ofScientific and Research Publications, 3*(2), 1–8.

Compton, J. E., & Boone, R. D., (2000). Long term impacts of agriculture on soil carbon and nitrogen in New England forests. *Ecology, 81*(8), 2314–2330.

Corwin, D. L., & Lesch, S. M., (2005). Apparent soil electrical conductivity measurements in agriculture. *Computers and Electronics in Agriculture, 46*(1–3), 11–43.

Cusack, D. F., Chadwick, O. A., Ladefoged, T., & Vitousek, P. M., (2013). Long term effects of agriculture on soil carbon pools and carbon chemistry along a Hawaiian environmental gradient. *Biogeochemistry, 112*(1–3), 229–243.

Czubaszek, A. W., & Czubaszek, R., (2014). Tillage erosion: The principles, controlling factors and main implications for future research. *Journal of Ecological Engineering, 15*(4), 150–159.

Dalal, R. C., (1998). Soil microbial biomass-what do the numbers really mean? *Australian J. Exp. Agric., 38,* 649–665.

De Souza, A. D., Colozzi-Filho, A., Balota, E. L., & Hungria, M., (2003). Long-term effects of agricultural practices on microbial community. In:Garcia-Torres, L., (ed.), *Conservation Agriculture* (pp. 301–306). Kluwer Academic Publishers, Dordrecht: Netherlands.

Deike, S., Pallutt, B., Melander, B., Strassemeyer, J., & Christen, O., (2008). Long-term productivity and environmental effects of arable farming as affected by crop rotation, soil tillage and strategy of pesticide use: A case study of two long-term field experiments in Germany and Denmark. *Eur. J. Agron., 29,* 191–199.

Dexter, A. R., (2004). Soil physical quality: Part I. Theory, effects of soil texture, density, and organic matter, and effects on root growth. *Geoderma, 120,* 201–214.

Dick, R. P., (1992). A review: Long-term effects of agricultural systems on soil biochemical and microbial parameters. *Agriculture, Ecosystems and Environment, 40*(1–4), 25–36.

Ding, G., Novak, J. M., Amarasiriwardena, D., Hunt, P. G., & Xing, B., (2002). Soil organic matter characteristics as affected by tillage management. *Soil Sci. Soc. Am. J., 66,* 421–429.

Doran, J. W., & Jones, A. J., (1996). *Methods for Assessing Soil Quality* (Vol. 49). SSSA Special Publication, Madison, WI: ASA.

Doran, J. W., & Parkin, T. B., (1994). Defining and assessing soil quality. In: Doran, J. W., Coleman, D. C., Bezdicek, D. F., & Stewart, B. A., (eds.), *Defining Soil Quality for a Sustainable Environment* (p. 410). Soil Science Society of America, Special Publication 35: Madison.

Doran, J. W., & Parkin, T. B., (1996). Quantitative indicators of soil quality: A minimum data set. In: Doran, J. W., & Jones, A. J., (eds.), *Methods for Assessing Soil Quality* (pp. 25–37). Soil Science Society of America, Special Publication 49: Madison.

Doran, J. W., & Zeiss, M. R., (2000). Soil health and sustainability: Managing the biotic component of soil quality. *Applied Soil Ecology, 15,* 3–11.

Doran, J. W., Elliott, E. T., & Paustian, K., (1998). Soil microbial activity, nitrogen cycling and long-term changes in organic carbon pools as related to fallow tillage management. *Soil and Tillage Research, 49,* 3–18.

Doran, J. W., Sarrantonio, M., & Lieberg, M. A., (1996). Soil health and sustainability. *Advances in Agronomy, 56,* 1–54.

Eivazi, F., Bayan, M., & Schmidt, K., (2003). Soil enzyme activities in the historic Sanborn field as affected by long term cropping systems. *Commun. Soil Sci. Plant Anal., 34*(15/16), 2259–2275.

El Titi, A., (2003). Implications of soil tillage for weed communities. In: El Titi, A., (ed.), *Soil tillage in Agro Ecosystems* (pp. 147–185). CRC Press, Boca Raton: FL.

Endo, T., Sadahiro, Y., Larrinaga, J. A., Fujiyama, H., & Honna, T., (2011). Status and causes of soil salinization of irrigated agricultural lands in Southern Baja California, Mexico. *Applied and Environmental Soil Science,* 1–12.

Entry, J. A., Fuhrmann, J. J., Sojka, R. E., & Shewmaker, G. E., (2004). Influence of irrigated agriculture on soil carbon and microbial community structure. *Environmental Management, 33,* S363–S373.

Erkossa, T., Itanna, F., & Stahr, K., (2007). Indexing soil quality: A new paradigm in soil science research. *Australian Journal of Soil Research, 45,* 129–137.

Fabrizzi, K. P., Garcia, F. O., Costa, J. L., & Picone, L. I., (2005). Soil water dynamics, physical properties and corn and wheat responses to minimum and no till systems in the southern Pampas of Argentina. *Soil andTillage Research, 81*(1), 576–569.

Farquharson, R. J., Schwenke, G. D., & Mullen, J. D., (2003). Should we manage soil organic carbon in vertosols in the northern grains region of Australia? *Aust. J. Exp. Agr., 43,* 261–270.

Filho, C. C., Lourenco, A., Guimaraesde, M. F., & Fonseca, I. C. B., (2002). Aggregate stability under different management systems in a red latosol in the State of Parana, Brazil. *Soil Tillage Research, 65,* 45–51.

Fliebach, A., Oberholzer, H. R., Gunst, L., & Mader, P., (2007). Soil organic matter and biological soil quality indicators after 21 years of organic and conventional farming. *Agriculture, Ecosystems and Environment, 118,* 273–284.

Flowers, T. J., (1999). Salinization and horticultural production. *Scientia. Hortic., 78,* 1–14.

Fontaine, S., & Barot, S., (2005). Size and functional diversity of microbe populations control plant persistence and long-term soil carbon accumulation. *Ecology Letters, 8,* 1075–1087.

Fontaine, S., Barot, S., Barre, P., Bdioui, N., Mary, B., & Rumpel, C., (2007). Stability of organic carbon in deep soil layers controlled by fresh carbon supply. *Nature, 450,* 277–280.

Gajda, A., & Martyniuk, S., (2005). Microbial biomass C and N and activity of enzymes in soil under winter wheat grown in different crop management systems. *Pol. J. Environ. Stud., 14,* 159–163.

Ge, T., Nie, S., Wu, J., Shen, J., Xiao, H., Tong, C., & Iwasaki, K., (2011). Chemical properties microbial biomass and activity differ between soils of organic and conventional horticultural systems under greenhouse and open field management: A case study. *J. Soils Sediments, 11,* 25–36.

Geisseler, D., Horwath, W. R., & Scow, K. M., (2011). Soil moisture and plant residue addition interact in their effect on extracellular enzyme activity. *Pedobiologia, 54,* 71–78.

Ghaemi, M., Astaraei, A. R., Emami, H., Mahalati, N. M., & Sanaeinejad, S. H., (2014). Determining soil indicators for soil sustainability assessment using principal component analysis of Astan Quds-east of Mashhad-Iran. *Journal of Soil Science and Plant Nutrition, 14*, 1005–1020.

Gianfreda, L., Roa, M. A., Piotrowksa, A., Palumbo, G., & Clombo, C., (2005). Soil enzymes as affected by anthropogenic alterations: Intensive agricultural practices and organic pollution. *Science of the TotalEnvironment, 341*(1–3), 265–279.

Gil, S. V., Meriles, J., Conforto, C., Figoni, G., Basanta, M., Lovera, E., & March, G. J., (2009). Field assessment of soil biological and chemical quality in response to crop management practices. *World J. Microbiol. Biotechnol., 25*, 439–448.

Gol, C., (2009). The effects of land use change on soil properties and organic carbon at Dagdami river catchment in Turkey. *Journal of Environmental Biology, 30*(5), 825–830.

Gomiero, T., Pimentel, D., & Paoletti, M. G., (2011). Environmental impact of different agricultural management practices: Conventional vs. organic agriculture. *Critical Reviews in Plant Sciences, 30*, 95–124.

Gosai, K., Arunachalam, A., & Dutta, B. K., (2010). Tillage effects on soil microbial biomass in a rain fed agricultural system of northeast India. *Soil and Tillage Research, 109*, 68–74.

Grace, P. R., Ladd, J. N., & Skjemstad, J. O., (1994). The effect of management practices on soil organic matter dynamics. In: Pankhurst, C. E., (ed.), *Soil Biota* (pp. 162–170). CSIRO Information Services, Melbourne: Australia.

Gregorich, E. G., Carter, M. R., Angers, D. A., Monreal, C. M., & Ellert, B. H., (1994). Towards a minimum dataset to assess soil organic matter quality in agricultural soils. *Canadian Journal of Soil Science, 74*, 367–385.

Gulzar, S., Khan, M. A., Ungar, I. A., & Liu, X., (2005). Influence of salinity on growth and osmotic relations of *Sporobolus ioclados*. *Pak. J. Bot., 37*(1), 119–129.

Gupta, V. V. S. R., Roper, M. M., Kirkegaard, J. A., & Angus, J. F., (1994). Changes in microbial biomass and organic matter levels during the first year of modified tillage and stubble management practices on a red earth. *Australian Journal Soil Research, 32*, 1339–1354.

Hakansson, I., & Lipiec, J., (2000). A review of the usefulness of relative bulk density values in studies of soil structure and compaction. *Soil Tillage Research, 53*, 71–85.

Hamblin, A., (1991). *Environmental Indicators for Sustainable Agriculture*. Report of a National Workshop, LWRRDC and GRDC.

Haynes, R. J., & Naidu, R., (1998). Influence of lime, fertilizer and manure applications on soil organic matter and soil physical conditions: A review. *Nutrient Cycling in Agro Ecosystems, 51*, 123–137.

Haynes, R. J., (2008). Soil organic matter quality and the size and activity of the microbial biomass: Their significance to the quality of agricultural soils. In: Huang, Q., & Huang, V. A., (eds.), *Soil Mineral-Microbe-Organic Interactions: Theories andApplications* (pp. 201–230). Springer: Berlin.

Heckrath, G., Djurhuus, J., Quine, T. A., Van, O. K., Govers, G., & Zhang, Y., (2005). Tillage erosion and its effect on soil properties and crop yield in Denmark. *J. Environ. Qual., 34*, 312–324.

Herrick, J. E., & Whitford, W. G., (1995). Assessing the quality of rangeland soils: Challenges and opportunities. *J. Soil Water Conserv., 50*, 237–242.

Hillel, D., (2000). *Salinity Management for Sustainable Agriculture: Integrating Science, Environment and Economics*. The World Bank, Washington: DC.

Holland, J., (2004). The environmental consequences of adopting conservation tillage in Europe: Reviewing the evidence. *Agriculture, Ecosystems and Environment, 103*, 1–25.

Houk, E., Frasier, M., & Schuck, E., (2006). The agricultural impacts of irrigation induced water logging and soil salinity in the Arkansas basin. *Agricultural Water Management, 85*(1), 175–183.

Idowu, O. J., Van, E. H. M., Abawi, G. S., Wolfe, D. W., Schindelbeck, R. R., Moebius-Clune, B. N., & Gugino, B. K., (2009). Use of an integrative soil health test for evaluation of soil management impacts. *Renew Agric. Food Syst., 24*, 214–224.

Islam, K. R., & Weil, R. R., (2000). Soil quality indicator properties in mid-Atlantic soils as influenced by conservation management. *Journal of Soil Conservation, 55*, 69–78.

Jacobs, A., Helfrich, M., Hanisch, S., Quendt, U., Rauber, R., & Ludwig, B., (2010). Effect of conventional and minimum tillage on physical and biochemical stabilization of soil organic matter. *Biology and Fertility of Soils, 46*(7), 671–680.

Jacobs, A., Rauber, R., & Ludwig, B., (2009). Impact of reduced tillage on carbon and nitrogen storage of two Haplic luvisols after 40 years. *Soil and Tillage Research, 102*, 158–164.

Jarecki, M. K., & Lal, R., (2003). Crop management for soil carbon sequestration. *Crit. Rev. Plant Sci., 22*(6), 471–502.

Jarvis, N. J., (2007). A review of non-equilibrium water flow and solute transport in soil macropores: Principles, controlling factors and consequences for water quality. *Eur. J. Soil Sci., 58*, 532–546.

Jordan, D., Kremer, R. J., Bergfield, W. A., Kim, K. Y., & Cacnio, V. N., (1995). Evaluation of microbial methods as potential indicators of soil quality in historical agricultural fields. *Biology and Fertility of Soils, 19*(4), 297–302.

Kandeler, E., Tscherko, D., & Spiegel, H., (1999). Long-term monitoring of microbial biomass, N mineralization and enzyme activities of a chernozem under different tillage management. *Biology and Fertility of Soils, 28*, 343–351.

Karlen, D. L., Doran, J. W., Andrews, S. S., & Wienhold, B. J., (2003). Soil Quality Humankinds' foundation for survival. *J. Soil Water Conserv., 58*, 171–179.

Karlen, D. L., Mausbach, M. J., Doran, J. W., Cline, R. G., Harris, R. F., & Schuman, G. E., (1997). Soil quality: A concept, definition, and framework for evaluation. *Soil Sci. Soc. Am. J., 61*, 4–10.

Karlen, D. L., Parkin, T. P., & Eash, N. S., (1996). Use of soil quality indicators to evaluate conservation reserve program sites in Iowa. In: Doran, J. W., & Jones, A. J., (eds.), *Methods for Assessing Soil Quality* (pp. 345–355). SSSA: Madison.

Kelley, K. W., Long, J. H., & Todd, T. C., (2003). Long-term crop rotations affect soybean yield, seed weight, and soil chemical properties. *Field Crops Research, 83*, 41–50.

Kennedy, A. C., & Papendick, R. I., (1995). Microbial characteristics of soil quality. *Journal of Soil and WaterConservation*, 243–248.

Kibblewhite, M. G., Ritz, K., & Swift, M. J., (2008). Soil health in agricultural systems. *Phil. Trans. R. Soc. B., 363*, 685–701.

Klose, S., & Ajwa, H. A., (2004). Enzyme activities in agricultural soils fumigated with methyl bromide alternatives. *Soil Biology and Biochemistry, 36*, 1625–1635.

Kushwaha, C. P., Tripathi, S. K., & Singh, K. P., (2001). Soil organic matter and water-stable aggregates under different tillage and residue conditions in a tropical dry land agro ecosystem. *Applied Soil Ecology, 16*, 229–241.

Laishram, J., Saxena, K. G., Maikhuri, R. K., & Rao, K. S., (2012). Soil quality and soil health: A review. *International Journal of Ecology and Environmental Sciences, 38*(1), 19–37.

Landers, J. N., Barros, G. S. C., Rocha, M. T., Manfrinato, W. A., & Weiss, J., (2001). Environmental impacts of zero tillage in Brazil- a first approximation. In: Garcia-Torres, L., Benitez, J., & Martinez-Villela, A., (eds.), *Conservation Agriculture: A Worldwide Challenge* (pp. 317–326). ECAF-Madrid: FAO-Rome.

Laudicina, V. A., Badalucco, L., & Palazzolo, E., (2011). Effect of compost input and tillage intensity on soil microbial biomass and activity under Mediterranean conditions. *Biology and Fertility of Soils, 47*(1), 63–70.

Lazarovtis, G., Tenuta, M., & Conn, K. L., (2001). Organic amendments as a disease control strategy for soil borne disease of high-value agricultural crops. *Australas. Plant Pathol., 30*, 111–117.

Liebig, M. A., Tanaka, D. L., & Wienhold, B. J., (2004). Tillage and cropping effects on soil quality indicators in the northern Great Plains. *Soil Tillage Research, 78*, 131–141.

Lindstrom, M. J., Nelson, W. W., & Schumacher, T. E., (1992). Quantifying tillage erosion rates due to moldboard plowing. *Soil and Tillage Research, 24*, 243–255.

Liu, X. B., Han, X. Z., Herbert, S. J., & Xing, B., (2003). Dynamics of soil organic carbon under different agricultural management systems in the black soil of China. *Commun. Soil Sci. Plant Anal., 34*, 973–984.

Liu, X. B., Liu, J. D., Xing, B., Herbert, S. J., & Zhang, X. Y., (2005). Effects of long-term continuous cropping, tillage, and fertilization on soil carbon and nitrogen in Chinese Mollisols. *Commun. Soil Sci. Plant Anal., 36*, 1229–1239.

Liu, X., Herbert, S. J., Hashemi, A. M., Zhang, X., & Ding, G., (2006). Effects of agricultural management on soil organic matter and carbon transformation: A review. *Plant Soil Environ., 52*(12), 531–543.

Livia, B., Uwe, L., & Frank, B., (2005). Microbial biomass, enzyme activities and microbial community structure in two European long-term field experiments. *Agric. Ecosyst. Environ., 109*, 141–152.

Lobb, D. A., Kochanoski, R. G., & Miller, M. H., (1995). Tillage translocation and tillage erosion on shoulder slope landscape positions measured using Cs-137 as a tracer. *Canadian Journal of Soil Science, 75*, 211–218.

Luo, Z., Wang, E., & Sun, O. J., (2010). Soil carbon change and its responses to agricultural practices in Australian agroecosystems: A review and synthesis. *Geoderma, 155*, 211–223.

Machado, R. M. A., & Serralheiro, R. P., (2017). Soil salinity: Effect on vegetable crop growth. Management practices to prevent and mitigate soil salinization. *Horticulturae, 3*(30), 1–13.

Mader, P., Fliessbach, A., Dubois, D., Gunst, L., Fried, P., Niggli, U., & Van, D. H. M. G. A.,(2002). Soil fertility and biodiversity in organic farming. *Science, 296*, 1694–1697.

Maharjan, M., Sanaullaha, M., Razavi, B. S., & Kuzyakova, Y., (2017). Effect of land use and management practices on microbial biomass and enzyme activities in subtropical top-and sub-soils. *Applied Soil Ecology, 113*, 22–28.

Mairura, F. S., Mugendi, D. N., Mwanje, J. I., Ramisch, J. J., Mbugua, P. K., & Chianu, J. N., (2007). Integrating scientific and farmer's evaluation of soil quality indicators in Central Kenya. *Geoderma, 139*, 134–143.

Malik, S. S., Chauhan, R. C., Laura, J. S., Kapoor, T., Abhilashi, R., & Sharma, N., (2014). Influence of organic and synthetic fertilizers on soil properties. *International Journal of Current Microbiology and Applied Sciences, 3*(8), 802–810.

Masto, R. E., Pramod, K., Singh, C. D., & Patra, A. K., (2009). Changes in soil quality indicators under long-term sewage irrigation in a sub-tropical environment. *Environmental Geology, 56*, 1237–1243.

Mathew, R. P., Feng, Y., Githinji, L., Ankumah, R., & Balkcom, K. S., (2012). Impact of no tillage and conventional tillage system on soil microbial communities. *Applied Environmental Soil Science,* 1–10.

McLauchlan, K., (2006). The nature and longevity of agricultural impacts on soil carbon and nutrients: A review. *Ecosystems, 9,* 1364–1382.

McLeod, M. K., Schwenke, G. D., Cowie, A. L., & Harden, S., (2013). Soil carbon is only higher in the surface soil under shallow tillage in vertosols and chromoslos of New South Wales north-west slopes and plains, Australia. *Soil Research, 51,* 680–694.

McVay, K. A., Budde, J. A., Fabrizzi, K., Mikha, M. M., Rice, C. W., Schlegel, A. J., Peterson, D. E., et al., (2006). Management effects on soil physical properties in long-term tillage studies in Kansas. *Soil Sci. Soc. Am. J., 70,* 434–438.

Miglierina, A. M., Iglesias, J. O., Landriscini, M. R., Galantini, J. A., & Rosell, R. A., (2000). The effects of crop rotation and fertilization on wheat productivity in the pampean semi-arid region of Argentina: Soil physical and chemical properties. *Soil and Tillage Research, 53,* 129–135.

Mikha, M. M., & Rice, C. W., (2004). Tillage and manure effects on soil and aggregate associated carbon and nitrogen. *Soil Science Society of America Journal, 68,* 809–816.

Mikhailova, E. A., Bryant, R. B., Vassenev, I. I., Schwager, S. J., & Post, C. J., (2000). Cultivation effects on soil carbon and nitrogen contents at depth in the Russian Chernozem. *Soil Science Society of America Journal, 64,* 738–745.

Moebius, B. N., Van, E. H. M., Schindelbeck, R. R., Idowu, O. J., Thies, J. E., & Clune, D. J., (2007). Evaluation of laboratory measured soil properties as indicators of soil physical quality. *Soil Science, 172,* 895–912.

Montgomery, D. R., (2007). Soil erosion and agricultural sustainability. *PNAS, 104,* 13268–13272.

Moraru, P. I., & Rusu, T., (2010). Soil tillage conservation and its effect on soil organic matter, water management and carbon sequestration. *Journal of Food, Agriculture and Environment, 8,* 309–312.

Murphy, B. W., (2015). Impact of soil organic matter on soil properties- a review with emphasis on Australian soils. *Soil Research, 53,* 605–635.

Murty, D., Kirschbaum, M. U. F., Mcmurtrie, R. E., & Mcgilvray, H., (2002). Does conversion of forest to agricultural land change soil carbon and nitrogen?: A review of literature. *Global Change Biology, 8,* 105–123.

Nayyar, A., Hamel, C., Lafond, G., Gossen, B. D., Hanson, K., & Germida, J., (2009). Soil microbial quality associated with yield reduction in continuous pea. *Applied Soil Ecology, 43,* 115–121.

Nguyen, X. H., & Pham, A. H., (2018). Assessing soil erosion by agricultural and forestry production and proposing solutions to mitigate: A case study in son la province, Vietnam. *Applied and Environmental Soil Science,* 1–10.

Nigigi, S. N., (2002). Review of irrigation development in Kenya. In: Blank, H. G., Mutero, H. G., & Murray, H., (eds.), *Opportunities of Anticipating Change in Eastern and Southern Africa.* Water Management Institute: Kenya.

Nogueira, M. A., Albino, U. B., Brandao-Junior, O., Braun, G., Cruz, M. F., Dias, B. A., Duarte, R. T. D., et al., (2006). Promising indicators for assessment of agro ecosystems alteration among natural, reforested and agricultural land use in southern Brazil. *Agriculture, Ecosystems and Environment, 115,* 237–247.

O Farrell, P. J. O., Donaldson, J. S., & Hoffman, M. T., (2010). Vegetation transformation, functional compensation, and soil health in a semi-arid environment. *Arid Land Res. Manage., 24*, 12–30.

Ogle, S. M., Breidt, F. J., & Paustian, K., (2005). Agricultural management impacts on soil organic carbon storage under moist and dry climatic conditions of temperate and tropical regions. *Biogeochemistry, 72*, 87–121.

Ogwo, P. A., & Ogu, O. G., (2014). Effects of biosolids application rates on bulk density and total porosity of degraded soils of Abia state, Nigeria. *Journal of Physical Sciences andEnvironmental Safety, 4*(1), 51–60.

Oliver, D. P., Bramley, R. G. V., Riches, D., Porter, I., & Edwards, J., (2013). A review of soil physical and chemical properties as indicators of soil quality in Australian viticulture. *Australian Journal of Grape and WineResearch, 19*, 129–139.

Osunbitan, J. A., Oyedele, D. J., & Adekalu, K. O., (2005). Tillage effects on bulk density, hydraulic conductivity and strength of a loamy sand soil in southwestern Nigeria. *Soil and Tillage Research, 82*(1), 57–64.

Pagliai, M., Pellegrini, S., Vignozzi, N., Rousseva, S., & Grasselli, O., (2000). The quantification of the effect of subsoil compaction on soil porosity and related physical properties under conventional to reduced management practices. *Adv. Geo. Ecology, 32*, 305–313.

Pandolfini, T., Gremigni, P., & Gabbrielli, R., (1997). Bio-monitoring of soil health by plants. In: Pankhurst, C. E., Doube, B. M., & Gupta, V. V. S. R., (eds.), *Biological Indicators of Soil Health* (pp. 325–347). New York, NY, USA: CAB International.

Pankhurst, C. E., Doube, B. M., & Gupta, V. V. S. R., (1997). *Biological Indicators of Soil Health* (p. 45)Wallingford: UK CAB International.

Pankhurst, C. E., McDonald, H. J., Hawke, B. G., & Kirkby, C. A., (2000). Effect of tillage and stubble management on chemical and microbiological properties and the development of suppression towards cereal root disease in soils from two sites in NSW, Australia. *Soil Biology and Biochemistry, 34*, 833–840.

Pardo, A., Amato, M., & Chiaranda, F. Q., (2000). Relationships between soil structure, root distribution and water uptake of chick-pea (*Cicer arietinum*). Plant growth and water distribution. *European Journal ofAgronomy, 13*, 39–45.

Patel, A. H., (2015). Electrical conductivity as soil quality indicator of different agricultural sites of Kheda District in Gujarat. *International Journal of Innovative Research in Science, Engineering and Technology, 4*(8), 7305–7309.

Patil, A., & Lamnganbi, M., (2018). Impact of climate change on soil health: A review. *International Journal ofChemical Studies, 6*(3), 2399–2404.

Pattison, A. B., Moody, P. W., Badcock, K. A., Smith, L. J., Armour, J. A., Rasiah, V., Cobon, J. A., et al., (2008). Development of key soil health indicators for the Australian banana industry. *Appl. Soil Ecol., 40*, 155–164.

Peigne, J., Ball, B. C., Roger-Estrade, J., & David, C., (2007). Is conservation tillage suitable for organic farming?: A review. *Soil Use and Management, 23*, 129–144.

Pekrun, C., Kaul, H. P., & Claupein, W., (2003). Soil tillage for sustainable nutrient management. In: El Titi, A., (ed.), *Soil Tillage inAgroecosystems* (pp. 83–113). CRC Press, Boca Raton, FL: USA.

Pimentel, D., & Burgess, M., (2013). Soil erosion threatens food production. *Agriculture, 3*(3), 443–463.

Poirier, V., Angers, D. A., Rochette, P., Chantigny, M. H., Ziadi, N., Tremblay, G., & Fortin, J., (2009). Interactive effects of tillage and mineral fertilization on soil carbon profiles. *Soil Science Society of America Journal, 73*, 255–261.

Ponting, C. A., (2007). *New Green History of the World: The Environment and the Collapse of Great Civilizations.* Vintage Books: London.

Pulleman, M., Jongmans, A., Marinissen, J., & Bouma, J., (2003). Effects of organic versus conventional arable farming on soil structure and organic matter dynamics in a marine loam in the Netherlands. *Soil Use and Management, 19*, 157–165.

Qi, Y., Darilek, J. L., Huang, B., Zhao, Y., Sun, W., & Gu, Z., (2009). Evaluating soil quality indices in an agricultural region of Jiangsu Province, China. *Geoderma, 149*, 325–334.

Rahman, L., Chan, K. Y., & Heenan, D. P., (2007). Impact of tillage, stubble management and crop rotation on nematode populations in a long-term field experiment. *Soil and Tillage Research, 95*, 110–119.

Rahman, M. H., Okubo, A., Sugiyama, S., & Mayland, H. F., (2008). Physical: Chemical and microbiological properties of an andisol as related to land use and tillage practice. *Soil Tillage Research, 101*, 10–19.

Rao, C. S., Grover, M., Kundu, S., & Desai, S., (2017). Soil enzymes. In: *Encyclopedia of Soil Science* (pp. 2100–2107). Taylor and Francis.

Raper, R. L., (2005). Agricultural traffic impacts on soil. *Journal of Terramechanics, 42*, 259–280.

Rasmussen, K. J., (1999). Impact of plough less soil tillage on yield and soil quality: A Scandinavian review. *Soil and Tillage Research, 53*, 3–14.

Reynolds, W. D., Bowman, B. T., Drury, C. F., Tan, C. S., & Lu, X., (2002). Indicators of good soil physical quality: Density and storage parameters. *Geoderma, 110*, 131–146.

Reynolds, W. D., Drury, C. F., Tan, C. S., Fox, C. A., & Yang, X. M., (2009). Use of indicators and pore volume function characteristics to quantify soil physical quality. *Geoderma, 152*, 252–263.

Riley, H., Pommeresche, R., Eltun, R., Hansen, S., & Korsaeth, A., (2008). Soil structure, organic matter and earthworm activity in a comparison of cropping systems with contrasting tillage, rotations, fertilizer levels and manure use. *Agriculture, Ecosystems, and Environment, 124*, 275–284.

Rimal, B. K., & Lal, R., (2009). Soil and carbon losses from five different land management areas under simulated rainfall. *Soil Tillage Research, 106*, 62–70.

Ritz, K., Black, H. I. J., Campbell, C. D., Harris, J. A., & Wood, C., (2009). Selecting biological indicators for monitoring soils: A framework for balancing scientific and technical opinion to assist policy development. *Ecological Indicators, 9*(6), 1212–1221.

Robinson, C. A., Cruse, R. M., & Kohler, K. A., (1994). Soil management. In: Hatfield, J. L., & Karlen, D. L., (eds.), *Sustainable Agricultural Systems* (pp. 109–134). Lewis Publishers, CRC Press, Boca Raton, FL: USA.

Romig, D. E., Garlynd, M. J., Harris, R. F., & Mcsweeney, K., (1995). How farmers assess soil health and quality (special issue on soil quality). *Journal of Soil and Water Conservation, 50*, 229–236.

Roscoe, R., & Burman, P., (2003). Tillage effects on soil organic matter in the density fractions of a cerrado oxisol. *Soil Tillage Research, 70*, 107–119.

Ruiz, G. R., Ochoa, V., Vinegla, B., Hinojosa, M. B., Pena-Santiago, R., Liebanas, G., Linares, J. C., & Carreira, J. A., (2009). Soil enzymes, nematode community and selected physico-chemical properties as soil quality indicators in organic and conventional olive oil farming: Influence of seasonality and site features. *Applied Soil Ecology, 41*(3), 305–314.

Rybicki, R., Obroslak, R., Mazur, A., & Marzec, M., (2016). Assessment of tillage translocation and tillage erosion on loess slope by contour mould board tillage. *Journal of Ecological Engineering, 17*(5), 247–253.

Saggar, S., Yeates, G. W., & Shepherd, T. G., (2001). Cultivation effects on soil biological properties, micro fauna and organic matter dynamics in eutric gleysol and gleyic luvisol soils in New Zealand. *Soil TillageResearch, 58*, 55–68.

Saha, N., & Mandal, B., (2009). Soil health-a precondition for crop production. In: Khan, M. S., Zaidi, A., & Musarrat, J., (eds.), *Microbial Strategies for CropImprovement* (pp. 161–168). Springer: Heidelberg.

Salem, H. M., Valero, C., Munoz, M. A., Rodriguez, G., & Silva, L. L., (2015). Short term effects of four tillage practices on soil physical properties, soil water potential and maize yield. *Geoderma, 237, 238*, 60–70.

Sanchez-Maranon, M., Siano, M., Delgado, G., & Delgado, R., (2002). Soil quality in Mediterranean mountain environments: Effects of land use change. *Soil Science Society of America Journal, 66*(3), 948–958.

Sarker, J. R., Singh, B. P., Dougherty, J. W., Fang, Y., Badgery, W., Hoyle, F. C., Dalal, R. C., & Cowie, A. L., (2018). Impact of agricultural management practices on the nutrient supply potential of soil organic matter under long-term farming systems. *Soil and Tillage Research, 175*, 71–81.

Sharma, P., Abrol, V., & Sharma, R. K., (2011). Impact of tillage and mulch management on economics, energy requirement and crop performance in maize-wheat rotation in rain-fed sub humid inceptisols, India. *European Journal of Agronomy, 34*(1), 46–51.

Sheoran, H. S., Phogat, V. K., Dahiya, R., & Gera, R., (2018). Long-term effect of organic and conventional farming practices on microbial biomass carbon, enzyme activities and microbial populations in different textured soils of Haryana State (India). *Applied Ecology and EnvironmentalResearch, 16*(3), 3669–3689.

Shukla, M. K., Lal, R., & Ebinger, M., (2006). Determining soil quality indicators by factor analysis. *Soil TillageResearch, 87*(2), 194–204.

Singh, A., (2015). Soil salinization and waterlogging: A threat to environment and agricultural sustainability. *Ecological Indicators, 57*, 128–130.

Six, J., Bossuyt, H., Degryze, S., & Denef, K., (2004). A history of research on the link between (micro) aggregates, soil biota, and soil organic matter dynamics. *Soil Tillage Research, 79*, 7–31.

Slawinski, C., Cymerman, J., Witkowska-Walczak, B., & Lamorski, K., (2012). Impact of diverse tillage on soil moisture dynamics. *Int. Agrophys., 26*, 301–309.

Smith, J. L., & Paul, E. A., (1990). The significance of soil biomass estimations. In: Bollog, J. M., & Stotzky, G., (eds.), *Soil Biochemistry* (pp. 357–359). Marcel Dekker: New York.

Smith, J. L., Halvoson, J. J., & Bolton, H. Jr., (2002). Soil properties and microbial activity across a 500 m elevation gradient in a semi-arid environment. *Soil Biol. Biochem., 34*, 1749–1757.

Soil and Water Conservation Society, (1995). Farming for a better environment: A white paper. *Soil Water Conserv. Soc.* Ankeny, IA.

Sousa, A. A. T., & Figueiredo, C. C., (2016). Sewage sludge biochar: Effects on soil fertility and growth of radish. *Biological Agriculture and Horticulture, 32*(2), 127–138.

Sparling, G. P., (1997). Soil microbial biomass, activity and nutrient cycling as indicators of soil health. In: Pankhurst, C. E., Doube, B. M., & Gupta, V. V. S. R., (eds.), *Biological Indicators of Soil Health* (pp. 97–119). CAB International: Wallingford, UK.

Sparling, G. P., Schipper, L. A., Bettjeman, W., & Hill, R., (2004). Soil quality monitoring in New Zealand: Practical lessons from a 6 year trial. *Agriculture, Ecosystems and Environment, 104*, 523–534.

Staben, M. L., Bezdicek, D. F., Smith, J. L., & Fauci, M. F., (1997). Assessment of soil quality in conservation reserve program and wheat-fallow soils. *Soil Sci. Soc. Am. J., 61*, 124–130.

Stark, C., Condron, L. M., Stewart, A., Di, H. J., & O'Callaghan, M., (2007). Effects of past and current crop management on soil microbial biomass and activity. *Biol. Fert. Soils, 43*, 531–540.

Sun, B., Hallett, P. D., Caul, S., Daniell, T. J., & Hopkins, D. W., (2010). Distribution of soil carbon and microbial biomass in arable soils under different tillage regimes. *Plant and Soil, 338*, 17–25.

Swiatkiewicz, D., & Gastol, M., (2012). Soil chemical properties under organic and conventional crop management systems in south Poland. *Biological Agriculture and Horticulture, 54*, 1–17.

Takoutsing, B., Weber, J., Aynekulu, E., Martin, J. A. R., Shepherd, K., Sila, A., Tchoundjeu, Z., & Diby, L., (2016). Assessment of soil health indicators for sustainable production of maize in smallholder farming systems in the highlands of Cameroon. *Geoderma, 276*, 64–73.

Tebrugge, F., & During, R. A., (1999). Reduced tillage intensity: A review of results from long term study in Germany. *Soil Tillage Research, 53*, 15–28.

Tejada, M., Gonzalez, J. L., Garcia-Martinez, A. M., & Parrado, J., (2008). Application of a green manure and green manure composted with beet vinasse on soil restoration: Effects on soil properties. *Bioresource Technology, 99*, 4949–4957.

Tejeda, M., Garcia, C., Gonzalez, J. L., & Hernandez, M. T., (2006). Organic amendment based on fresh and composted beet vinasse: Influence on soil physical, chemical and biological properties and wheat yield. *Soil Sci. Soc. Am. J., 70*, 900–908.

Thierfelder, C., & Wall, P. C., (2010). Rotation in conservation agriculture systems of Zambia: Effects on soil quality and water relations. *Expl. Agric., 46*(3), 309–325.

Tilman, D., (1999). Global environmental impacts of agricultural expansion: The need for sustainable and efficient practices. *Proc. Natl. Acad. Sci., 96*, 5995–6000.

Tomich, T. P., Brodt, S., Ferris, H., Galt, R., Horwath, W. R., Kebreab, E., Leveau, J. H. J., et al., (2011). Agroecology: A review from a global-change perspective. *Annu. Rev. Env. Resour., 36*, 193–222.

Tsiafouli, M. A., Thebault, E., Sgardelis, S. P., De Ruiter, P. C., Van, D. P. W. H., Birkhofer, K., Hemerik, L., et al., (2014). Intensive agriculture reduces soil biodiversity across Europe. *Global Change Biology, 21*, 973–985.

Turkelboom, F., Poesen, J., Ohler, I., Van, K. K., Ongprasert, S., & Vlassak, K., (1999). Reassessment of tillage erosion rates by manual tillage on steep slope in northern Thailand. *Soil Tillage Research, 51*, 245–259.

Utobo, E. B., & Tewari, L., (2015). Soil enzymes as bioindicators of soil ecosystem status. *Applied Ecology andEnvironmental Research, 13*(1), 147–169.

Uzoma, K. C., Inoue, M., Andry, H., Fujimaki, H., Zahoor, A., & Nishihara, E., (2011). Effect of cow manure biochar on maize productivity under sandy soil condition. *Soil Use Manage, 27*, 205–212.

Vakali, C., Zaller, J. G., & Kopke, U., (2011). Reduced tillage effects on soil properties and growth of cereals and associated weeds under organic farming. *Soil and Tillage Research, 111*, 133–141.

Van, M. W., Govers, G., & Van, O. K., (2002). Identification of important factors in the process of tillage erosion: The case of mouldboard tillage. *Soil and Tillage Research, 65*, 77–93.

Van, O. K., Van, M. W., Govers, G., Deckers, J., & Quine, T. A., (2005). From water to tillage erosion dominated landform evolution. *Geomorphology, 72*, 193–203.

Van-Camp, L., Bujarrabal, B., Gentile, A. R., Jones, R. J. A., Montanarella, L., Olazabal, C., & Selvaradjou, S. K., (2004). *Reports of the Technical Working Groups Established Under the Thematic Strategy for Soil Protection.* EUR 21319- EN/5, Luxembourg, Europe: European Communities.

Varvel, G. E., (2000). Crop rotation and nitrogen effects on normalized grain yields in a long-term study. *Agron. J., 92*, 938–941.

Visser, S., & Parkinson, D., (1992). Soil biological criteria as indicators of soil quality: Soil microorganisms. *American Journal of Alternative Agriculture, 7*(1/2), 33–37.

Vogeler, I., Horn, R., Wetzel, H., & Kruemmelbein, J., (2006). Tillage effects on soil strength and solute transport. *Soil Tillage Research, 88*, 193–204.

Volk, L. B. S., Cogo, N. P., & Streck, E. V., (2004). Water erosion influenced by surface and subsurface soil physical conditions resulting from its management, in the absence of vegetal cover. *Brazilian Journal of Soil Science, 28*, 763–774.

Wang, W. J., Dalal, R. C., & Moody, P. W., (2004). Soil carbon sequestration and density distribution in a Vertosol under different farming practices. *Australian Journal of Soil Research, 42*, 875–882.

Weil, R. R., & Magdoff, F., (2004). Significance of soil organic matter to soil quality and health. In:Weil, R. R., & Magdoff, F., (eds.), *Soil Organic Matter in Sustainable Agriculture* (pp. 1–43). CRC Press: Florida.

Weil, R. R., (1992). Inside the heart of sustainable farming. *The New Farm*, 43–48.

West, T. O., & Marland, G., (2002). A synthesis of carbon sequestration, carbon emissions and net carbon flux in agriculture: Comparing tillage practices in the United States. *Agric. Ecosys. Environ., 91*, 217–232.

West, T. O., & Post, W. M., (2002). Soil organic carbon sequestration rates by tillage and crop rotation: A global data analysis. *Soil Science Society of America Journal, 66*, 1930–1946.

Whipps, J. M., (2004). Prospects and limitations for mycorrhizas in bio control of root pathogens. *Canadian Journal of Botany, 82*, 1198–1227.

Williams, N. D., & Petticrew, E. L., (2009). Aggregate stability in organically and conventionally farmed soils. *Soil Use Manage, 25*, 284–292.

Witt, C., Cassman, K. G., Olk, D. C., Biker, U., Liboon, S. P., Samson, M. I., & Ottow, J. C. G., (2000). Crop rotation and residue management effects on carbon sequestration, nitrogen cycling and productivity of irrigated rice system. *Plant Soil, 225*, 263–278.

World Resource Institute, (1998). *World Resources: A Guide to the Global Environment.* Oxford University Press: New York, USA.

Yang, X. M., & Kay, B. D., (2001). Rotation and tillage effects on soil organic carbon sequestration in a typic Hapludalf in Southern Ontario. *Soil Tillage Research, 59*, 107–114.

Yazdanpanah, N., Mahmoodabadi, M., & Cerda, A., (2016). The impact of organic amendments on soil hydrology, structure and microbial respiration in semi arid lands. *Geoderma, 266*, 58–65.

Yemefack, M., Jetten, V. G., & Rossiter, D. G., (2006). Developing a minimum data set for characterizing soil dynamics in shifting cultivation systems. *Soil Tillage Research, 86,* 84–98.

Yusuf, A. A., Abaidoo, R. C., Iwuafor, E. N. O., Olufajo, O. O., & Sanginga, N., (2009). Rotation effects of grain legumes and fallow on maize yield, microbial biomass and chemical properties of an Alfisol in the Nigerian savanna. *Agric. Ecosyst. Environ., 129,* 325–331.

Zalidis, G., Stamatiadis, S., Takavakoglou, V., Eskridge, K., & Misopolinos, N., (2002). Impacts of agricultural practices on soil and water quality in the Mediterranean region and proposed assessment methodology. *Agriculture, Ecosystems and Environment, 88,* 137–146.

Zhang, P., Li, L., Pan, G., & Ren, J., (2006). Soil quality changes in land degradation as indicated by soil chemical, biochemical and microbiological properties in a Karst area of southwest Guizhou, China. *Environmental Geology, 51,* 609–619.

Zhao, G., Bryan, B. A., King, D., Luo, Z., Song, E. X., & Yu, Q., (2013). Impact of agricultural management practices on soil organic carbon: Simulation of Australian wheat systems. *Global Change Biology, 19,* 1585–1597.

Zhou, X., Yu, G., & Wu, F., (2011). Effects of intercropping cucumber with onion or garlic on soil enzyme activities, microbial communities and cucumber yield. *European Journal of Soil Biology, 47*(5), 279–287.

Zibilske, L. M., Bradford, J. M., & Smart, J. R., (2002). Conservation tillage induced changes in organic carbon, total nitrogen and available phosphorus in a semi-arid alkaline subtropical soil. *Soil Tillage Research, 66,* 153–163.

Zornoza, R., Acosta, J. A., Bastida, F., Dominguez, S. G., Toledo, D. M., & Faz, A., (2015). Identification of sensitive indicators to assess the interrelationship between soil quality, management practices and human health. *Soil, 1,* 173–185.

Zuber, S. M., & Villamil, M. B., (2016). Meta-analysis approach to assess effect of tillage on microbial biomass and enzyme activities. *Soil Biology and Biochemistry, 97,* 176–187.

CHAPTER 4

Global Scenario of Remediation Techniques to Combat Pesticide Pollution

REZWANA ASSAD, IQRA BASHIR, IFLAH RAFIQ, IRSHAD AHMAD SOFI, SHOWKAT HAMID MIR, ZAFAR AHMAD RESHI, and IRFAN RASHID

Department of Botany, University of Kashmir, Srinagar – 190006, Jammu and Kashmir, India, Phone: +91-7889481075, E-mail: rezumir@gmail.com (R. Assad)

ABSTRACT

Globally, in the process of agricultural development, the initial use of pesticides was an indispensable tool that improved food security by protecting economically important crops. Nevertheless, incessant and indiscriminate utilization of pesticides due to modern agricultural and healthcare practices contaminated the environment and has become a serious ecological and public health hazard due to their persistent nature, multifaceted toxicity, and recalcitrance, long-range environment transport potential, bioaccumulation, and biomagnification knack. To combat pesticide pollution in agricultural and non-agricultural settings, diverse techniques categorized as physical, chemical, physicochemical, biochemical, and biological remediation techniques have been employed worldwide either independently or in the alliance. Selection of effective remediation technique through proper screening method is crucial for restoration of pesticide-polluted environs. This chapter appraises global ecological and health hazards of pesticides, global scenario of pesticide production and usage, selection of remediation technique through screening approaches, global scenario of pesticide remediation techniques, indicators of sustainable remediation, and economics of global remediation sector. This chapter particularly summarizes the current global scenario of all the existing and emerging pesticide remediation techniques used to combat the menace of pesticide pollution hitherto. Moreover, this chapter focuses on knowledge gaps and future research directions for developing novel strategies for minimizing pesticide-related ecological problems.

4.1   INTRODUCTION

Environmental pollution and global health impacts caused by incessant use of pesticides is a major global concern. As stated by Environmental Protection Agency (EPA) and Food and Agriculture Organization (FAO), pesticide is any substance or concoction of substances premeditated for preventing, repelling, eradicating, killing or mitigating any pest (weeds, insects, pathogens, bacteria, fungi, viruses, nematodes, snails, mites, and/or rodents) menacing animal/human health and agricultural/industrial production (Li and Jennings, 2017; Morillo and Villaverde, 2017). Pesticides comprise a variety of chemicals including algicides, antifoliants, avicides, bactericides, disinfectants, fungicides, herbicides, insect, and animal repellents, insecticides, miticides, molluscicides, nematicides, piscicides, rodenticides, and viricides (Frazar, 2000; Marican and Durán-Lara, 2017; Kumar et al., 2019). On the basis of chemical nature, pesticides can be categorized into carbamates, chlorophenols, organochlorines, organophosphorus, and synthetic pyrethroids (Hamza et al., 2016; Marican and Durán-Lara, 2017).

Some conventional pesticides like red pepper, salt, sulfurous rock, tobacco extract, wooden ash, etc. were used in ancient times. These were then replaced by chemical pesticides due to their instant and effective crop protection ability (Pandey et al., 2018), which have become an integral part of the contemporary world. Nowadays, pesticides are globally used as important tools in agricultural, domestic, industrial, and public health sectors to boost crop production, protect agricultural land, stored food crops, wood, gardens, urban plantations, sustain agro-productivity and to combat disease-transmitting harmful pests (Gill and Garg, 2014; Pariatamby and Kee, 2016; Bhandari, 2017). It has been estimated that if pesticides are not applied, around 40% of agricultural crops would be lost, infectious diseases would amplify manifold and invasive species will replace native ones (Rice et al., 2007; Kumar et al., 2018).

Spraying of pesticides is still regarded as the most effectual means for crop protection and mitigation of pests (Sun et al., 2018). Although novel biological management tools are being developed as alternatives to chemical pesticides, yet the use of pesticides endure to be an indispensable tool in integrated pest management (IPM) (Rice et al., 2007). Thus, pesticides have grown to be an inevitable element of modern agricultural practices. However, incessant use of pesticides pose serious threats to human, animal, plant, and environmental health, thus pesticide usage must be regularly monitored and controlled (Rani and Dhania, 2014; Pariatamby and Kee, 2016; Dixit et al.,

2019; Kumar et al., 2019). This chapter provides current global scenario of all the existing and emerging pesticide remediation techniques used to combat menace of pesticide pollution hitherto.

4.2 GLOBAL ECOLOGICAL AND HEALTH HAZARDS OF PESTICIDES

Pesticides belong to the group 'persistent organic pollutants' (POPs). POPs are the toxic organic compounds that resist degradation and persist for a long time period in the environment, migrate to long distances with high stability and mount up to levels harmful to the environmental and public health (Pariatamby and Kee, 2016; Bharat, 2018). The characteristics that make POPs source of major global concern are: Persistency, long-range environmental transport potential, bioaccumulation ability, and high toxicity even at small concentrations (Pariatamby and Kee, 2016; Bharat, 2018). At Stockholm convention (1995), twelve important toxic POPs (mostly pesticides) were listed and named as 'Dirty dozen' (Bharat, 2018). However, more chemicals were added to this list later.

The environmental risk assessment of pesticide is not an easy task. The environmental fate of pesticides depends on the nature and quantity of pesticide sprayed, ecotoxicity, half-life of pesticide, physicochemical characteristics of soil/water, and the site (Pandey et al., 2018). Assessing pesticide residues' environmental fate and their impending exposure risks to environment and public health is crucial (Rice et al., 2007). Levitan et al. (1995) tried to find a holistic method for assessing the environmental impacts of pesticides by grouping various approaches of environmental risk assessment into different categories like anecdotal assessments, directory-format, and tabular databases, single, and multiple-parameter assessments, composite environmental impact rating systems, economic assessments, site-specific assessment tools, and holistic assessments. However, after properly analyzing these categories, they found that a universal all-inclusive pesticide impact assessment system is still lacking (Levitan et al., 1995).

Of all applied pesticides, just 0.1% reaches their target life form and the remaining 99.9% penetrate and contaminate the environment (air, soil, and water) as pesticide residues, enter the food chain and strike non-target organisms including humans (Pimentel, 1995; Bhandari, 2017; Sun et al., 2018; Dixit et al., 2019). This results in development of pest resistance, elimination of parasites, predators, pollinators, and decomposers, resurgence of new pests, destruction of useful insects, and damage ecosystem stability (Tewari et al., 2012; Gill and Garg, 2014).

Pesticide residues have been found in soil and aquatic ecosystems across the world (Moschet et al., 2014). Pesticides are ubiquitous substances that persist and contaminate soil and result in contamination of surface water through runoff and percolate into groundwater via leaching (Hamza et al., 2016). Thus, water plays a key role in the widespread transport of pesticides. Offsite transport of pesticides to surface water and groundwater is a grim global environmental concern (Rice et al., 2007). This results in the spatial and temporal inconsistency of pesticide occurrence and hurdles the formulation of effective mitigation measures restoration practice of contaminated areas (US-EPA, 2019).

On the one hand, pesticides improve food security by preventing pest attack on economically important crops. Alternatively, indiscriminate use of pesticides causes serious impacts on environment and public health (Baldissarelli et al., 2019). Pesticides are extremely noxious, destroying biodiversity and environment and causing chronic abnormalities in humans (Marican and Durán-Lara, 2017). Pesticides reach human body through direct and/or indirect multi-pathway exposure (Kim et al., 2017). Owing to their stability and lipophilicity, pesticides pass through food chain and accumulate in non-target living organisms' tissues and results in bioaccumulation and biomagnification (FAO, 2019). Pesticides exposure can cause range of acute/chronic public health problems like flu, headache, fatigue, dizziness, nausea, mood swings, cramps, anxiety, skin diseases, skin blistering, lacrimation, acute poisoning, neurological disorders, reduced intelligence, learning disability, memory loss, depression, behavioral disorders, cardiopulmonary disorders, hypertension, gastrointestinal problems, obesity, diabetes, respiratory diseases, asthma, bronchitis, bronchospasm, hypoxemia, pulmonary edema, geno, and cytotoxicity, suppression, and dysfunctioning of immune system, blindness, kidney failures, abnormal bone growth, fatal deformities, paralysis, endocrine hormone disruption, reproductive disorders, infertility, miscarriages, stillbirth, cancer, and even immediate mortality (Kamel and Hoppin, 2004; Abhilash and Singh, 2009; Gill and Garg, 2014; Weiss et al., 2016; Bhandari, 2017; Bharat, 2018; Bhat et al., 2019; Boudh and Singh, 2019; Dixit et al., 2019; McLellan et al., 2019). World Health Organization (WHO) estimated that globally, there are 3,000,000 cases of pesticide poisoning, resulting in approximately 220,000 deaths annually (Lah, 2011; Dixit et al., 2019). Pesticides are currently rightly termed as 'economic poisons' in the USA (Sun et al., 2018). Although pesticide persistence causes serious global environmental security, socio-economic, and public health concerns (Chaussonnerie et al., 2016) yet, their consumption cannot be stopped abruptly and wholly.

Pesticide pollution results in the loss of biodiversity (Boudh and Singh, 2019). Several species are on the verge of extinction risk due to pesticide contamination (Jalees and Vemuri, 1980). Significant population reduction has been observed in amphibians, birds, earthworms, fishes, insects (bees, beetles, spiders, wasps), nematodes, and non-target plants as a consequence of pesticide pollution (Boudh and Singh, 2019; US-EPA, 2019). A decline in insect population, in turn, result in decreased crop production, since $1/3^{rd}$ of crops are dependent on insects for pollination (Boudh and Singh, 2019). In non-target plants, pesticides affect photosynthesis, destroy plant cell membrane, promote oxidative stress by producing reactive oxygen species, inhibit cell division, growth, and survival (Weiss et al., 2016). Furthermore, long term pesticide contamination destructs microbial diversity of soil and aquatic ecosystems, alter growth activities of microbes, reduce soil fertility, influence nutrient cycling, affect operative soil/water enzyme systems, destroy earthworms (bioindicators of soil toxicity), cause adverse effects and eliminate beneficial micro-flora and fauna, (Megharaj et al., 2000; Bhandari, 2017; Sun et al., 2018). Hence, for the maintenance of environment and public health safeguard, sustainable remediation of pesticide-contaminated sites is prerequisite.

4.3   GLOBAL SCENARIO OF PESTICIDE PRODUCTION AND USAGE

Generally, acquiring comprehensive and reliable data regarding global production and usage of pesticides is a very difficult task.

4.3.1   HISTORY OF PESTICIDE PRODUCTION AND USAGE

Some natural products like plant extracts, red pepper, salt, sulfurous rock, a tobacco extract, wooden ash, etc. were used as pesticides in ancient times. In 1600, honey and arsenic mixture was used for insect management (History of pesticide use 1998). In 1800, the farmer community of the USA commenced the usage of some particular chemicals such as calcium arsenate, nicotine sulfate, and sulfur for agricultural pest management purposes (Delaplane, 2000). Following World War II, in 1939, several efficient chemical pesticides like aldrin, DDT, dieldrin, and endrin and several other pesticides were successfully produced and used in large quantities (Jalees and Vemuri, 1980; Jabbar and Mallick, 1994; Delaplane, 2000; Castelo-Grande et al., 2010; Dixit et al., 2019). DDT was the first organic chemical to be used as a pesticide (insecticide), as it protected several economical crops and eradicated malaria from numerous parts of the world (Kumar et al., 2018).

Period between 1940 and 1950 has been rightly termed as 'pesticide era' due to global incessant pesticide usage (Graeme, 2005; Bhandari, 2017). Although, in 1961, pesticide usage peaked, with the rising public awareness of the environmental ill effects of chemical pesticides, there was a significant decrease in new pesticide production 1962 afterwards.

A marine biologist and writer, Rachel Carson penned down a book 'Silent Spring' in 1962, wherein she highlighted direct and indirect toxicological effects of pesticides on non-target organisms in such a manner that it created mass public awareness and outcry, which made politicians and policymakers to formulate environmental regulations against indiscriminate use of harmful pesticides (Jabbar and Mallick, 1994; Delaplane, 2000; Dixit et al., 2019). Although, IPM based on use of biological predators for pest control was introduced in 1969, but that time it failed to be an effective alternative of chemical pesticides (Delaplane, 2000; Dixit et al., 2019). In 1972, production and use of DDT was entirely banned and use of other harmful pesticides like dieldrin, endrin was restricted in the USA. Till 1975, around only 1175 pesticides were registered in the United States (Pandey et al., 2018). However, scores of new pesticides were introduced from 1975–1990 (History of pesticide use 1998) and around 23,400 pesticides were registered with US-EPA (Environmental Protection Agency) by 1991 (Singhvi et al., 1994).

In 1995, a convention held in Stockholm proposed to ban 12 toxic POPs (dirty dozen), to which more chemicals were added afterwards. Later in 2001, an international legal treaty named as 'Stockholm Convention' was opened for ratification and was finally signed by 179 countries, which then came into power in 2004, around after a decade after UNEP's (United Nations Environment Programme) call for global action on POPs in 1995 (UNEP, 2009). This convention proposed to ban or restrict the production, release, and use of selected harmful chemicals including polychlorinated biphenyls (PCBs) and a range of organochlorine pesticides (OCPs) in all participant countries (UNEP, 2009; Bharat, 2018). In the year 1997, 1.2 billion pounds of pesticides were utilized globally, of which 77% was applied in the agriculture sector, 12% by government and industrial organizations, and 11% by households (US-EPA, 2000).

4.3.2 CURRENT GLOBAL SCENARIO OF PESTICIDE PRODUCTION AND USAGE

Nowadays, a wide range of chemical pesticides, composed of diverse active ingredients, is available globally (Schwarzenbach et al., 2010; US-EPA,

2019). Globally, 7 million tons of chemical pesticides are manufactured annually (Tilman et al., 2001). Around 4.6 million tons of pesticides are applied annually across the world (Zhang et al., 2011), of which 45% is applied in Europe, 24% in the USA only, just 3.75% in India and remaining 27% in rest of the world (Pathak and Dikshit, 2011), signifying that Europe is rampant pesticide consumer and is followed by Asia (Zhang et al., 2011). Brazil, China, France, Japan, and the United States are the leading pesticide producers globally (Zhang et al., 2011; Pandey et al., 2018). In Europe, 99% of pesticide emanation is from the agriculture sector; however, industrial and household pesticide usages have a negligible impact (EEA, 2016; Spina et al., 2018). Recently, it has been estimated that in the global pesticide market, bactericides represent 26%, insecticides 30%, and herbicides represent 30–40% of all the agricultural applied pesticide (Zhao et al., 2017; Sun et al., 2018).

Several workers provided different estimates of pesticide usage per hectare; however, such estimates have proven to be inexact, as the quantity of chemicals requisite to manage pests depends on several factors such as type of pesticide applied, method of application, crop to be treated and climatic conditions of that region (Schwarzenbach et al., 2010). For instance, (a) newly developed pesticides work even at lower concentrations as compared to conventional ones (Schwarzenbach et al., 2010); (b) pesticides are applied throughout the year in tropical and subtropical regions, as there is no winter in such regions and agricultural practices continue throughout the year. This continuous application considerably increases pesticide usage and raises the pesticide concentration in environment (Weiss et al., 2016).

World population has been estimated to reach nearly 9.4 billion by 2050, subsequently growing population will necessitate increase in the productivity of crops, which will further intensify the use of diverse pesticides to increase crop production, so as to warrant food security (Tilman et al., 2001; Alexandratos and Bruinsma, 2012; Gill and Garg, 2014). Although, large-scale usage of pesticides contributed a lot to the progress and success of green revolution and is still essential means of sustaining agriculture and global food security (Carvalho, 2006; Rani and Dhania, 2014). However, despite pesticide application, around 1/3rd of the global agricultural production is still lost annually as a consequence of attack by resistant pests (Rani and Dhania, 2014).

4.3.3   SCENARIO OF PESTICIDES PRODUCTION AND USAGE IN INDIA

In India, the use of synthetic pesticides commenced in 1948, and their production commenced in 1952 from BHC technical plant in Kolkata (Abhilash and Singh, 2009). Then, Union Carbide constructed a pesticide production plant UCIL (Union Carbide India Ltd.) in Bhopal in 1969 and continued production of pesticides till 1984's Bhopal Tragedy. Currently, India is the largest pesticide producer in Asia and the 12th largest pesticide user globally (Abhilash and Singh, 2009; Dixit et al., 2019).

In India, the consumption of pesticides is 0.5 Kg/hectare (Abhilash and Singh, 2009; Boudh and Singh, 2019), which is far lower as compared to developed countries; however, the problem of persistent pesticide residues is steep in India since 1960 (Pathak and Dikshit, 2011). As against the global pattern of highest herbicide consumption, insecticides (76%) are the most used pesticide in India, followed by herbicides and fungicides (Abhilash and Singh, 2009), mainly due to increased insect outbreaks by warm humid climatic conditions in India (Rani and Dhania, 2014). Currently, India has adopted IPM system for pest management, as a substitute to synthetic pesticides (Rani and Dhania, 2014).

4.4   SELECTION OF REMEDIATION TECHNIQUE THROUGH SCREENING APPROACHES

Globally, numerous techniques are available for decontamination of pesticide polluted environments (Sun et al., 2017), which differ in terms of efficiency, environmental impact and economic benefit, thus posing difficulties in the selection of suitable remediation technology for a particular situation. In addition, remediation techniques to combat pollutants vary for air, soil, and water decontamination. Nature of contaminated area, type, and concentration of contaminant, and cost-benefit involved in treatment also plays a crucial role in the selection of appropriate remediation technique (Castelo-Grande et al., 2010; Bhawana and Fulekar, 2012; Morillo and Villaverde, 2017; Kumar et al., 2018).

In the present era of science and technology, most apposite remediation technology for a particular circumstance is not selected based on familiarity; instead, it is selected based on rational screening methods (Mardani et al., 2015; Tian et al., 2019). Selection of appropriate remediation technology through proper screening methods is crucial for the restoration of pesticide-contaminated ecosystems (Bhawana and Fulekar, 2012; Tian et al., 2019). Several factors like environmental factors, economic factors, social factors, and technical factors are used as screening indexes (Zhang et al., 2012; Søndergaard et al., 2018; Tian et al., 2019). Application of screening method also proves to be economical in terms of time and cost. Tian et al. (2019) reported that several screening approaches are available for selection of suitable remediation technique viz., AHP (analytic hierarchy process), OWA (ordered weighted average), PROMETHEE (Preference Ranking Organization Method for Enrichment Evaluation), SAW (simple additive weighting), and TOPSIS (technique for order preference by similarity to ideal solution). However, using a hybrid approach, formed by combining several approaches, provided results similar to practical remediation (Tian et al., 2019). Furthermore, this chapter presents an overview of cost-benefit assessment of the existing and emerging remediation technologies, which will facilitate the selection of the most promising remediation technology.

4.5 GLOBAL SCENARIO OF PESTICIDE REMEDIATION TECHNIQUES

Abandoning pesticide-contaminated agricultural areas for a long time can reduce the contamination to some extent (Kardol and Wardle, 2010). In nature, several physical, chemical, physicochemical, biological, and biochemical degradation processes enact on pesticide residues; however, they still persist till centuries and stockpile in the ecosystem because of their highly stable nature and slow degrability (Morillo and Villaverde, 2017). Therefore, it becomes pertinent to search for remediation technologies for significant decontamination of pesticide-contaminated environments. Nevertheless, remediation technologies face numerous challenges due to onsite simultaneous presence of many pollutants and currently, very few remediation technologies have been tested and applied commercially at full-scale. Only those techniques that uphold the concept of "green sustainable restoration," i.e., reduce the secondary pollution while remediating is preferable.

Across the globe, several techniques are used to decontaminate pesticide-polluted environments (Sun et al., 2017). To combat pesticide pollution in agricultural and non-agricultural settings, diverse techniques categorized as physical, chemical, physicochemical, biochemical, and biological remediation techniques have been employed worldwide either independently or in alliance. This chapter does not describe methodological details of each pesticide remediation technique, instead it provides global scenario of various remediation techniques employed hitherto. The outline of all the remediation technologies employed for pesticide-contaminated environs is presented in Figure 4.1.

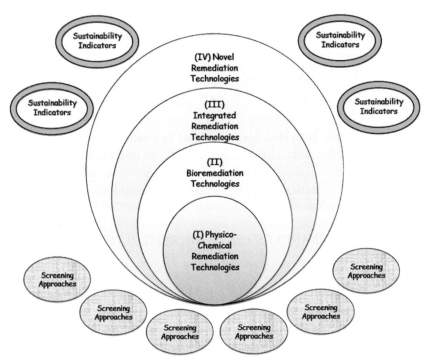

FIGURE 4.1   Pesticide remediation technologies selected through screening approaches for sustainable remediation of pesticide-contaminated environs.

Pesticide treatment technologies are of two types: (a) contaminant immobilization technologies (b) treatment technologies. Processes involved in treatment technologies are separation and destruction (Morillo and Villaverde, 2017). Remediation treatments can be applied in three ways viz., (i) *In situ* (treatment under natural setting on the place where the contamination occurred), (ii) Onsite (contaminated soil is mined, cleaned onsite, and then returned to the original spot), or (iii) *Ex situ* (contaminated soil is mined and carried to other sites for treatment) (Morillo and Villaverde, 2017).

4.5.1   PHYSICO-CHEMICAL REMEDIATION TECHNOLOGIES

Globally, physical and chemical remediation technologies are most often used in alliance in order to increase remediation efficiency (Marican and Durán-Lara, 2017), and are collectively termed as physicochemical remediation technologies. These are presently considered as 'conventional technologies.' Pesticide remediation through physicochemical technologies involves the application of physical means and chemical agents for the removal of hazardous pesticides (Marican and Durán-Lara, 2017).

To tackle the problem of pesticide pollution across the world, several phyico-chemical remediation techniques were employed over the last few decades viz., (a) land-filling (dig and dump for soil and pump and treat for water), (b) adsorption (using activated carbon, clays, cyclodextrins, dendrimers, hyper-crosslinked polymers, polymeric material, synthetic surfactants and zeolites), (c) thermal techniques (incineration, desorption, pyrolysis, and vitrification), (d) soil washing and flushing, (e) chemical oxidation-reduction technology (fenton reaction, dehalogenation, and ozonation by zero valent iron, UV-H_2O_2 and UV-ozone), (f) photocatalysis and photodegradation, (g) air sparging, (h) ultra sound/ultrasonic assisted remediation, (i) chemical leaching repair technology, (j) chemical/solvent extraction, (k) solidification/stabilization, (l) soil vapor extraction, (m) ion exchange, (n) supercritical fluids extraction, (o) plasma technology, and (p) electrokinetic remediation (Frazar, 2000; Castelo-Grande et al., 2010; Dadrasnia et al., 2013; Jin et al., 2017; Marican and Durán-Lara, 2017; Morillo and Villaverde, 2017; Baldissarelli et al., 2019; Boudh and Singh, 2019; Xing-Lu et al., 2019). Among the studied physicochemical remediation techniques, Fenton technology is the most promising technique for remediation of pesticide-contaminated environs (Baldissarelli et al., 2019).

Each technique has its merits and demerits. These conventional remediation techniques are less efficient, labor-intensive, expensive, cause secondary pollution, and thus can damage the environment (Bhandari, 2017; Marican and Durán-Lara, 2017). The costs involved in carrying out different remediation techniques are mentioned in the subsequent section.

4.5.2 BIOREMEDIATION TECHNOLOGIES

Bioremediation or biological remediation technologies involving the use of microbes, plants or merely their enzymes for degradation of pollutants, have been investigated for decades as an alternative sustainable environmental clean-up technology (Bhandari, 2017; Morillo and Villaverde, 2017; Kumar et al., 2018). In contrast to conventional remediation techniques, bioremediation is efficient, economic, and environmentally safe tactic (Zhang and Qiao, 2002; Jin et al., 2017; Marican and Durán-Lara, 2017; Kumar et al., 2018; Parween et al., 2018; Sun et al., 2018; Boudh and Singh, 2019; Xing-Lu et al., 2019), and has emerged as a powerful practical technology for the remediation of diverse pollutants (Spina et al., 2018).

Globally, various bioremediation techniques are employed viz., (a) phytoremediation (phytodegradation, phytoextraction, phytostabilization, phytostimulation, phytotransformation, phytovolatilization, and rhizoremediation), (b) microbioremediation(bacterial bioremediation, mycoremediation, and phycoremediation), (c) land farming, (d) composting, (e) animal repair technology (vermiremediation), (f) biosurfactants, (g) constructed wetlands, (h) bioattenuation, (i) bioaugmentation, (j) use of biobeds, (k) biofiltration, (l) bioleaching, (m) biopiling, (n) bioreactors, (o) biorehabilitation, (p) biosparging, (q) biostimulation, and (r) bioventing (Castelo-Grande et al., 2010; Fulekar, 2010; Dadrasnia et al., 2013; Sharma and Pandey, 2014; Tripathia et al., 2015; Jin et al., 2017; Marican and Durán-Lara, 2017; Morillo and Villaverde, 2017; Kumar et al., 2018, 2019; Parween et al., 2018; Sun et al., 2018; Boudh and Singh, 2019).

Phytoremediation is an emerging remediation technology that has attracted lots of attention since the last two decades (Sharma and Pandey, 2014; Koelmel et al., 2015; Kumar et al., 2019). Likewise, microbioremediation has been found to be a safe and sustainable means of decontaminating the environment (Kumar et al., 2018; Pandey et al., 2018; Spina et al., 2018; Boudh and Singh, 2019). New emerging reliable techniques like bioremediation is the most preferred and widely used sustainable remediation technique for mitigation of detrimental pesticides nowadays (Bhandari, 2017; Kumar et al., 2018; Parween et al., 2018; Boudh and Singh, 2019). However, it still demands further research for its full-scale implementation commercially.

4.5.3 INTEGRATED REMEDIATION TECHNOLOGIES

The sole aim of each individual remediation technology is to eliminate pollutants from contaminated areas; however, no attention is paid to the improvement of soil quality, community structure and function of microbes, and restoration and reuse of contaminated areas (Tripathi et al., 2014). Integrated remediation technology is the most effective remediation technology, since it overcome the demerits of every individual remediation technique (Sun et al., 2017), by undertaking dual function of removing pollutants by systematic and sequential integration of several remediation techniques and simultaneously improving the structural and functional attributes of soil and aquatic ecosystems (Tripathi et al., 2019). Various cases wherein integrated remediation technology was successfully employed are:

1. **Sequential and Integrated Remediation Approach (SIRA):** Tripathi et al. (2019) recommended a novel approach, namely SIRA, which integrates several remediation techniques systematically in step-wise comportment, starting from addition of organic residues and other organic amendments, application of pesticide degrading microbes (microbioremediation), chemical alterations, followed by phytoremediation using herbs, shrubs, and trees for rapid, economic, and successful removal of pesticides and sustainable recuperation and reuse of polluted soil.

2. **Integrated Biochemical Remediation Techniques:** Sun et al. (2017) suggested that preeminent techniques to combat multiple pollutants are integrated biochemical remediation techniques such as surfactant-enhanced-bioremediation (SERB). Surfactants accelerate pesticide biodegradation by facilitating their trans-membrane movement into microbial cells (Zhang and Zhu, 2012).

3. **Plant-Microbe-Associated Bioremediation:** Vergani et al. (2017) used rhizospheric synergistic interactions of microbial and plant communities for bioremediation of recalcitrant pesticides in agricultural ecosystems. Root exudates augment the bioavailability of pesticides for efficient degradation by microbes (Javorska et al., 2009).

4. **Electrokinetic-Bioremediation:** Recently, Barba et al. (2017) developed an integrated remediation technique by combining electrokinetic and bioremediation techniques for *in situ* remediation of pesticides.

However, additional research is prerequisite for designing and evaluation of such integrated remediation technologies, through pilot studies under varied climatic conditions, for their successful field implementation.

4.5.4 NOVEL REMEDIATION TECHNOLOGIES

Under the changing climate scenario, incessant modernization of human life, and limitations of the existing remediation technologies, it is imperative to develop novel remediation technologies for combating continuously evolving pollutants. Many countries across the globe, invested resources for the development of advanced and innovative remediation technologies

(Kuppusamy et al., 2016). Additionally, in order to assess the reliability of innovative technologies, various laboratory and pilot-scale experiments have been conducted by numerous research groups. Superfund innovative technology evaluation (SITE) and Federal Remediation Technologies Roundtable (FRTR) are some important programs amongst several global programs that promote novel remediation technologies (Frazar, 2000). Various novel efficient remediation techniques have been developed so far:

1. **Nanotechnology-based Remediation Technologies:** In the emerging field of environmental nanotechnology, remediation techniques based on nanotechnology are novel, innovative, and promising strategies for decontamination of pesticide-contaminated air, soil, and water ecosystems, with the least negative environmental impacts (Fulekar, 2010;Abhilash et al., 2012; Das et al., 2015, 2018; Rani et al., 2017; Singh et al., 2020). Nanomaterials and nanoparticles act as superb adsorbents, transformers, catalysts, and sensors and possess tremendous application potentials for the detection, deterrence, monitoring, and remediation of contaminated ecosystems owing to their novel physical, chemical, biological, and electronic properties like large specific surface area, surface multi-functionality, super-reactivity, and lack of toxicity (Bhawana and Fulekar, 2012; Zhan and Jiang, 2015; Das et al., 2018; Singh et al., 2020). A potent pesticide, 2,4-dicholorophenoxyacetic acid (2,4-D) has been successfully remediated by using magnetic nanoparticles. Metal coated algae, bacteria, fungi, viruses, and yeasts act as 'biological nanofactories' and effectively help in environmental remediation (Das et al., 2018). Furthermore, nanomaterials' integrated application for augmenting plant-microbe-associated bioremediation has intense future prospects (Abhilash et al., 2012; Sun et al., 2017).

   Alternatively, nanocides can also be used as a substitute to the commercially available chemical pesticides, which will subsequently reduce the chemical pollutant load. For example, surface-functionalized nanosilica has been used as an effective insecticide (Debnath et al., 2012; Das et al., 2015). These nanocides are target specific, non-toxic, and environmentally least detrimental (Debnath et al., 2012). With the progress in environmental nanotechnology, more nanotools for environmental remediation will be on hand; however, all those nanotools must be monitored continuously for their possible detrimental environmental effects (Das et al., 2015).

2. **Bio-Electrochemical Remediation Systems:** The application of electrochemically active microorganisms called bio-electrochemical systems (BES), for bioremediation is a novel innovative facet in the field of electromicrobiology (Fernando et al., 2018). Electrochemically active microorganisms are formed by extracellular electron transfer (EET) from/to an electron donor/acceptor located on the microbial cell membrane. Domínguez-Garay et al. (2016) recently conducted an innovative study and reported that by using BES, atrazine degrade 20-fold faster in comparison to non-BES control. BES is one of the most promising, eco-friendly, and sustainable pesticide remediation technology (Fernando et al., 2018); however, only a few studies have been conducted in this perspective globally.

3. **Role of Biochar in Pesticide Remediation:** Biochar is carbon-rich material produced by combustion of biological residues under oxygen limiting conditions (Beesley et al., 2011; Carey et al., 2015; Hadi et al., 2015; Varjania et al., 2019). Biochar enhances the water holding capacity of soil, improves aeration, enhances growth of soil microbes, and alleviates their metabolic activity, which in turn facilitates pesticide degradation (Varjania et al., 2019). Biochar acts as a prospective sink for pesticides and reduces pesticide mobility due to their large surface areas and high sorption capability (Beesley et al., 2011; Cederlund et al., 2016; Varjania et al., 2019).

Globally, amongst all the pesticide remediation technologies, substantial research has been done on the application of biochar during the past decade (Alvarez et al., 2017; Varjania et al., 2019). Furthermore, several researchers reported that soil biochar amendment suppresses greenhouse gases emissions and, thus, mitigates global climate change (Cederlund et al., 2016; Varjania et al., 2019). Therefore, to comprehend the full potential of biochar in pesticide remediation and the mitigation of global climate change, an in-depth research is requisite.

4. **Genetically Modified Organism (GMO)-Based Remediation Technology:** The prospective of genetic engineering and the role of GMOs in enhancing pesticide remediation was recognized in early 1980, with a preliminary focus mainly on engineered microorganisms (Macek et al., 2007). Genetically modified plants (GMP) reduce environmental contamination either by diminishing the dependence on large quantities of pesticides or by raising the bioremediation

efficiency. For example, propagation of a genetically engineered pest-resistant cotton called Bt-cotton (gene coding for Bt toxin from *Bacillus thuringiensis* bacterium inserted in cotton to make it pest resistant) eventually results in the reduction of pesticide usage in cotton fields (Macek et al., 2007; Abhilash et al., 2012; Marican and Durán-Lara, 2017). Furthermore, specific microorganisms' bioremediation efficiency can be raised by applying gene-editing tools like CRISPR Cas, TALEN, and ZFNs (Rani and Dhania, 2014; Jaiswal et al., 2019). Thus, genetic engineering can be employed to reduce the environmental pollutant load.

5. **Role of Multi-Omics Technologies:** Multi-omics, i.e., genomics, metabolomics, metagenomics, proteomics, and transcriptomics approaches, can be employed for discovering the potential pesticide degrading microbes, functional genes, their biodegradation pathways and for enhancing their pesticide biodegradation potential (Jeffries et al., 2018; Kumar et al., 2018; Jaiswal et al., 2019). Especially, metagenomics is deemed as a ladder for moving up and progressing in the bioremediation field (Guerra et al., 2018; Jeffries et al., 2018; Roy et al., 2018; Jaiswal et al., 2019). Thus, rapidly developing multi-omics approaches equips global environmental scientists with vast opportunities to efficiently remediate pesticide-contaminated areas.

6. **Integrated Pest Management (IPM) System:** Based on use of biological control agents for pest control, IPM system acts as an effectual substitute of synthetic chemical pesticides (Tewari et al., 2012; Dixit et al., 2019). These biocontrol agents are also recognized as bio-rational pesticides. Furthermore, biopesticides are important as future pesticide agents (Bhat et al., 2019). In order to curb the menace of pesticide pollution, IPM system has been adopted globally for pest management (Dixit et al., 2019).

4.6 INDICATORS OF SUSTAINABLE REMEDIATION

Foundation of sustainable remediation is based on three pillars of sustainability viz., economy, environment, and society (SuRF-UK, 2010; Sparrevik et al., 2011, 2012; Rosén et al., 2015), which serve as sustainable indicators for remediation technologies (Harclerode et al., 2015; Tripathia et al., 2015).

Sustainable remediation implies efficiently remediating the contaminated site with minimum negative economic, environmental, and social impacts (Søndergaard et al., 2018). Workers from various academia, organizations, and committees are currently in the hunt for development of valid evaluation methods, that will function as indicators to reveal amendments in bioavailability, concentration, and toxicity of target contaminant, for ensuring that the remediation goals are accomplished, including sustainability criteria (Harclerode et al., 2015; Tripathia et al., 2015; Song et al., 2017). However, till date, only a few studies have reported valid and reliable indicators of sustainable remediation.

The amount of remaining pollutants, ecological threats, and soil/sediment characteristics double up as indicators of *in situ* remediation sustainability (Song et al., 2017). Bioassays (like biotoxicity tests) are employed to assess the success of the bioremediation process (Prokop et al., 2016; Garcia-Carmona et al., 2017). Yang et al. (2016) developed and used a mathematical modeling technique based on *Caenorhabditis elegans* bioindicator for evaluating the threat of nanoparticles produced while *in situ* remediation (Song et al., 2017). Søndergaard et al. (2018) selected a multi-criteria assessment (MCA) technique for assessing the sustainability of remediation alternatives for Groyne 42 site, one of the major polluted sites in Denmark. As the name indicates, this technique permits combined assessment of array of indicators like remediation effect, remediation time, economic, environmental, and social indicators, which may be assessed either quantitatively or qualitatively.

Besides being an imperative component of the entire remediation project, these indicators also assist in further reuse and management of contaminated site (Apitz et al., 2005; Tripathia et al., 2015; Kuppusamy et al., 2016). However, most of the indicators developed so far lack practicability and are not feasible. Therefore, while developing new remediation technologies, it is crucial to focus on development of reliable and valid post-remediation evaluation/monitoring methods for determining the remediation effectiveness, success, and sustainability (Garcia-Carmona et al., 2017; Song et al., 2017).

4.7 ECONOMICS OF GLOBAL REMEDIATION SECTOR

Globally, US$505.5 billion are contributed annually to revenue by the remediation sector indicating that this sector is economically significant (Kuppusamy et al., 2016). The United States is the leading global contributor to environmental income through remediation sector, followed by Western Europe (Singh et al., 2009; EBI, 2013; US-ITC, 2013). Global remediation sector will show further substantial growth in the future across the world (US-EPA, 2019).

In the process of decision-making, comparative cost-benefit appraisal of the possible technological options is of utmost importance (Marican and Durán-Lara, 2017). The cost-benefit appraisal is an element of complex spatial planning that reduces remediation costs and upgrades the benefits (Kuppusamy et al., 2016). The cost of remediation depends on the remediation technique selected, which is in turn determined by type of contaminant, site characteristics (*in situ* or *ex situ*), and volume of material to be remediated (Kuppusamy et al., 2016; Marican and Durán-Lara, 2017). However, such specific studies concerning the accurate and reliable cost-benefit assessment of pesticide remediation techniques are not easy and are still lacking.

Scores of remediation technologies are currently available. Globally, physical, chemical, physicochemical, biochemical, and biological remediation techniques, employed either independently or in alliance, are the only options for the remediation of contaminated sites. However, modern research is determined on developing novel remedial methods by integrating the principles of promising existing technologies. Moreover, it becomes easier to choose and implement the most suitable remediation technology, only when proper cost-benefit appraisal of the existing and emerging remediation technologies is easily accessible. In this regard, this chapter is of great substance.

Comparing the exact costs involved in different remediation technologies accurately is difficult, as the costs vary widely according to the type of pollutant remediated (Frazar, 2000; Castelo-Grande et al., 2010). So, estimated costs are considered while comparing different remediation techniques. It has been reported that conventional *ex situ* remediation technologies are much more expensive than the *in situ* technologies (Marican and Durán-Lara, 2017). Quarry of contaminated soil and disposal as landfills is the most expensive remediation technique and costs around US$ 2200–2400/ton (Weber et al., 2011). *In situ* soil washing costs around $50–$80/m³, while as *ex situ* soil flushing costs $150–$200/m³ (Bini, 2009; Kuppusamy et al., 2016). Field-scale chemical dehalogenation treatment costs around US$200–$500/ton (FRTR, 2012; Kuppusamy et al., 2016). The estimated cost of cyclodextrins-based remediation technology is about $220/ton (Gruiz et al., 2009). When a specific type of contaminated site is selected for remediation, incineration costs US$0.74–$1.25/m³ (Kuppusamy et al., 2016). US$115–$205/ton is the operational cost of thermal desorption technique

(Frazar, 2000; Castelo-Grande et al., 2010). Pyrolysis costs nearly about US$300/ton (FRTR, 2012; Kuppusamy et al., 2016). US$31.1/m^3 is the estimated cost of ultrasonic thermal desorption technique upon using 31 kHz frequency for onsite remediation (Kuppusamy et al., 2016). Solidification/ stabilization costs nearly US$132–$263/m^3 (FRTR, 2012; Kuppusamy et al., 2016). Air sparging costs less than $1/1000 L and falls among the cheapest remediation techniques (Dadrasnia et al., 2013). The estimated cost involved in ion exchange remediation technology is nearly about US$ 0.3–$ 0.8/4000 L of water (FRTR, 2012; Kuppusamy et al., 2016).

Gavrilescu (2005) estimated that phytoremediation costs about $60/m^3, however, the maintenance costs of phytoremediation are very low (Gerhardt et al., 2009). *In situ* microbioremediation costs about $6.4–$150/m^3, while as *ex-situ* microbioremediation costs $150–$500/m^3 (Gavrilescu, 2005; Bini, 2009). Bioremediation, by using aerobic/anaerobic cycling system costs in the range of $80–$120/ton (Frazar, 2000). Land farming is simple, economical, and eco-friendly technique and costs around US$30–$60/ton of soil (US-EPA, 2012; Kuppusamy et al., 2016). It has been extensively implemented in most polluted sites across the world (Hejazi et al., 2003; Martin et al., 2005). Composting costs around US$314–$458/m^3 (FRTR, 2012; Kuppusamy et al., 2016). Biosurfactants' production is quite expensive and costs between 1 to 60 USD/Kg (Mao et al., 2015). The estimated cost of constructed wetland treatment is around US$ 1.36/1000 gal water over a period of ten years (FRTR, 2012; Kuppusamy et al., 2016). Biopiling is easy and effective remediation technique for diverse pollutants and costs around US$30–$90/ton of polluted soil (Kuppusamy et al., 2016). Remediation cost by means of bioreactors/bioslurries varies from US$130–200/m^3 (US-EPA, 2012; Kuppusamy et al., 2016).

4.7.1   RELATIONSHIP BETWEEN ECONOMY AND ENVIRONMENTAL REMEDIATION

Erakhrumen (2011) hypothesized that globally there exist a relationship between country's economic status (per capita income) and their echelon of environmental concerns, and remediation technologies. Initially, with increase in per capita income, environmental degradation increases till a turning point from which environment starts improving through the availability of suitable resources, constructive policies, increased public awareness and institutional interferences, ensuring the desired improvement (Erakhrumen, 2011).

Developing countries focus more on instantaneously discernible environmental contamination like soil contamination, surface-water pollution, etc. Alternatively, developed countries, besides this, deal continually with background-concealed pollution like groundwater pollution, effects of which might come to surface after a protracted time period (Elekwachi et al., 2014). Thus developing nations are at extra jeopardy of critical environmental problems. Furthermore, developed nations reserve huge funds (from taxes and polluter money) for remediation of contaminated sites and environmental regulations are established and implemented through robust institutions. However, such institutions either do not exist or are weak in developing countries which establish unproductive environmental regulations, often followed with delayed enforcement (Lee and George, 2000; Koelmel et al., 2015).

Moreover, in developing countries, data on exposure scenarios and toxicity risk factor of poor populations residing mostly in direct contact of various pollution sources are lacking (Maantay, 2002; Koelmel et al., 2015; Weiss et al., 2016). In such areas, the issue of pollution does not receive high precedence due to lack of awareness about pollution impacts, limited number of monitoring systems to oversee the production of hazardous pollutants, lack of sound data and good risk assessments, limited budget, lack of appropriate policy and analytical technologies, weak environmental regulations, and lack of government enforcements, which result in huge production and release of noxious pollutants into the environment, which further lead to higher exposures and risks (Blacksmith Institute and Green Cross, 2012; Weiss et al., 2016). For example, improper and profuse pesticide application near drinking water sources further increase the risk of exposure to pesticide pollution (Pesticide Action Network Asia and the Pacific, 2010; Weiss et al., 2016).

In comparison to developing nations, developed ones have officially documented number of contaminated sites, along with substantial technological frameworks, policies, and management measures for clean-up of those contaminated sites (Tripathia et al., 2015). In addition, developed countries practice sorting municipal and industrial waste methods, which reduce the cost of remediation. However, no such practice is followed by developing nations which further enhance the cost of remediation (Koelmel et al., 2015). Consequently, there is an immediate necessity to devise apposite technological and policy frameworks based on green technologies to remediate and restore contaminated sites in developing countries (Tripathia et al., 2015).

4.8 CONCLUSION

With the advent of green revolution conventional natural pesticides were replaced by synthetic chemical pesticides. Nowadays, a wide range of chemical pesticides composed of diverse active ingredients are available in the global market, which are currently an imperative part of modern agriculture and healthcare. However, environmental contamination by pesticides has emerged as a major threat globally, which destroys biodiversity, threatens global environment security, and causes chronic abnormalities in humans. For restoration of pesticide-contaminated ecosystems, selection of appropriate remediation technology through proper screening methods is crucial. This chapter presents an overview of the cost-benefit assessment of the existing and emerging remediation technologies, facilitating the selection of the most promising remediation technology. Very few remediation technologies have been tested and applied commercially at full-scale. This chapter provides global scenario of pesticide remediation techniques employed to clean pesticide pollutants from contaminated environments. In order to combat pesticide pollution in agricultural as well as non-agricultural settings, diverse techniques categorized as physical, chemical, physicochemical, biochemical, and biological remediation techniques have been employed worldwide either independently or in alliance. Physico-chemical remediation technologies are the conventional techniques and are less efficient, labor-intensive, expensive, cause secondary pollution, and thus can damage the environment. New emerging reliable techniques like bioremediation are the most preferred and widely used sustainable remediation technique for mitigation of detrimental pesticides. However, it demands further research for its full-scale implementation commercially. Integrated remediation technologies are gaining attention globally as the most effective remediation technology for pesticide removal and restoration and reuse of contaminated areas. Many countries across the globe invested resources for the development of advanced and innovative remediation technologies. The use of novel remediation technologies will help in the safeguard of biodiversity and will prevent environmental pollution. While developing new remediation technologies, it is crucial to focus on the development of reliable and valid post-remediation evaluation/monitoring methods which will serve as indicators for determining the remediation effectiveness, success, and sustainability. Moreover, it becomes easier to choose and implement the most suitable remediation technology, only when proper cost-benefit appraisal of the existing and emerging remediation technologies is easily accessible. In this regard, this chapter is of great substance. In developing countries, there is an immediate necessity to devise appropriate technological and policy frameworks based on green technologies to remediate and restore contaminated sites. Modern research is determined on developing novel remedial methods by integrating the principles of promising existing technologies. In order to curb the menace of pesticide pollution, IPM system has been adopted globally, as an effectual substitute to synthetic pesticides.

KEYWORDS

- **bioremediation**
- **global remediation sector**
- **integrated remediation technology**
- **persistent organic pollutants**
- **pesticide pollution**
- **sustainable remediation technologies**

REFERENCES

Abhilash, P. C., & Singh, N., (2009). Pesticide use and application: An Indian scenario. *J. Hazard Mater., 165*, 1–12.

Abhilash, P. C., Powell, J. R., Singh, H. B., & Singh, B. K., (2012). Plant-microbe interactions: Novel applications for exploitation in multipurpose remediation technologies. *Trends. Biotechnol., 30*, 416–420.

Alexandratos, N., & Bruinsma, J., (2012). *World Agriculture Towards 2030/2050: The 2012 Revision*. ESA Working Paper. Rome: FAO.

Alvarez, A., Saez, J. M., Costa, J. S. B., Colin, V. L., Fuentes, M. O., Cuozzo, S. A., Benimeli, C. S., et al., (2017). Actinobacteria: Current research and perspectives for bioremediation of pesticides and heavy metals. *Chemosphere, 166*, 41–62.

Apitz, S. E., Davis, J. W., Finkelstein, K., Hohreiter, D. W., Hoke, R., & Jensen, R. H., (2005). Assessing and managing contaminated sediments: Part I, developing an effective investigation and risk evaluation strategy. *Integr. Environ. Assess Manag., 1*, 2–8.

Baldissarelli, D. P., Vargas, G. D. L. P., Korf, E. P., Galon, L., Kaufmann, C., & Santos, J. B., (2019). Remediation of soils contaminated by pesticides using physicochemical processes: A brief review. *Planta Daninha, 37*, e019184975.

Barba, S., Villasenor, J., Rodrigo, M. A., & Canizares, P., (2017). Effect of the polarity reversal frequency in the electrokinetic-biological remediation of oxyfluorfen polluted soil. *Chemosphere, 177*, 120–127.

Beesley, L., Moreno-Jiménez, E., Gomez-Eyles, J. L., Harris, E., Robinson, B., & Sizmur, T., (2011). A review of biochars' potential role in the remediation, revegetation and restoration of contaminated soils. *Environ. Pollut, 159*, 3269–3282.

Bhandari, G., (2017). Mycoremediation: An eco-friendly approach for degradation of pesticides. In: Prasad, R., (ed.), *Mycoremediation and Environmental Sustainability, Fungal Biology* (pp. 119–131). Springer International Publishing AG.

Bharat, G. K., (2018). *Persistent Organic Pollutants in Indian Environment: A Wake-Up Call for Concerted Action.* Policy Brief, The Energy and Resources Institute (TERI). www. teriin.org (accessed on 30 October 2020).

Bhat, R. A., Beigh, B. A., Mir, S. A., Dar, S. A., Dervash, M. A., Rashid, A., & Lone, R., (2019). Biopesticide techniques to remediate pesticides in polluted ecosystems. In: *IGI Global* (pp. 387–407).

Bhawana, P., & Fulekar, M. H., (2012). Nanotechnology: Remediation technologies to clean up the environmental pollutants. *Res. J. Chem. Sci., 2*, 90–96.

Bini, C., (2009). From soil contamination to land restoration. In: Steimberg, R. V., (ed.), *Contaminated Soils: Environmental Impact, Disposal and Treatment* (pp. 97–137). Nova Science Publishers, Inc., New York.

Boudh, S., & Singh, J. S., (2019). Pesticide contamination: Environmental problems and remediation strategies. In: Bharagava, R. N., & Chowdhary, P., (eds.), *Emerging and Eco-Friendly Approaches for Waste Management* (pp. 245–269). Springer Nature Singapore Pte Ltd.

Carey, D. E., McNamara, P. J., & Zitomer, D. H., (2015). Biochar from pyrolysis of biosolids for nutrient adsorption and turfgrass cultivation. *Water Environ. Res., 87*, 2098–2106.

Carvalho, F. P., (2006). Agriculture, pesticides, food security and food safety. *Environ. Sci. Policy, 9*, 685–692.

Castelo-Grande, T., Augusto, P. A., Monteiro, P., Estevez, A. M., & Barbosa, D., (2010). Remediation of soils contaminated with pesticides: A review. *Int. J. Environ. Anal. Chem., 90*, 438–467.

Cederlund, H., Börjesson, E., Lundberg, D., & Stenström, J., (2016). Adsorption of pesticides with different chemical properties to a wood biochar treated with heat and iron. *Water Air Soil Pollut., 227*, 1–12.

Chaussonnerie, S., Saaidi, P. L., Ugarte, E., Barbance, A., Fossey, A., Barbe, V., & Fouteau, S., (2016). Microbial degradation of a recalcitrant pesticide: Chlordecone. *Front Microbiol., 7*, 2025.

Dadrasnia, A., Shahsavari, N., & Emenike, C. U., (2013). Remediation of contaminated sites. In: *Hydrocarbon* (pp. 65–82). Intech Open.

Das, S., Chakraborty, J., Chatterjee, S., & Kumar, H., (2018). Prospects of biosynthesized nanomaterials for the remediation of organic and inorganic environmental contaminants. *Environ. Sci. Nano, 5*, 2784–2808.

Das, S., Sen, B., & Debnath, N., (2015). Recent trends in nanomaterials applications in environmental monitoring and remediation. *Environ. Sci. Pollut. Res.*

Debnath, N., Das, S., Patra, P., & Mitra, S., (2012). Toxicological evaluation of entomotoxic silica nanoparticles. *Toxicol. Environ. Chem., 94*, 944–951.

Delaplane, K. S., (2000). *Pesticide Usage in the United States: History, Benefits, Risks, and Trends* (p. 1121). Cooperative Extension Service, The University of Georgia, College of Agricultural and Environmental Sciences, Bulletin.

Dixit, S., Srivastava, M. P., & Sharma, Y. K., (2019). Pesticide and human health: A rising concern of the 21st century. In: *IGI Global*, 85–104.

Domínguez-Garay, A., Boltes, K., & Esteve-Núñez, A., (2016). Cleaning-up atrazine-polluted soil by using microbial electro remediating cells. *Chemosphere, 161*, 365–371.

EBI, (2013). US remediation industry generates $8.07 billion in revenues. *Environmental Business International.* San Diego, USA.

EEA, (2016). *EMEP/EEA Air Pollutant Emission Inventory Guidebook: 3. Df, 3.1 Agriculture Other Including Use of Pesticides.* Luxembourg: Office for Official Publications of the European Communities.

Elekwachi, C. O., Andresen, J., & Hodgman, T. C., (2014). Global use of bioremediation technologies for decontamination of ecosystems. *J. Bioremed. Biodegrad., 5*, 1.

Erakhrumen, A. A., (2011). Research advances in bioremediation of soils and groundwater using plant-based systems: A case for enlarging and updating information and knowledge in environmental pollution management in developing countries. In: *Bio-Management of Metal-Contaminated Soils* (pp. 143–166). Springer, Netherlands.

Fernando, E. Y., Keshavarz, T., & Kyazze, G., (2018). *The Use of Bio-Electrochemical Systems in Environmental Remediation of Xenobiotics: A Review.*

Frazar, C., (2000). *The Bioremediation and Phytoremediation of Pesticide-Contaminated Sites.* U.S. Environmental Protection Agency, office of solid waste and emergency response technology innovation office, Washington, DC.

FRTR, (2012). *Remediation Technologies Screening Matrix and Reference Guide Version 4.0-Remediation Technology.* Federal Remediation Technologies Roundtable, Washington, DC. www.frtr.gov (accessed on 30 October 2020).

Fulekar, M. H., (2010). Global status of environmental pollution and its remediation strategies. In: Fulekar, M. H., (ed.), *Bioremediation Technology: Recent Advances.* Capital Publishing Company.

Garcia-Carmona, M., Romero-Freire, A., Sierra-Aragon, M., Martinez-Garzon, F. J., & Martin-Peinado, F. J., (2017). Evaluation of remediation techniques in soils affected by residual contamination with heavy metals and arsenic. *J. Environ. Manage, 191*, 228–236.

Gavrilescu, M., (2005). Fate of pesticides in the environment and its bioremediation. *Engineer in Life Sci., 5*, 497–526.

Gerhardt, K. E., Huang, X. D., Glick, B. R., & Greenberg, B. M., (2009). Phytoremediation and rhizoremediation of organic soil contaminants: Potential and challenges. *Plant Sci., 176*, 20–30.

Gill, H. K., & Garg, H., (2014). Pesticides: Environmental impacts and management strategies. In:*Pesticides- Toxic Aspects* (pp. 187–230). Intech. Open.

Graeme, M., (2005). *Resistance Management: Pesticide Rotation.* Ontario Ministry of Agriculture, Food and Rural Affairs.

Gruiz, K., Molnár, M., & Fenyvesi, E., (2009). Verification tool for in situ soil remediation. *Land Contam. and Reclam., 17*, 339–362.

Guerra, A. B., Oliveira, J. S., Silva-Portela, R. C., Araujo, W., Carlos, A. C., & Vasconcelos, A. T. R., (2018). Metagenome enrichment approach used for selection of oil-degrading bacteria consortia for drill cutting residue bioremediation. *Environ. Pollut., 235*, 869–880.

Hadi, P., Xu, M., Ning, C., Lin, C. S. K., & McKay, G., (2015). A critical review on preparation, characterization and utilization of sludge-derived activated carbons for wastewater treatment. *Chem. Eng. J., 260*, 895–906.

Hamza, R. A., Iorhemen, O. T., & Tay, J. H., (2016). Occurrence, impacts and removal of emerging substances of concern from wastewater. *Environ. Tech Innov., 5*, 161–175.

Harclerode, M., Ridsdale, D. R., Darmendrail, D., Bardos, P., Alexandrescu, F., Nathanail, P., Pizzol, L., & Rizzo, E., (2015). Integrating the social dimension in remediation decision-making: State of the practice and way forward. *Remediat. J., 26*, 11–42.

Hejazi, R. F., Husain, T., & Khan, F. I., (2009). Landfarming operation of oily sludge in arid region- Human health risk assessment. *J. Hazard. Mater., 99*, 287–302.

Jabbar, A., & Mallick, S., (1994). *Pesticides and Environment Situation in Pakistan (Working Paper Series No. 19).* Available from Sustainable Development Policy Institute (SDPI).

Jaiswal, S., Singh, D. K., & Shukla, P., (2019). Gene editing and systems biology tools for pesticide bioremediation: A review. *Front Microbiol., 10*, 87.

Jalees, K., & Vemuri, R., (1980). Pesticide pollution in India. *Int. J. Environ. Stud., 15*, 49–53.

Javorska, H., Tlustos, P., & Kaliszova, R., (2009). Degradation of polychlorinated biphenyls in the rhizosphere of rape, *Brassica napus* L. *B. Environ. Contam. Tox., 82.*, 727–731.

Jeffries, T. C., Rayu, S., Nielsen, U. N., Lai, K., Ijaz, A., Nazaries, L., & Singh, B. K., (2018). Metagenomic functional potential predicts degradation rates of a model organophosphorus xenobiotic in pesticide contaminated soils. *Front Microbiol., 9*, 147.

Jin, X., Xing, C., & Wei, Z., (2017). Present situation and evaluation of contaminated soil disposal technique. *4th International Conference on Environmental Systems Research (ICESR 2017) IOP Conf. Series:Earth and Environmental Science, 178*, 012029.

Kamel, F., & Hoppin, J. A., (2004). Association of pesticide exposure with neurologic dysfunction and disease. *Environ. Health Perspect., 112*, 950–958.

Kardol, P., & Wardle, D. A., (2010). How understanding aboveground-belowground linkages can assist restoration ecology. *Trends Ecol. Evol., 25*, 670–679.

Kim, K. H., Kabir, E., & Jahan, S. A., (2017). Exposure to pesticides and the associated human health effects. *Sci. Total Environ., 575*, 525–535.

Koelmel, J., Prasad, M. N. V., & Pershell, K., (2015). Bibliometric analysis of phytotechnologies for remediation: Global scenario of research and applications. *Int. J. Phytorem., 17*, 145–153.

Kumar, P. S., Carolin, C. F., & Varjani, S. J., (2018). Pesticides Bioremediation. In: Varjani, S. J., et al., (eds.), *Bioremediation: Applications for Environmental Protection and Management, Energy, Environment, and Sustainability* (pp. 197–222). Springer Nature Singapore Pte Ltd.

Kumar, S., Singh, R., Behera, M., Kumar, V., Sweta, R. A., Kumar, N., & Bauddh, K., (2019). Restoration of pesticide-contaminated sites through plants. In: *Phytomanagement of Polluted Sites* (pp. 313–327). Elsevier Inc.

Kuppusamy, S., Palanisami, T., Megharaj, M., Venkateswarlu, K., & Naidu, R., (2016). *Ex-situ* remediation technologies for environmental pollutants: A critical perspective. In: De Voogt, P., (ed.), *Reviews of Environmental Contamination and Toxicology* (Vol. 236, pp. 117–192). Springer International Publishing Switzerland.

Lee, N., & George, C., (2000). *Environmental Assessment in Developing and Transitional Countries-Principles, Methods, and Practice.* Hoboken (NJ): Wiley.

Levitan, L., Merwin, I., & Kovach, J., (1995). Assessing the relative environmental impacts of agricultural pesticides: The quest for a holistic method. *Agri. Ecosyst. Environ., 55*, 153–168.

Li, Z., & Jennings, A., (2017). Worldwide regulations of standard values of pesticides for human health risk control: A review. *Int. J. Environ. Res. Public Health, 14*, 826.

Maantay, J., (2002). Mapping environmental injustices: Pitfalls and potential of geographic information systems in assessing environmental health and equity. *Environ. Health Perspect., 110*, 161–171.

Macek, T., Kotrba, P., Svatos, A., Novakova, M., Demnerova, K., & Mackova, M., (2007). Novel roles for genetically modified plants in environmental protection. *Trends Biotechnol., 26*, 146–152.

Mao, X., Jiang, R., Xiao, W., & Yu, J., (2015). Use of surfactants for the remediation of contaminated soils: A review. *J. Hazard. Mater.*, *285*, 419–435.

Mardani, A., Jusoh, A., & Zavadskas, E. K., (2015). Fuzzy multiple criteria decision-making techniques and applications: Two decades review from 1994 to 2014. *Expert. Syst. Appl.*, *42*, 4126–4148.

Marican, A., & Durán-Lara, E. F., (2017). A review on pesticide removal through different processes. *Environ. Sci. Pollu. Res.*

Martin, J. A., Hernandez, T., & Garcia, C., (2005). Bioremediation of oil refinery sludge by landfarming in semiarid conditions: Influence on soil microbial activity. *Environ. Res.*, *98*, 185–195.

McLellan, J., Gupta, S. K., & Kumar, M., (2019). Feasibility of using bacterial-microalgal consortium for the bioremediation of organic pesticides: Application constraints and future prospects. In: *Application of Microalgae in Wastewater Treatment* (pp. 341–362). Springer, Cham.

Megharaj, M., Kantachote, D., Singleton, I., & Naidu, R., (2000). Effects of long-term contamination of DDT on soil microflora with special reference to soil algae and algal transformation of DDT. *Environ. Pollu.*, *109*, 35–42.

Morillo, E., & Villaverde, J., (2017). Advanced technologies for the remediation of pesticide-contaminated soils. *Sci. Total Environ.*, *586*, 576–597.

Moschet, C., Wittmer, I., Simovic, J., Junghans, M., Piazzoli, A., Singer, H., Stamm, C., et al., (2014). How a complete pesticide screening changes the assessment of surface water quality. *Environ. Sci. Technol.*, *48*, 5423–5432.

Pandey, C., Prabha, D., & Negi, Y. K., (2018). Mycoremediation of common agricultural pesticides. In: Prasad, R., (ed.), *Mycoremediation and Environmental Sustainability, Fungal Biology*, 155–179.

Pariatamby, A., & Kee, Y. L., (2016). Persistent organic pollutants management and remediation. *Proc. Environ. Sci.*, *31*, 842–848.

Parween, T., Bhandari, P., Sharma, R., Jan, S., Siddiqui, Z. H., & Patanjali, P. K., (2018). Bioremediation: A sustainable tool to prevent pesticide pollution. In: Oves, M., et al., (eds.), *Modern Age Environmental Problems and their Remediation*. Springer International Publishing AG.

Pathak, R. K., & Dikshit, A. K., (2011). Various techniques for atrazine removal. *International Conference on Life Science and Technology (IPCBEE)*, *3*. IACSIT Press, Singapore.

Pimentel, D., (1995). Amounts of pesticides reaching target pests: Environmental impacts and ethics. *J. Agric. Environ. Ethics.*, *8*, 17–29.

Prokop, Z., Nečasova, A., Klanova, J., & Čupr, P., (2016). Bioavailability and mobility of organic contaminants in soil: New three-step ecotoxicological evaluation. *Environ. Sci. Pollut. Res.*, *23*, 4312–4319.

Rani, K., & Dhania, G., (2014). Bioremediation and biodegradation of pesticide from contaminated soil and water: A novel approach. *Int. J. Curr. Microbiol. App. Sci.*, *3*, 23–33.

Rani, M., Shanker, U., & Jassal, V., (2017). Recent strategies for removal and degradation of persistent and toxic organochlorine pesticides using nanoparticles: A review. *J. Environ. Manage*, *190*, 208–222.

Rice, P. J., Rice, P. J., Arthur, E. L., & Barefoot, A. C., (2007). in pesticide environmental fate and exposure assessments. *J. Agric. Food Chem.*, *55*, 5367–5376.

Rosén, L., Back, P. E., Söderqvist, T., Norrman, J., Brinkhoff, P., Norberg, T., Volchko, Y., et al., (2015). SCORE: A novel multi-criteria decision analysis approach to assessing the sustainability of contaminated land remediation. *Sci. Total Environ.*, *511*, 621–638.

Roy, A., Dutta, A., Pal, S., Gupta, A., Sarkar, J., & Chatterjee, A., (2018). Biostimulation and bioaugmentation of native microbial community accelerated bioremediation of oil refinery sludge. *Bioresource Technol.*, *253*, 22–32.

Schwarzenbach, R. P., Egli, T., Hofstetter, T. B., Von, G. U., & Wehrli, B., (2010). Global water pollution and human health. *Annu. Rev. Environ. Resour.*, *35*, 109–136.

Sharma, P., & Pandey, S., (2014). Status of phytoremediation in world scenario. *Int. J. Environ. Bioremed. Biodegrad.*, *2*, 178–191.

Singh, A., Kuhad, R. C., & Ward, O. P., (2009). Biological remediation of soil: An overview of global market and available technologies. In: Singh, A., et al., (eds.), *Advances in Applied Bioremediation, Soil Biology*. Springer-Verlag Berlin Heidelberg.

Singh, D. S., Bhat, R. A., Dervash, M. A., Qadri, H., Mehmood, M. A., Dar, G. H., Hameed, M., & Rashid, N., (2020). Wonders of nanotechnology for remediation of polluted aquatic environs. In: Qadri, H., et al., (eds.), *Fresh Water Pollution Dynamics and Remediation* (pp. 319–339). Springer Nature Singapore Pte Ltd.

Singhvi, R., Koustas, R. N., & Mohn, M., (1994). *Contaminants and Remediation Options at Pesticide Sites*. EPA/600/R-94/202, US EPA, Office of Research and Development, Risk Reduction Engineering Laboratory. Cincinnati, OH.

Søndergaard, G. L., Binning, P. J., Bondgård, M., & Bjerg, P. L., (2018). Multi-criteria assessment tool for sustainability appraisal of remediation alternatives for a contaminated site. *J. Soils and Sediments*, *18*, 3334–3348.

Song, B., Zeng, G., Gong, J., Liang, J., Xu, P., Liu, Z., Zhang, Y., et al., (2017). Evaluation methods for assessing effectiveness of in situ remediation of soil and sediment contaminated with organic pollutants and heavy metals. *Environ. Int.*, *105*, 43–55.

Sparrevik, M., Barton, D. N., Bates, M. E., & Linkov, I., (2012). Use of stochastic multi-criteria decision analysis to support sustainable management of contaminated sediments. *Environ. Sci. Technol.*, *46*, 1326–1334.

Sparrevik, M., Barton, D. N., Oen, A. M. P., Sehkar, N. U., & Linkov, I., (2011). Use of multicriteria involvement processes to enhance transparency and stakeholder participation at Bergen Harbor, Norway. *Integr. Environ. Assess Manag.*, *7*, 414–425.

Spina, F., Cecchi, G., Landinez-Torres, A., Pecoraro, L., Russo, F., Wu, B., Cai, L., et al., (2018). Fungi as a toolbox for sustainable bioremediation of pesticides in soil and water. *Plant Biosyst.*, *152*, 474–488.

Sun, J., Pan, L., Tsang, D. C. W., Zhan, Y., Zhu, L., & Li, X., (2017). Organic contamination and remediation in the agricultural soils of China: A critical review. *Sci. Total Environ.*, *615*, 724–740.

Sun, S., Sidhu, V., Rong, Y., & Zheng, Y., (2018). Pesticide pollution in agricultural soils and sustainable remediation methods: A review. *Current Pollution Reports*.

SuRF-UK, (2010). *A Framework for Assessing the Sustainability of Soil and Groundwater Remediation*. Sustainable Remediation Forum UK, Published by Contaminated Land: Applications in Real Environments (CL: AIRE), London.

Tewari, L., Saini, J., & Arti, (2012). Bioremediation of pesticides by microorganisms: General aspects and recent advances. In: *Bioremediation of Pollutants*. I. K. International Publishing House Pvt. Ltd. New Delhi.

Tian, J., Huo, Z., Ma, F., Gao, X., & Wu, Y., (2019). Application and selection of remediation technology for OCPs-contaminated sites by decision-making methods. *Int. J. Environ. Res. Public Health, 16,* 1888.

Tilman, D., Fargione, J., Wolff, B., D'Antonio, C., & Dobson, A., (2001). Forecasting agriculturally driven global environmental change. *Science, 292,* 281–284.

Tripathi, V., Dubey, R. K., Edrisi, S. A., Narain, K., Singh, H. B., Singh, N., & Abhilash, P. C., (2014). Towards the ecological profiling of a pesticide contaminated soil site for remediation and management. *Ecol. Eng., 71,* 318–325.

Tripathi, V., Edrisi, S. A., Chaurasia, R., Pandey, K. K., Dinesh, D., Srivastava, R., Srivastava, P., & Abhilash, P. C., (2019). Restoring HCHs polluted land as one of the priority activities during the UN-International Decade on Ecosystem Restoration (2021–2030): A call for global action. *Sci. Total Environ., 689,* 1304–1315.

Tripathia, V., Fracetob, L. F., & Abhilasha, P. C., (2015). Sustainable clean-up technologies for soils contaminated with multiple pollutants: Plant-microbe-pollutant and climate nexus. *Ecol. Eng., 82,* 330–335.

UNEP (United Nations Environment Programme), (2009). Global monitoring report under the global monitoring plan for effectiveness evaluation. In: *Proceedings of the UNEP/POPS/ COP.4/33 Conference of the Parties of the Stockholm Convention on Persistent Organic Pollutants, Fourth Meeting.* Geneva, Switzerland.

US-EPA (US Environment Protection Agency), (2000). *Pesticides Industry Sales and Usage: 1996 and 1997 Market Estimates.* Office of Pesticide Programs, Washington, DC.

US-EPA (US Environment Protection Agency), (2004). *Cleaning Up the Nation's Waste Sites: Markets and Technology Trends* (pp. 3–9). Office of solid waste and emergency response, Washington, DC.

US-EPA (US Environment Protection Agency), (2012). *About Remediation Technologies.* US EPA Office of Superfund Remediation and Technology Innovation (CLU-IN), Washington, DC.

US-ITC, (2013). *US-ITC Finds Few Trade Barriers Specific to core Environmental Science.* US International Trade Commission, Washington, DC.

Varjani, S., Kumar, G., & Rene, E. R., (2019). Developments in biochar application for pesticide remediation: Current knowledge and future research directions. *J. Environ. Manag., 232,* 505–513.

Vergani, L., Mapelli, F., Zanardini, E., Terzaghi, E., Di, Guardo, A., Morosini, C., & Borin, S., (2017). Phyto-rhizoremediation of polychlorinated biphenyl contaminated soils: An outlook on plant-microbe beneficial interactions. *Sci. Total Environ., 575,* 1395–1406.

Weber, R., Watson, A., Forter, M., & Oliaei, F., (2011). Persistent organic pollutants and landfills: A review of past experiences and future challenges. *Waste Manag. Res., 29,* 107–121.

Weiss, F. T., Leuzinger, M., Zurbrügg, C., & Eggen, R. I. L., (2016). *Chemical Pollution in Low and Middle-Income Countries.* Swiss Federal Institute of Aquatic Science and Technology, Switzerland.

Xing-Lu, P., Feng-Shou, D., Xiao-Hu, W., Jun, X., Xin-Gang, L., & Yong-Quan, Z., (2019). Progress of the discovery, application, and control technologies of chemical pesticides in China. *J. Integ. Agri., 18,* 840–853.

Yang, Y. F., Cheng, Y. H., & Liao, C. M., (2016). In situ remediation-released zero-valent iron nanoparticles impair soil ecosystems health: A *C. elegans* biomarker-based risk assessment. *J. Hazard. Mater., 317,* 210–220.

Zhan, H., & Jiang, Y., (2015). Metal oxide nanomaterials for the photodegradation of phenol. *Anal. Lett., 49*, 855–866.

Zhang, D., & Zhu, L. Z., (2012). Effects of tween 80 on the removal, sorption and biodegradation of pyrene by *Klebsiella oxytoca* PYR-1. *Environ. Pollut., 164*, 169–174.

Zhang, J. L., & Qiao, C. L., (2012). Novel approaches for remediation of pesticide pollutants. *Int. J. Environ. Pollut., 18*, 423–433.

Zhang, Q., Jiang, D., Gu, Q., Li, F., Zhou, Y., & Hou, H., (2012). Selection of remediation techniques for contaminated sites using AHP and TOPSIS. *Acta Pedol. Sin., 49*, 1087–1094.

Zhang, W., Jiang, F., & Ou, J., (2011). Global pesticide consumption and pollution: With China as a focus. *Proc. Int. Acad. Ecol. Environ. Sci., 1*, 125.

Zhao, X., Reitz, S. R., Yuan, H., Lei, Z., Paini, D. R., & Gao, Y., (2017). Pesticide-mediated interspecific competition between local and invasive thrips pests. *Sci. Rep., 7*, 40512. doi: 10.1038/srep40512.

WEBLINKS

Blacksmith Institute, Green Cross, (2012). *The World's Worst Pollution Problems: Assessing Health Risks at Hazardous Waste Sites.* www.worstpolluted.org (accessed on 30 October 2020).

FAO (Food and Agriculture Organization), (2019). *Prevention and Disposal of Obsolete Pesticides.* Retrieved from: http://www.fao.org/agriculture/crops/obsolete-pesticides/what-dealing/obs-pes/en/ (accessed on 30 October 2020).

History of Pesticide Use, (1998) Accessible from: http://www2.mcdaniel.edu/Biology/eh01/pesticides/historyofpesticidesuse.html (accessed on 30 October 2020).

Lah, K., (2011). *Effects of Pesticides on Human Health.* In: Toxipedia. www.toxipedia.org (accessed on 30 October 2020).

Pesticide Action Network Asia and the Pacific (2010). *Communities in Peril: Asian Regional Report on Community Monitoring of Highly Hazardous Pesticide Use.* www.panap.net (accessed on 30 October 2020).

US-EPA (US Environment Protection Agency), (2019). *Pesticides.* http://www.epa.gov/pesticides/ (accessed on 30 October 2020).

CHAPTER 5

Woodchip Bioreactors for Nitrate Removal in Agricultural Land Drainage

MEHRAJ U. DIN DAR,[1] AAMIR ISHAQ SHAH,[2] SYED ROUHULLAH ALI,[3] and SHAKEEL AHMAD BHAT[3]

[1] Department of Soil and Water Engineering,
Punjab Agricultural University, Ludhiana – 141004, Punjab, India,
E-mail: mehrajudindar24@gmail.com

[2] Department of Hydrology, Indian Institute of Technology,
Roorkee – 247667, Uttarakhand, India

[3] College of Agricultural Engineering and Technology,
SKUAST K – 190025, Srinagar, Jammu and Kashmir, India

ABSTRACT

Bioreactors are getting a major boost over the recent past for improving the water qualities in regions affected by waterlogging and poor drainage problems. In order to decrease the quantity of nitrate in agricultural drainage, Woodchip or denitrification bioreactors have come up as an innovative, engineering-based technology. Bioreactors form an active area of research, though still addressing the problems related to drainage water quality, concerns do not seem magical solution to a complicated problem. There is a need for meticulous analysis of several features of these systems, besides increasing the number of bioreactor installations among practitioners. This chapter examines the plan and establishment of bioreactors, gives an overall view towards improved denitrification treatment of agricultural drainage, and explains various factors impacting the nitrate removal. This chapter offers various ideas towards the monitoring and management of agricultural drainage bioreactors. In order to improve drainage water quality bioreactors have become a promising technology, although greater emphasis has to be laid on understanding and optimizing their performance. A more comprehensive and clear image of these frameworks' potential commitment will be produced with extra appraisal and assessment of bioreactors.

5.1   INTRODUCTION

Nitrogen (N) forms a chief component of proteins and nucleotides and hence is very crucial for life on the planet earth (Robertson and Vitousek, 2009). Although nitrogen is the most abundant gas in the atmosphere, still it is limiting for more than 99% on organisms living on the earth. The reason for the unavailability of nitrogen is that more than 99% of nitrogen is present as nitrogen gas (N_2) (Galloway et al., 2003). As a result available nitrogen forms a limiting factor in crop production and affects ecosystem structure (Robertson and Vitousek, 2009). In order to meet the demands of growing global population, there is a need to surmount the N limitation in agricultural food production. This has led to the growth of the Haber-Bosch process, which changes the nitrogen gas into ammonia (NH_3) and augmented cultivation of N fixing plants (Galloway et al., 2003; Seitzinger et al., 2006). The increased nitrogen inputs results in increased agricultural production. While there are notable benefits, excess nitrogen from the agricultural system usually ends up in rivers, groundwater, and other freshwater sources. The presence of excess nitrogen in aquatic systems has led to many difficulties including eutrophication, N_2O production (greenhouse gas) and associated hypoxic zones, which adversely affect human health and aquatic ecosystem vigor (Camargo and Alonso, 2006). In order to avoid many adverse consequences as N moves to downstream water bodies, it becomes imperative to treat the nitrogen at the source of origin (Galloway et al., 2003). Denitrification process involves reduction of nitrate (NO_3^-) molecules into to inert N_2 gas (Figure 5.1) by the action of microbes (Seitzinger et al., 2006) and mechanism of the process is as shown in the flowchart (Figure 5.2). It may be said to be principal mechanism of removal of N from various constituents of ecosystem (Burgin and Hamilton, 2007), and therefore forms key towards maintaining water quality. Reactive N within the aquatic or terrestrial system is maintained by all other transformation processes (Myrold, 2004). Absence of oxygen (O_2), availability of NO_3^- and unstable C to act as an energy source are the primary controls of denitrification process (Teidje, 1988; Seitzinger et al., 2006). Approaches for improving denitrification in drainage water and agricultural groundwater has become critical since all these modern agricultural practices result into high levels of NO_3^- leaching into drainage water and groundwater.

Installation of denitrification walls is one approach for raising denitrification in groundwater (Jaynes et al., 2008). A mixture of organic soil and sawdust and sand/soil mixture below the soil and even 100% woodchips are used to construct denitrification walls (Cherry and Robertson, 1995; Vojvodic-Vukovic and Schipper, 1998). Walls for denitrification process require a large quantity of groundwater to flow through the walls, avoid groundwater re-routing below the wall and hence are designed to maintain high value hydraulic conductivities (Schipper et al., 2004, 2010). With the use of denitrification walls the nitrate removal rate usually ranges from 0.014–3.6 gNm^{-3} day^{-1} (Schipper et al., 2010). The working life of denitrification walls depends on two important factors (1) maintenance of hydraulic conductivity (2) Carbon supply (Schipper et al., 2010). The carbon in denitrification walls decays slowly, which is why no denitrification walls suffered failure because of C depletion (Schipper and Vojvodic-Vukovic, 2001; Moorman et al., 2010). Nevertheless, with regards to the sustainability of denitrification walls, the subject has been poorly interpreted. For a Single-family septic system it was estimated using a stoichiometric technique that a denitrification wall with 20% of sawdust would have enough carbon to support 200 years of denitrification, expecting 100% carbon exhausted by denitrifiers (Robertson and Cherry, 1995). The permanence of NO_3^- removal in walls of denitrification could be greatly reduced by periodic exposure to aerobic conditions and could enhance emission of N_2O (Moorman et al., 2010; Bock et al., 2018). A 9 years successful working of denitrification walls with regards to NO_3^- removal was reported by Moorman et al. (2010) in denitrification walls installed at central Iowa, USA (Jaynes et al., 2008). The only decadal study on determining the continued effectiveness in NO_3-removal was performed in Canada which indicated successful performance in relation to NO_3^- removal following 15 years after installation (Robertson et al., 2008). Direct field sampling technique of changes in groundwater NO_3^- concentrations was not used rather laboratory column tests of nearly 15-year old wall material rather was done in this study. Long-term studies related towards evaluating the effectiveness of denitrification walls still remain sparse and hence more efforts are to be made in this regard. The average annual drainage NO_3^- losses in New Zealand, under dairy pastures for grazing are 25 to 30 kg N ha^{-1}, which is almost same to concentration from an agricultural area under row crop in the US Midwest (Monaghan et al., 2002). Midwestern drainage differs from New Zealand drainage system in the respect that at a given site over a drainage season, NO_3^- concentrations are relatively consistent while as there is a significant trend of decreasing NO_3^- concentrations over the season in New Zealand drainage systems with the first 100 to 150 mm of drainage showing highest concentrations (Monaghan et al., 2002; Houlbrooke et al., 2004). Changing flow rates results in changing bioreactor retention times which results from uncontrolled and irregular pulsed drainage flow rates which is presently a challenge for bioreactor treatment (Gottschall et al., 2016; Husk et al., 2017). At peak flow rates low bioreactor retention time may cause too high dissolved oxygen concentrations which are high for denitrifiers to reduce NO_3^-. Decreased bioreactor NO_3^- removal at higher flow rates have been documented by several past works (Woli et al., 2010; Christianson et al., 2011). Short duration, intensive, and also short duration flows present design issues and hence an impractically large bioreactor volume would be required for designing a system for 100% of the peak flow. Bioreactors in the Midwest are currently designed only for a part of peak flow rate indicating that not all of the total annual flow volume receives bioreactor treatment (Christianson et al., 2009; USDA-NRCS, 2009; Moorman et al., 2015). Early season drainage water have highest NO_3^- concentrations and in New Zealand, drainage water NO_3^- mitigation could focus on capturing and treating early season drainage water (i.e., first 100–150 mm of drainage). Early season drainage water has highest NO_3^- concentration and hence requires longer retention time of drainage water. In order to retain the drainage water, temporary diversions may be constructed in paddock gullies. The impounded drainage can be released in a controlled manner into a denitrification bioreactor which would allow more consistent retention time and treatment for longer duration. Almost all the drainage volume in the early season can be treated with this two-stage treatment system setup and effective treatment can be ensured by maintaining a proper retention time. A two-stage denitrification bioreactor design in the US Midwest design is a major shift from current denitrification bioreactor design. The containment of drainage prior to treatment provides two major benefits including: (1) it is possible to treat early-season drainage, which usually contains the highest NO_3^- concentrations (2) treatment at a longer time, stabilization of flow rate variability, and more constant bioreactor retention time. Consistent long-term NO_3 removal rates varying from 5–15 years were noted for woodchips and sawdust (wood particle media) besides requiring little maintenance (Robertson et al., 2000, 2008, 2009; Schipper and Vojvodic-Vukovic, 2001). A variety of immiscible liquids and carbonaceous solids have been successfully tested such as straw, cotton burr, leaf compost, newspaper, cardboard, corn stalks, vegetable oil, sawdust, etc., (Greenan et al., 2006). Finer wood particle media have been used in 'denitrification walls' using finer wood and have proven barrier bypass because of reduced permeability of reactive mixture (Schipper et al., 2004). As a result, for its use in nitrate bioreactors coarser wood particle media with greater permeability are being used (Van Driel et al., 2006; Greenan et al., 2006; Jaynes et al., 2008; Schipper et al., 2004). Deeper groundwater flow can be captured with the use of PRBs (permeable reactive barrier) high permeability media allows in shallow-installed PRBs (Robertson et al., 2009). First-order kinetics would apply where NO_3 concentration affects NO_3 mass removal rate. Zero-order kinetics applies to the reactions where NO_3^- removal is unaffected by NO_3 concentrations and control is unaffected by the availability of usable organic carbon (Bekins et al., 2005). In denitrifying bioreactors both zero-order (Schipper et al., 2004) and first-order approaches have been considered in early studies related to the evaluation of reaction rates. Bioreactors have been found to have successfully reduced the nitrate concentration near shallow ground and drainage water at numerous installation sites around the world (Volokita et al., 1996; Robertson and Cherry, 1995; Robertson et al., 2000; Blowes et al., 1994; Schipper and Vojvodic-Vukovic, 1998). With the use of wood chips, tree bark and leaf compost as a source of carbon in bioreactors designed by Blowes et al. (1994) a decrease from 3–6 mg NO_3^--N L^{-1} to <0.2 mgL^{-1} in the agricultural drainage water was reported. NO_3^--N removal from agricultural groundwater using a denitrification wall composed of a mixture of sawdust and soil was obtained by Schipper and Vojvodic-Vukovic (1998, 2000). The NO_3^--N removal was ascribed to enhanced denitrification enzymatic activity inside denitrification wall as compared with soil outside the wall where amendments are not used. Jaynes et al. (2008) used woodchip-based denitrification walls placed on either side of subsurface drainage lines in order to remove nitrate from corn-soybean drainage water. Flow-weighted nitrate concentrations over 5 years of operation as compared to a value of 22 mg NO_3^--N L^{-1} in untreated drainage water had an average value of 8.8 mg NO_3^--N L^{-1} in the treated drainage water. The nitrate concentration in drainage water from a golf course and cornfield was reduced by 53% and 32%, respectively, over 4 years using two small wood chip bioreactors in Ontario State (Van Driel et al., 2006). The results indicate that although there are certain important issues towards nitrate removal from drainage water that need to be addressed, long term nitrate removal can still be achieved. The carbon supply duration to the denitrification organisms within the soil determines the longevity and effectiveness of installation of long-term denitrification walls. Denitrification like other processes such as sulfate reduction, fermentation, and aerobic respiration also requires a constant supply of substrate. Extensive investigation regarding knowing the bioreactor media and associated

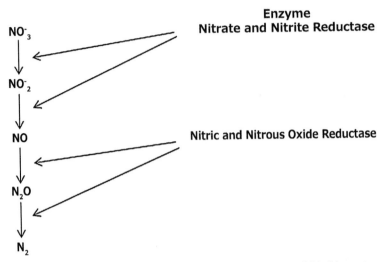

FIGURE 5.1 Nitrate conversion into atmospheric nitrogen in a woodchip bioreactor.

FIGURE 5.2 Woodchip bioreactor installation steps: (A) woodchips of a particular size, (B) construction of a trench, (C) woodchips being filled in the trench, (D) covering of woodchips with a plastic cover (Source: Reprinted with permission from Christianson et al., 2011).

Source: Christianson, Laura

denitrifier activity associated with the dynamics of wood substrate decay and populations have to be carried out. Increased emission of N_2O which is a byproduct of the denitrification reaction is another big concern related to the denitrification process. The greenhouse gas N_2O production accounts for <0.03% of the nitrate denitrified in short-term laboratory experiments (Greenan et al., 2009). Efficiency of bioreactors in converting nitrate to N_2 still remains to be investigated in the areas where fluctuation of hydrology, nitrate fluxes, and weather is recorded.

5.2 ALTERNATIVE BIO-FILTER MEDIA

Worldwide, there has been a rapid expansion of Aquaculture in the recent past as compared to any other animal food-producing sector (FAO, 2002). There has been greater emphasis on producing more fish yield using less water to avoid negative environmental and land impacts. Recirculating aquaculture systems (RAS) were adopted by Lepine et al., (2018) in order to effectively use the available soil and water resources. An effective recirculating system design must have a separate unit to control nitrate and ammonia concentrations and pathogenic bacteria while keeping a pace with their generation rates. It must also have a separate solid waste removal unit (Losordo et al., 1998). A concentration of as high 100 mg/L in waste streams was reported due to water reuse from waste stream by Chen et al. (2002). There is a high potential for toxic algal blooms and eutrophication in receiving waters due to augmented levels of nitrate in the wastewater stream. Methemoglobinemia among infants may result from consumption of groundwater contaminated with elevated nitrate-nitrogen concentrations. The problems related to the discharge of nitrate have led to concerns within RAS designers to control and treat the discharge within or at the end of pipe treatment of wastewater leaving the system. Denitrification process involves conversion of nitrate-nitrogen to di-nitrogen (N_2) gas with the involvement of bacteria. Due to reduced space requirements in aquaculture applications, using attached growth processes such as bio-filter with structured media has been favored instead of suspended growth processes. The relatively high cost (often in excess of US$ 1000/m^3) of plastic media may be said to be one of the major limitations. There have been numerous studies on testing the use of wood and agricultural byproduct as structured biofilter media for use in denitrification of residential wastewater, agricultural runoff, and drinking water. Wheat straw was also used for denitrification of nitrogen rich and turbid water by Lowengart et al. (1993). The use of wood chips as a bio-filter for the treatment of irrigation and runoff water has also been successfully demonstrated by Blowes et al. (1994). Examinations identified with nitrate removal with the utilization of wheat straw or 100% wood chips have been led by Kim et al. (2003). Nitrate-loaded water treatment with the use of sawdust and leaf fertilizers has been reported by Robertson et al. (2000). Nitrate-nitrogen removal from pretreated septic tank outflow using a wood-based biofilter has also been reported by Robertson et al. (2005). Volokita et al. (1996) suggested the use of shredded paper as a source of carbon and use as a bio-filter in denitrification sections. However, removal rates for nitrate-nitrogen in the groundwater studies listed above were reported to be low due to limited carbon availability and low substrate (NO_3^-N) concentrations for fueling the denitrification process. The availability of carbon dissolved in residential wastewater result in a high rate of nitrate removal (Robertson et al., 2005). In both cases, limited carbon was available for denitrification from the agrarian or potentially wood items media. Greenan et al. (2006) reported denitrification rates of just 66.0 mg N/kg in subsurface wastewater using hardwood chips as a carbon source in a study relating carbon-based substrates.

5.3 SOLID CARBON SOURCE USE FOR ENHANCED DENITRIFICATION TREATMENT

The results for influent indicated a possible potential of organic media for upgrading NO_3^- removal and showed a decrease of NO_3^- from 2–6 mg NO_3^-N/L to below 0.02 mg NO_3^-N/L. Similar investigations based on this work relating to the septic treatment of drainage water with the use of Nitrex Tm, a University of Waterloo trademark. The product is reactive flow through the barrier and well suited for passive drainage water treatment (Robertson et al., 2005a). Upgraded denitrification permeable reactive boundaries have been a result of ongoing improvements in the field of remediation of water. The addition of a strong carbon source leads to aerobic respiration and a decrease of dissolved oxygen. It also offers a continuous source of carbon for use in denitrification (Schipper et al., 2005; Hashemi et al., 2011).

5.3.1 PERFORMANCE FACTORS

Numerous variables can influence bioreactor NO_3^- removal ability, including microbiology, temperature, and time of retention. Schipper et al. (2010a) provided a total overview of longevity and kinetics on denitrification and not discussed in detail. Designs for enhanced-denitrification of NO_3^- that could use zero-order kinetics were reported through a review by Schipper et al. (2010a). Several factors, such as flow characteristics, including the amount and type of carbon source, consistency, and level of saturation have been found having an impact on longevity on bioreactor (Schipper et al., 2010a). The performance evaluation of denitrification systems is usually for the longer time scale of the order of several decades with the use of empirical data. The empirical data should show performance evaluation for a time span of at least 10 years. The fluctuations in the levels of saturation and flow depths may complicate the performance evaluation of life of drainage bioreactors. The woodchips placed in the deeper layers of soil will stay longer while those at surface would tend to degrade more quickly (Moorman et al., 2010). The bioreactors for treatment of drainage waters are currently designed to last for a minimum 10 years (USDA NRCS, 2009).

5.3.2   HYDRAULICS AND RETENTION

Design factors related to flow volume and media porosity determine the retention time of wastewaters in a bioreactor. The liquid to be treated must remain in the reactor for an appreciable time until fully treated. Hence, the solute and liquids' retention time is centric towards design of bioreactors. The objective to decrease influent drainage dissolved oxygen to an appreciably low level may be strenuous with small retention times as experienced in bioreactors (Damaraju et al., 2015). Greater retention time gives appreciable NO_3^- removal yet additionally the potential for oxidation decrease possibilities (ORPs) characteristic of bothersome procedures, similar to sulfate decrease mercury methylation. In denitrification drainage systems, higher NO_3^- removal is typically correlated with a greater retention time. Chun et al. (2009) studied a reduction of 10 to 40% with a time of retention timeless than 5 hours with 100% removal at retention times of 15.6 and 19.2 hours. On a longer time scale, removal efficiencies of 30% to 100% were reported by Greenan et al. (2009) for retention times varying from 2.1 to 9.8 days, respectively. In Iowa and Illinois, the correlation of NO_3^- removal with retention time at field scale has been done (Woli et al., 2010). Closer management of retention times in Midwestern bioreactors has been made possible with the usage of outflow and inflow control structures (Chun et al., 2010). The diversion structure allows the water to by-pass and escape at high flow events, besides it also routes water into the bioreactor. The structure requires most of the in-field management and hence control of retention time can also be achieved with the outflow structure (Figure 5.3) (Chun et al., 2010).

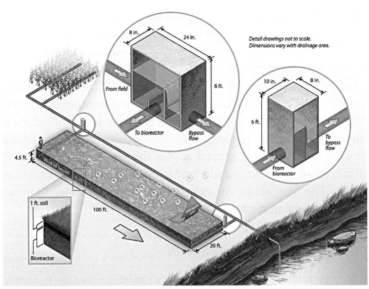

FIGURE 5.3 Schematic diagram locations of pressure transducers of a denitrification bioreactor for agricultural drainage (Christianson et al., 2011).

Stoplogs can be used for management of bioreactor where during low flows the log height can be lowered in the control structure. Roberson and Merkely (2009) used moveable pipes as a low cost alternative for controlling retention time, head, and rate of flow. The reduction of sulfate is a major concern since it can be closely linked with (1) methylation of mercury (2) essential carbon loss for denitrifiers (3) Noxious gas production such as hydrogen sulfide. The management and design of bioreactors for minimizing sulfate reduction is possible through the retention of NO_3^- in the effluent in very low concentrations (Robertson and Merkley, 2009). The height of the stop-log should be decreased in order to allow unrestricted flow of water through the reactor, if hydrogen sulfide is observed (Christianson and Helmers, 2011a). Hydrologically, majority of drainage systems have been found to experience dry or periods of low flow even during the period of active drainage season. However, the periods of low flow have not been problematic since bioreactors startup once the flow resumes (Van Driel et al., 2006a). In general, decreased Nitrogen removal performance and decreased retention times result from a drainage event hydrograph advancing via a bioreactor. As compared to steady-state bioreactors for equal time of retention, Christianson et al., (2011c) observed that the ones that experience fluctuating flow rate may have reduced performance.

5.3.3   TEMPERATURE

Christianson et al. (2012a) reported varying temperature of drainage water entering a bioreactor with late summer temperatures over 15°C and early spring temperature being just above freezing. Even though NO_3^- removal has been reported at water temperatures 2°C–4°C, the temperature factor still plays a very important role in denitrification process in a bioreactor (Merkley and Robertson, 2009). Several studies (Volokita et al., 1996; Diaz et al., 2003; Cameron and Schipper, 2010) documented an increase in NO_3^- removal at higher temperatures. The Q_{10} value, i.e., the factor owing to which rate of reactor changes for every 10°C change in temperature usually varies from 1 to 3 with majority of values around 2 (Van Driel et al., 2006a; Robertson and Merkley, 2009; Cameron and Schipper, 2010; Warneke et al., 2011a). A reported increase of having higher retention times at low temperatures of drainage water using the Van't Hast-Arrhenius Law was observed by Cooke et al. (2001). Since this temperature dependence of denitrification is so imperative, Cameron, and Schipper (2010) experimented increasing the temperature using passive solar heating at a denitrification in New Zealand, no appreciable increase in N removal rate was observed as the experiment resulted in temperature increase of bioreactor by just 3.4°C mean and no huge increment in N expulsion rate. Randall and Goss (2001) observed that temperature in combination with several other factors has an impact on the design of bioreactor. For instance, waste N loads are mostly high in the spring when the temperatures remain low, while as during the summers maximum temperatures are reached which presents a major challenge for optimization of bioreactor treatment plants. Although the bioreactor performance is greatly affected by temperature, it is practically possible to reduce the sensitivity of bioreactor on temperature by effective seasonal retention time management and better understand operational parameters (Soupir et al, 2018). The problem of higher flow rates associated with low temperatures in the Midwest can be solved by control structure management (Volokita et al., 1996; Robertson et al., 2005a).

5.3.4   MICROBIOLOGY

Environment is abundant of denitrifiers which has not let their inoculation till date (Schipper et al., 2010a) besides the inclusion of soil, in low amounts, e.g., 1 L by Christianson et al. (2011c) or 1 kg by Blowes et al. (1994). Nevertheless, bioreactor slowly starts-up, attributed to the slow growing microbial community after one early spring installation (Wildman, 2001). Appleford et al. (2008) contemplated that woodchip surfaces and in bioreactor arrangement denitrifiers are plentifully present, and Moorman et al. (2010) revealed that encompassing soil have less denitrifiers than woody media dividers/ walls which underpins more elevated amounts of denitrifiers. Denitrification locales are not restricted by the chips' outside surface; Robertson et al. (2000) found that wood particles got stretched out a few mm because of dull shading, indicating that water mixed wood was denitrified also (Robertson et al., 2005b). There is a change in bioreactor microbial networks throughout the year with depth and towards the flow (Andrus, 2010). In the recent past, Warneke et al. (2011b) considered that the bacterial network in a little scale woody media bioreactor was higher than the network in a maize cob bioreactor, showing the greater plausibility of carbon to be utilized by non-denitrifiers in the maize reactor. It is additionally recorded by Warneke et al., (2011b) that the traits required for denitrification were almost four times the concentration at 27.1°C versus 16.1°C. Andrus (2010) proposed that due deliberation is given to expanded populace of the microbial colonies inside a bioreactor, which should be possible either through natural administration or immunization of ideal species to change the network to high performing NO_3^- removers, which could ameliorate the capability of a bioreactor. Other than denitrifiers, microbes, which are seen as biofilms, have been accounted for at bioreactor installation sites (Chun et al., 2009). Such biofilms may obstruct the control structures or the lines, so flushing (through stop log control) or blending can be prescribed as the best administration decision (Wildman, 2001; Van Driel et al, 2006a). At the same time, there may be issues when denitrifiers clean flow at high rates (Volokita et al., 1996); however, this has never been perceived in a drainage bioreactor at field-scale.

5.4   DESIGN OF BIOREACTOR

One of the greatest challenges that design and the performance of drainage denitrification system poses is its variability. Flow rates are inherent to this drainage system, and some are often unidentified (Woli et al., 2010; Christianson et al., 2009). Flow rate ranges from zero to maximum within a given time period (or greater, since this is theoretical) depending upon precipitation patterns with high and low flow periods interspersed. Christianson et al. (2011a) developed a method by taking into account flow rate and retention time with the reactor taking in only a certain percentage of peak flow for a designated retention time. Although it has been argued by different reports, that this type of reduction in the peak flow may not be economical for the performance of a bioreactor (Van Driel et al., 2006a). An alternative has been adopted by the USDA NRCS in (Iowa NRCS, 2010) to structure bioreactors which are looking for cost-share financing by means of the environmental quality incentives program (EQIP). The technique comprises correlating treatment region and bioreactor surface region (L × W) on the performance curve recommended by Illinois. For instance, roughly for each 1.2–1.4 hectares of drainage area (100 ft^2/3.0 to 3.5 acres) to achieve 60% load reduction, a 9.3 m^2 of bioreactor surface area would be required (Verma et al., 2010). Concerning drainage frameworks where coefficient and exact drainage area are not known, in spite of the fact that a design table from Wildman (2001) permits estimation of a required bioreactor volume dependent on drainage coefficient and drainage area. UMN Extension and MN Department of Agriculture (2011) from the Midwest suggested an informal method of using approximately 3.3 meters length of bioreactor length for every 0.4 hectares of drainage area. The designs developed by the University of Illinois and Christianson et al. (2011a) are most widely adopted designs available in the Midwest, currently used in Illinois and Iowa, respectively. A range of bioreactor sizes has been suggested by various works in this field. However, there has been no agreement on the development of standard sizes for different drainage areas of bioreactors and several studies are underway to address these issues. Different design methods and alternative configurations of denitrification setups are being examined in different studies. An in-stream bioreactor was installed by Robertson and Merkley (2009) in a drainage ditch. Christianson et al. (2010) using varying designs of bioreactors reported no significant effect on denitrification rate on varying cross-sections. Treatment of drainage water can be maximized by designing bioreactors in parallel or series or addition of baffles within the bioreactors (Cooke et al., 2001). In order to improve the water quality of various natural water sources and sumps, the bioreactors can be linked with wetlands along with other in-situ conservation practices (Robertson and Merkley, 2009). These thoughts might seem to be enthusiastic for the exploration domain, but for practical applicability and adoption at the farm scale lot of work has to be done in this direction with keeping economical cost and performance in mind. The removal of metals, trace organic contaminants and metals from bio-char amended woodchip reactor was performed on a pilot scale and urban runoff in the field was used for preconditioning the bioreactors (Figure 5.4). It was found that the woodchip bioreactors could successfully remove nitrate and

4 of the 5 metals tested. It could successfully remove Pb, Ni, Cu, and Cd, except Zn. No negative effect on removal of nitrate and metals was observed after addition of bio-char at 33% of dry weight. The addition of bio-char, however, substantially improved the removal of trace organic contaminants (Hua et al., 2016; Ashoori et al., 2019).

FIGURE 5.4   Biochar-amended woodchip bioreactors to remove nitrate, metals, and trace organic contaminants from urban stormwater runoff.

Source: Reprinted with permission from Ashoori et al. (2019). © Elsevier.

5.5 CONCLUSIONS

Enhanced-denitrification systems may be said to be a promising new technology to reduce NO_3^- concentrations from drainage output of agricultural fields. The drainage water quality enhancement although requires a total refinement in existing techniques and technologies, besides it gives a variety of edge-of-field and in-field approaches. However, several limitations come with this new water quality improvement option and more research needs to be done in this regard especially towards handling peak flows of drainage water. In order to develop better management procedures and optimize the performance of bioreactors, more field scale data is needed. The data can be used for evaluating design methods and quantifying deleterious effects. Professional and landowners can benefit from a more practice-oriented document that will help them monitor, understand, and manage the installed denitrification bioreactors for treating drainage from agriculture.

KEYWORDS

- **agricultural drainage**
- **bioreactors**
- **denitrification**
- **nitrogen**
- **water quality**
- **woodchips**

REFERENCES

Andrus, J. M., Rodriguez, L. F., Porter, M., Cooke, R. A. C., Zhang, Y., Kent, A. D., & Zilles, J. L., (2010). Microbial community patterns in tile drain biofilters in Illinois (extended abstract). *21ˢᵗ Century Watershed Technology: Improving Water Quality and Environment.* Universidad.

Appleford, J. M., Rodriguez, L. F., Cooke, R. A. C., Zhang, Y., Kent, A. D., & Zilles, J. L., (2008). Characterization of microorganisms contributing to denitrification in tile drain biofilters in Illinois. *ASABE Annual International Meeting, Providence.* RI. Paper Number: 084583.

Ashoori, N., Teixido, M., Spahr, S., LeFevre, G. H., Sedlak, D. L., & Luthy, R. G., (2019). Evaluation of pilot-scale biochar-amended woodchip bioreactors to remove nitrate, metals, and trace organic contaminants from urban stormwater runoff. *Water Res.*

Aslan, S., & Turkman, A., (2003). Biological denitrification of drinking water using various natural organic solid substrates. *Water Sci. Technol., 48*(11/12), 489–495.

Bedessem, M. E., Edgar, T. V., & Roll, R., (2005). Nitrogen removal in laboratory model leach fields with organic-rich layers. *J. Environ. Qual., 34*, 936–942.

Bekins, B. A., Warren, E., & Godsy, E. M., (2005). A comparison of zero-order, first-order, and Monod biotransformation models. *Ground Water, 36*(2), 261–268.

Blowes, D. W., Ptacek, C. J., Benner, S. G., McRae, C. W. T., Bennett, T. A., & Puls, R. W., (2000). Treatment of inorganic contaminants using permeable reactive barriers. *J. Contam. Hydrol., 45*, 123–137.

Blowes, D. W., Robertson, W. D., Ptacek, C. J., & Merkley, C., (1994). Removal of agricultural nitrate from tile drainage effluent water using in-line bioreactors. *J. Contam. Hydrol., 15*, 207–221.

Bock, E. M., Coleman, B. S., & Easton, Z. M., (2018). Effect of biochar, hydraulic residence time, and nutrient loading on greenhouse gas emission in laboratory-scale denitrifying bioreactors. *Ecol. Engg., 120*, 375–383.

Boussaid, F., Martin, G., & Mowan, J., (1988). Denitrification in-situ of groundwater with solid-carbon matter. *Environ. Technol. Lett.*, *9*(8), 803–816.

Bowler, D. G., (1980). *The Drainage of Wet Soils*. Hodder and Stoughton, Auckland, New Zealand.

Burgin, A. J., & Hamilton, S. K., (2007). Have we overemphasized the role of denitrification in aquatic ecosystems? a review of nitrate removal pathways. *Front Ecol. Environ.*, *5*(2), 89–96.

Camargo, J. A., & Alonso, A., (2006). Ecological and toxicological effects of inorganic nitrogen pollution in aquatic ecosystems: A global assessment. *Environ. Int.*, *32*, 831–849.

Cameron, S. G., & Schipper, L. A., (2010). Nitrate removal and hydraulic performance of organic carbon for use in denitrification beds. *Ecol. Engg.*, *36*, 1588–1595.

Chen, S., Summerfelt, S., Losordo, T., & Malone, R., (2002). Recirculating systems effluents, and treatment. In: Tomasso, J., (ed.), *Aquaculture and the Environment in the United States* (pp. 119–140). US Aquaculture Society, Baton Rouge, LA.

Christianson, L. E., Bhandari, A., & Helmers, M. J., (2011b). Pilot-scale evaluation of denitrification drainage bioreactors: Reactor geometry and performance. *J. Environ. Engg.*, *137*, 213–220.

Christianson, L. E., Hanly, J., & Hedley, M., (2011c). Optimized denitrification bioreactor treatment through simulated drainage containment. *Agric. Water Manage*, *99*(1), 85–92.

Christianson, L., & Matthew, H., (2011). *Woodchip Bioreactors for Nitrate in Agricultural Drainage*.

Christianson, L., & Tyndall, J., (2011). Seeking a dialog: A targeted technology for sustainable agricultural systems in the US Corn Belt. *Sustainability: Science, Practice, and Policy, 7*(2), Published online Sep. 02, 2011.

Christianson, L., Bhandari, A., & Helmers, (2011a). *Potential Design Methodology for Agricultural Drainage Denitrification Bioreactors*. EWRI Congress, Palm Springs, California.

Christianson, L., Bhandari, A., & Helmers, M. J., (2011). Pilot-scale evaluation of denitrification drainage bioreactors: Reactor geometry and performance. *J. Environ. Engg.*, *137*, 213–220.

Christianson, L., Bhandari, A., & Helmers, M., (2009). Emerging technology: Denitrification bioreactors for nitrate reduction in agricultural waters. *J. Soil and Water Conserv.*, *64*, 139A–141A.

Chun, J. A., Cooke, R. A., Eheart, J. W., & Cho, J., (2010). Estimation of flow and transport parameters for woodchip-based bioreactors. II. Field-scale bioreactor. *Biosys. Engg.*, *105*, 95–102.

Chun, J. A., Cooke, R. A., Eheart, J. W., & Kang, M. S., (2009). Estimation of flow and transport parameters for woodchip-based bioreactors: I. laboratory-scale bioreactor. *Biosys. Engg.*, *104*, 384–395.

Cooke, R. A., Doheny, A. M., & Hirschi, M. C., (2001). Bio-reactors for edge of field treatment of tile outflow. *2001 ASAE Annual International Meeting, Sacramento*. CA. Paper Number, 01-2018.

Damaraju, S., Singh, U. K., Sreekanth, D., & Bhandari, A., (2015). Denitrification in biofilm configured horizontal flow woodchip bioreactor: Effect of hydraulic retention time and biomass growth. *Ecohydro. Hydrobio.*, *15*(1), 39–48.

Diaz, R., Garcia, J., Mujeriego, R., & Lucas, M., (2003). A quick, low-cost treatment method for secondary effluent nitrate removal through denitrification. *Environ. Engg. Sci.*, *20*, 693–702.

FAO (Food and Agriculture Organization), (2002). *The State of World Fisheries and Aquaculture.* Food and Agriculture Organization of the United Nations, Rome.

Galloway, J. N., Aber, J. D., Erisman, J. W., Seitzinger, S. P., Howarth, R. W., Cowling, E. B., & Cosby, B. J., (2003). The nitrogen cascade. *Bioscience, 53*(4), 341–356.

Gierczak, R. F. D., Devlin, J. F., & Rudolph, D. L., (2007). Field test of a cross injection scheme for stimulating in situ denitrification near a municipal water supply well. *J. Contam. Hydrol., 89*, 48–70.

Goolsby, D. A., Battaglin, W. A., Aulenbach, B. T., & Hooper, R. P., (2001). Nitrogen input to the Gulf of Mexico. *J. Environ. Qual., 30*, 329–336.

Goss, M. J., Barry, D. A. J., & Rudolph, D. L., (1998). Contamination in Ontario farmstead domestic wells and it's associated with agriculture: 1. Results from drinking water wells. *J. Contam. Hydrol., 32*, 267–293.

Gottschall, N., Edwards, M., Craiovan, E., Frey, S. K., Sunohara, M., Ball, B., & Lapen, D. R., (2016). Amending woodchip bioreactors with water treatment plant residuals to treat nitrogen, phosphorus, and veterinary antibiotic compounds in tile drainage. *Ecol. Engg., 95*, 852–864.

Greenan, C. M., Moorman, T. B., Kaspar, T. C., Parkin, T. B., & Jaynes, D. B., (2006). Comparing carbon substrates for denitrification of subsurface drainage water. *J. Environ. Qual., 35*, 824–829.

Greenan, C. M., Moorman, T. B., Kaspar, T. C., Parkin, T. B., & Jaynes, D. B., (2009). Denitrification in wood chip bioreactors at different water flows. *J. Environ. Qual., 38*, 1664–1671.

Hashemi, S. E., Heidarpour, M., & Mostafazadeh-Fard, B., (2011). Nitrate removal using different carbon substrates in a laboratory model. *Water Sci. Technol., 63*(11), 2700–2706.

Hiscock, K. M., Lloyd, J. W., & Lerner, L. N., (1991). Review of natural and artificial denitrification of groundwater. *Water Res., 25*(9), 1099–1111.

Houlbrooke, D. J., Horne, D. J., Hedley, M. J., Hanly, J. A., & Snow, V. O., (2004). A review of literature on the land treatment of farm-dairy effluent in New Zealand and its impact on water quality. *New Zealand J. Agric. Res., 47*, 499–511.

Hua, G., Salo, M. W., Schmit, C. G., & Hay, C. H., (2016). Nitrate and phosphate removal from agricultural subsurface drainage using laboratory woodchip bioreactors and recycled steel byproduct filters. *Water Res., 102*, 180–189.

Hunter, W. J., (2001). Use of vegetable oil in a pilot-scale denitrifying barrier. *J. Contam. Hydrol., 53*, 119–131.

Husk, B. R., Anderson, B. C., Whalen, J. K., & Sanchez, J. S., (2017). Reducing nitrogen contamination from agricultural subsurface drainage with denitrification bioreactors and controlled drainage. *Biosys. Engg, 153*, 52–62.

Iowa NRCS, (2010). *Iowa Environmental Quality Incentives Program (EQIP) List of Eligible Practices and Payment Schedule FY2011.* United States Department of Agriculture Natural Resources Conservation Service.

Iowa State University Extension Publication, (2011). PMR 1008. Available at: https://store. extension.iastate.edu/ItemDetail.aspx?ProductID=13691 (accessed on 30 October 2020).

Jaynes, D. B., Kaspar, T. C., Moorman, T. B., & Parkin, T. B., (2008). In situ bioreactors and deep drain-pipe installation to reduce nitrate losses in artificially drained fields. *J. Environ. Qual., 37*, 429–436.

Kim, H., Seagren, E. A., & Davis, A. P., (2003). Engineered bioretention for removal of nitrate from storm water runoff. *Water Environ. Res., 75*(4), 355–367.

Ledgard, S. F., Penno, J. W., & Sprosen, M. S., (1999). Nitrogen inputs and losses from clover/grass pastures grazed by dairy cows, as affected by nitrogen fertilizer application. *J. Agric. Sci., 132*, 215–225.

Lepine, C., Christianson, L., Davidson, J., & Summerfelt, S., (2018). Woodchip bioreactors as treatment for recirculating aquaculture systems' wastewater: A cost assessment of nitrogen removal. *Aquacul. Engg., 83*, 85–92.

Losordo, T. M., Masser, M. P., & Rakocy, J., (1998). *Recirculating Aquaculture Tank Production Systems: An Overview of Critical Considerations.* SRAC Publication No. 451. Southern Regional Aquaculture Center, Stoneville, Mississippi.

Lowengart, A., Diab, S., Kochba, M., & Avnimelech, Y., (1993). Development of a bio filter for turbid and nitrogen-rich irrigation water. A: Organic carbon degradation and nitrogen removal processes. *Bioresour. Technol., 44*, 131–135.

Monaghan, R. M., Paton, R. J., & Drewery, J. J., (2002). Nitrogen and phosphorus losses in mole and tile drainage from a cattle-grazed pasture in Eastern Southland. *New Zealand J. Agric. Res., 45*, 197–205.

Moorman, T. B., Parkin, T. B., Kaspar, T. C., & Jaynes, D. B., (2010). Denitrification activity, wood loss, and N_2O emissions over 9 years from a wood chip bioreactor. *Ecol. Engg., 36*, 1567–1574.

Moorman, T. B., Tomer, M. D., Smith, D. R., & Jaynes, D. B., (2015). Evaluating the potential role of denitrifying bioreactors in reducing watershed-scale nitrate loads: A case study comparing three Midwestern (USA) watersheds. *Ecol. Engg., 75*, 441–448.

Myrold, D. D., (2004). Microbial nitrogen transformations. In: Sylvia, D. M., Fuhrmann, J. J., Hartel, P. G., & Zuberer, D. A., (eds.), *Principles and Applications of soil Microbiology* (2nd edn., pp. 333–372). Prentice Hall, Upper Saddle River, NJ.

Rabalais, N. N., Turner, R. E., & Wiseman, Jr. W. J., (2001). Hypoxia in the Gulf of Mexico. *J. Environ. Qual., 30*, 320–329.

Randall, G. W., & Goss, M. J., (2001). Chapter 5: Nitrate losses to surface water through subsurface, tile drainage. In: Hatfield, R. F. F., (ed.), *Nitrogen in the Environment: Sources, Problems, and Management* (pp. 23–35). Elsevier Science.

Robertson, G. P., & Vitousek, P. M., (2009). Nitrogen in agriculture: Balancing the cost of an essential resource. *Annu. Rev. Environ. Resour., 34*, 97–125.

Robertson, W. D., & Cherry, J. A., (1995). In situ denitrification of septic-system nitrate using reactive porous media barriers: Field trials. *Ground Water, 33*, 99–111.

Robertson, W. D., & Merkley, L. C., (2009). In-stream bioreactor for agricultural nitrate treatment. *J. Environ. Quality, 38*, 230–237.

Robertson, W. D., Blowes, D. W., Ptacek, C. J., & Cherry, J. A., (2000). Long-term performance of in situ reactive barriers for nitrate remediation. *Ground Water, 38*, 689–695.

Robertson, W. D., Ford, G. I., & Lombardo, P. S., (2005a). Wood-based filter for nitrate removal in septic systems. *Transactions of the ASAE, 48*, 121–128.

Robertson, W. D., Ford, G. I., & Lombardo, P. S., (2005b). Wood-based filter for nitrate removal in septic systems. *Trans. ASAE, 48*, 1–8.

Robertson, W. D., Vogan, J. L., & Lombardo, P. S., (2008). Nitrate removal rates in a 15-yearold permeable reactive barrier treating septic system nitrate. *Ground Water Monit. Remediat., 28*(3), 65–72.

Robertson, W. D., Yeung, N., Van, D. P. W., & Lombardo, P. S., (2005b). High-permeability layers for remediation of ground water; go wide, not deep (in English). *Ground Water, 43*, 574–581.

Royer, T. V., Tank, J. L., & David, M. B., (2004). Transport and fate of nitrate in headwater agricultural streams in Illinois. *J. Environ. Qual., 33*, 1296–1304.

Schipper, L. A., & Vojvodic-Vukovic, M., (1998). Nitrate removal from groundwater using a denitrification wall amended with sawdust: Field trial. *J. Environ. Qual., 27*, 664–668.

Schipper, L. A., & Vojvodic-Vukovic, M., (2000). Nitrate removal from groundwater and denitrification rates in a porous treatment wall amended with sawdust. *Ecol. Eng., 14*, 269–278.

Schipper, L. A., & Vojvodic-Vukovic, M., (2001). Five years of nitrate removal, denitrification, and carbon dynamics in a denitrification wall. *Water Res., 35*(14), 3473–3477.

Schipper, L. A., Barkle, G. F., & Vojvodic-Vukovic, M., (2005). Maximum rates of nitrate removal in a denitrification wall. *J. Environ. Qual., 34*, 1270–1276.

Schipper, L. A., Barkle, G. F., Hadfield, J. C., Vojvodic-Vukovic, M., & Burgess, C. P., (2004). Hydraulic constraints on the performance of a groundwater denitrification wall for nitrate removal from shallow groundwater. *J. Contam. Hydrol., 69*, 263–279.

Schipper, L. A., Cameron, S. C., & Warneke, S., (2010a). Nitrate removal from three different effluents using large-scale denitrification beds. *Ecol. Engg., 36*, 1552–1557.

Schipper, L. A., Robertson, W. D., Gold, A. J., Jaynes, D. B., & Cameron, S. C., (2010b). Denitrifying bioreactors: An approach for reducing nitrate loads to receiving waters. *Ecol. Engg., 36*, 1532–1543.

Seitzinger, S., Harrison, J. A., Bohlke, J. K., Bouwman, A. F., Lowrance, R., Peterson, B., Tobias, C., & Van, D. G., (2006). Denitrification across landscapes and waterscapes: A synthesis. *Ecol. Appl., 16*(6), 2064–2090.

Soares, M. I. M., & Abeliovich, A., (1998). Wheat straw as a substrate for water denitrification. *Water Res., 32*(12), 3790–3794.

Soupir, M. L., Hoover, N. L., Moorman, T. B., Law, J. Y., & Bearson, B. L., (2018). Impact of temperature and hydraulic retention time on pathogen and nutrient removal in woodchip bioreactors. *Ecol. Engg., 112*, 153–157.

Spalding, R. F., & Exner, M. E., (1993). Occurrence of nitrate in groundwater: A review. *J. Environ. Qual., 22*(3), 392–402.

Teidje, J. M., (1988). Ecology of denitrification and dissimilatory nitrate reduction to ammonium. In: Zehnder, J. B., (ed.), *Biology of Anaerobic Microorganisms* (pp. 179–244). Wiley, New York, New York.

Tomer, M. D., Meek, D. W., Jaynes, D. B., & Hatfield, J. L., (2003). Evaluation of nitrate nitrogen fluxes from a tile-drained watershed in Central Iowa. *J. Environ. Qual., 32*, 642–653.

Twarowska, J. G., Westerman, P. W., & Losordo, T. M., (1997). Water treatment and wastewater characterization evaluation of an intensive recirculation fish production system. *Aquacult. Eng., 16*(3), 133–147.

UMN Extension, and MN Department of Ag, (2011). *Woodchip Bioreactors.* University of Minnesota Extension and MN Department of Agriculture: https://bbe.umn.edu/sites/bbe. umn.edu/files/the_agricultural_best_management_practices_handbook_for_minnesota.pdf (accessed on 30 October 2020).

USDA-NRCS, (2009). *Natural Resources Conservation Service Conservation Practice Standard Denitrifying Bioreactor (Ac.) Interim Code,* 747.

Van, D. P. W., Robertson, W. D., & Merkley, L. C., (2006). Denitrification of agricultural drainage using wood-based reactors. *Trans. ASAE, 48*, 121–128.

Van, D. P. W., Robertson, W. D., & Merkley, L. C., (2006a). Up flow reactors for riparian zone denitrification. *J. Environ. Qual.*, 35, 412–420.

Van, D. P. W., Robertson, W. D., & Merkley, L. C., (2006b). Denitrification of agricultural drainage using wood-based reactors. *Transactions of the ASABE, 49*, 565–573.

Volokita, M., Abeliovich, A., & Soares, M. I. M., (1996b). Denitrification of groundwater using cotton as energy source. *Water Sci. Technol., 34*(1/2), 379–385.

Volokita, M., Belkin, S., Abeliovich, A., & Soares, M. I. M., (1996a). Biological denitrification of drinking water using newspaper. *Water Res., 30*(4), 965–971.

Warneke, S., Schipper, L. A., Bruesewitz, D. A., McDonald, I., & Cameron, S., (2011a). Rates, controls and potential adverse effects of nitrate removal in a denitrification bed. *Ecol. Engg., 37, 511–522.*

Warneke, S., Schipper, L. A., Matiasek, M. G., Scow, K. M., Cameron, S., Bruesewitz, D. A., & McDonald, I., (2011b). Nitrate removal, communities of denitrifiers and adverse effects in different carbon substrates for use in denitrification beds. *Water Res., 45*, 5463–5475.

Wildman, T. A., (2001). *Design of Field-Scale Bioreactors for Bioremediation of Nitrate in tile Drainage Effluent.* M.S. thesis, University of Illinois at Urbana-Champaign, Urbana-Champaign.

Woli, K. P., David, M. B., Cooke, R. A., McIsaac, G. F., & Mitchell, C. A., (2010). Nitrogen balance in and export from agricultural fields associated with controlled drainage systems and denitrifying bioreactors. *Ecol. Engg., 36*, 1558–1566.

CHAPTER 6

Biocontrol Agents in Organic Agriculture

AJAZ AHMAD KUNDOO,[1] MOONISA ASLAM DERVASH,[2]
ROUF AHMAD BHAT,[2] BARKAT HUSSAIN,[1] and MUNTAZIR MUSHTAQ[3]

[1] Division of Entomology, Sher-e-Kashmir University of
Agricultural Sciences and Technology, Kashmir Shalimar,
Jammu and Kashmir – 190025, India

[2] Division of Environmental Sciences, Sher-e-Kashmir
University of Agricultural Sciences and Technology, Shalimar,
Jammu and Kashmir – 190025, India,
E-mail: rufi.bhat@gmail.com (R. A. Bhat)

[3] Division of Biotechnology, Sher-e-Kashmir University of Agricultural
Sciences and Technology, Jammu and Kashmir – 190025, India

ABSTRACT

Bio-agents are among the well-known components in the integrative pest management (IPM) program. These are quite efficient in keeping the pest population below economic injury levels. Because of environmental issues, pesticide resistance, and health hazards of conventional pesticides, bio-agents are set to become a vital factor in managing the pest population. A remarkable feature of bio-agents is that of environmental safety, reduction in pesticide resistance, and health hazards. Bio-control agents will generate considerable interest in terms of organic agriculture and environmental safety. The most frequent "bio-control" agents used in pest management include predators, parasitoids, and "microbial antagonists" (bacteria, virus, fungi, and nematodes). Mode of accomplishment of bio-control agents includes contest, antibiosis, "mycoparasitism"/hyperparasitism, production of lytic enzymes, induced systemic resistance (ISR), and plant growth promotion. There are three important biocontrol *viz* approaches importation (traditional biological control), conservation, and augmentation (inoculation and inundation).

6.1 INTRODUCTION

In organic farming (organic agricultural industry), food is produced, which is devoid of any harmful chemical consortium. The cultivation strategies incorporate methods and techniques which take into cognizance ecological background and novel technologies in consonance with conventional farming strategies dependent on natural niche of organisms. Biological pest control approach possesses high potential to reduce the usage of synthetic pesticides which actually relies on capability of organisms (the natural enemies) through predation, parasitism, grazing, etc. Through this approach, the proliferation of pests is halted in an eco-friendly manner. Sometimes, humans deliberately play a vital role in introducing any natural enemy in any pest population to maintain the intricate balance in ecology. In addition, currently this approach is gaining recognition among stakeholders and scientific community to deal with pest populations on nominal economic investment as the biological control will sustain for generations and, moreover, without polluting aquatic as well as terrestrial ecosystems.

Biological control (bio-control) is counted as an effectual eco-friendly approach in extenuating pest populations through involvement of natural enemies without leaving off any chemical leftovers that are detrimental to other organisms and ecosystems.

The usage of chemical pesticides in traditional farming practices are persistent and cast negative impacts on linkages and cross-linkages of food chains and food webs, respectively, which eventually stem out like a phantom ecological problem termed as 'biological magnification.' Besides, traditional agrochemicals are detrimental to human health; thus, there are enormous fears among society's health-conscious sect. It has revolutionized the outlook of living with considerable transformation in people's thoughts in the direction of synthetic pesticide utilization and advocate shift towards sustainable agricultural pesticides. Due to the bulk application of synthetic pesticides, pest resistance boosts up exponentially resulting in pest resistance towards a particular pesticide. In contrast, biological control is an eco-friendly approach devoid of any ecological constraints and must be coupled with other control measures (management, escalation of and importation of normal enemies) on the basis of spatial and temporal aspects of a given site (Sharma et al., 2013). Management of normal enemies is one of the vital biological control practices that basically stresses on conserving the gene pool of bio-control agents. In augmentation, there is an incorporation of any trait (intrinsic or extrinsic) to boost the vigor and vitality of natural enemies, whereas, importation of natural enemies focuses on the spatial and temporal shifting to the dominant pest sites (Oztemiz, 2008).

On the basis of the mode of action of bio-control agents, there are broadly three approaches; as:

1. **Classical/Traditional Bio-Control:** In this approach, the target is either a single or multi-species exotic pest of bio-control mediators from the pest's native array (Newman et al., 1998).

2. **Novel Association or Neo-Classical Strategy:** In this scheme, targets are original resident pests with exotic biological control agents (Pimentel, 1963).

3. **Conservation (Ehler, 1998), Augmentation, and Inundation Approaches:** In its broader context, this mutually maintains or amplifies the profusion and effect of "bio-control agents" which are already there and in various cases inhabitant to the province.

The scope of bio-control agents has escalated over many decades (Bailey et al., 2009) due a large thrust for ecological stewardship between various stakeholders (McCaffery, 1998) and to nip the unchecked population of resistant pests (McCaffery, 1998). The consumer demands for organic agriculture have also been escalated (Dabbert et al., 2004); therefore, it is mandatory for agricultural stakeholders to pacify the research in this domain.

6.2 BIO-CONTROL AGENTS

They are normal adversaries and can be defined as creatures employed to manage pest groups, such as insect pests or plant diseases. These comprise of predators, parasitoids, and pathogens (Figure 6.1). The "biocontrol agents" against plant diseases are usually called "antagonists." Furthermore, the resistant managers of weeds include primary consumers and phytopathogens.

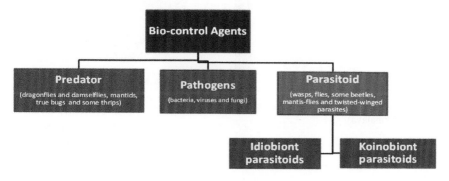

FIGURE 6.1   Various categories of biocontrol agents for degradation of pests.

6.2.1 PREDATOR

Generally, these organisms belong to arthropods (mites) and crustaceans (spiders), which usually eat other insects. They usually choose to hunt, kill, and steadfastly consuming them by varied activities. They may attack the pest in its juvenile and fully developed forms. It has been observed that more than one prey entity is needed for the "predator" to complete its "life cycle." Even though the majority of insect families (more than 100) contain predator species, among, twelve of which hold the foremost "bio-control agents" of crop pests. Key forms of predatory insect are "dragonflies and damselflies," mantids, "true bugs," "some thrips," etc. Spiders are also counted among the chief predators of arthropods (Figure 6.2).

Coccinalla septumpunctata   *Hippodamia variegata*

Syrphid fly   *Aiolacaria hexaspilota*

FIGURE 6.2 Representation of some predominant predators collected from North India (Kashmir Himalaya).

6.2.2 PARASITOID

These are the organisms which parasitize and execute other arthropods, miniature than its horde. They are parasitic in its juvenile phase and "free-living" as a mature, squander a considerable part of its life attached to or within a "host organism" in an association that is in soul freeloading (Consoli, 2010). However, parasitoids in due course of action clean thoroughly or kill, and occasionally swallow the "host." After completing its life cycle, they become free-living creatures and devoid of any enslavement on "host" (Consoli, 2010). As in the juvenile phase, they acquire nourishment by eating in or outside the host body (Consoli, 2010). Parasitoids have been exploited in bio-control on broader context as compared to its contemporaries. The key variety of insects that are "parasitoids" include: "wasps, flies, some beetles, mantis-flies, and twisted-winged parasites" (Consoli, 2010). They attacked varied life periods of pests mostly on their egg stage. Correspondingly, parasitoids that feed on eggs of pests usually extract nutrients and inhabitable within pests eggs. However, these organisms can be introverted state or colonies, however, in all cases avert the host eggs from "hatching," and employ merely a distinct host entity to fully fulfill their progression (van Lenteran, 2000). This distinguishes true egg parasitoids (*Trichogramma* spp. checks the proliferation of corn borer, i.e., *Ostrinia* spp.) from other pseudoparasitoids (such as "egg-pre-pupal parasitoids") that assault the host eggs but impede growth to destroy the host ahead of pupation (van Lenteran, 2000) and egg "predators" that devour numerous eggs inside ovisac (van Lenteran, 2000). On the basis of effects on host, parasitoids are classified as discussed in subsections.

6.2.2.1 IDIOBIONT PARASITOIDS

These organisms avert progressive growth of the host after primarily put out of their action, and approximately devoid of any exemption and grow outside the host (Consoli, 2010).

6.2.2.2 KOINOBIONT PARASITOIDS

Such kind of organisms permits the "host" to maintain its growth process while being trophic dependent and may "parasitize" its "host" at any instant. Furthermore, "koinobiont" can be classified into endo-parasitoids; which build up within the host's body, and "ecto-parasitoids"; which build up outside the "host" body. However, the "parasitoids" recurrently are fastened or entrenched in the host's tissues (Consoli, 2010). It is well-established fact, mature female parasitoids put their eggs within the host (the host arthropod is usually in its immature stage) by either piercing the body wall with their ovipositor, or by fastening their eggs to the outside of the host's body (Consoli, 2010).

6.2.3 PATHOGENS

Pathogens are microorganisms that include a consortium of various life forms (particularly microbes) that can transmit disease and exterminate the specific "host" and entirely undisruptive to other living creatures. This target specific pathogenicity labels numerous pathogens as precious bio-control agents of insect pests. Populations of several aphids, caterpillars, mites, and other invertebrate are seldom radically condensed by natural pathogens, typically under prolonged high humidity or crowded pest populations. Exploitation of miniscule pathogens has become an incredibly promised means of pest management. Parasitic disease causing agents (bacteria, viruses, fungi) will often kill their host and then release millions of spores or 'resting stages,' which are dispersed to infect other host individuals. Their comparative pathogenicity, towering growth rate, and in some, easiness of culturing ensure their use in both augmentation and inundating releases. Pathogens may also act as bio-control agents through competitive exclusion or through the production of antibiotics. Known as antagonists, they are predominantly valuable in the biological control of phytopathogens.

6.3 ROLE OF VARIOUS ORGANISMS IN BIOLOGICAL CONTROL OF PESTS

6.3.1 BACTERIA

Bacteria have been the natural source of toxins for pests that are nowadays prepared as "biopesticides." Remarkable examples include:

1. ***Bacillus thuringensis* (Bt.):** It is a dowel shaped "gram-positive bacteria" that develops a spore originate in soil environs. The US was the first country to employ it as a commercial "bio-pesticide"

in 1958 (Ring, 2017). This organism is lethal to "caterpillars," some flies, and "beetle larvae"; however, nontoxic to other living creatures. Various scientific investigations reported that Bt. var. kurstaki is poisonous to "butterfly," skipper and "moth larvae" (Ring, 2017). Furthermore, *Bt.* var. *aizawai* is deadly to wax "moth larvae" (Ring, 2017). The particular bacterium is offered to organic farmers in sachets of dried spores which are mixed with water and sprayed onto susceptible plants such as Brassicaceae and fruit trees (Ring, 2017).

2. *"Actinobacteria"* and *"Streptomycetes"*: These are acknowledged to comprise an outsized component of the rhizo soil "microbiota." They survive "saprophytically" and "endophytically" in both ordinary and as well as agri-environs and grow around root sphere higher plants (Saleem et al., 2016). Thus, taking into account their plant development promoting commotion, *"Streptomycetes"* symbolize an exceptional option for recuperating nutrient accessibility to crops and upholding improvement and sustainability in farming environs (Figueiredo et al., 2010). 'Plant growth-promoting streptomycetes" (PGPS) arouse and augment plentiful nonstop and tortuous "biosynthetic pathways" in plants. For instance, inorganic "phosphate solubilization," "biosynthesis of chelating compounds," "phytohormones production," "inhibition of plant pathogens," and improvement of assorted abiotic hassle (Vardharajula et al., 2017).

6.3.2 BACULOVIRUSES

These are outsized assemblage of "double-stranded DNA" viruses that attack insects and other arthropods. These organisms have been used as insecticide in an extensive array of circumstances from woodlands and crop fields, stashes, and "greenhouses" (Kost et al., 2005). They are helpful in biological pest control and have no unconstructive effects on non-target organisms (Clem and Passarelli, 2013). These transmit a disease to insects of the orders (Lepidoptera, Diptera, and Hymenoptera) and it is widely acknowledged that Baculoviruses can be as effective as chemical pesticides in wiping out specific insect pests. Certainly, Baculoviruses have numerous recompenses over traditional insecticides that formulate them extremely adequate control representatives (Kost et al., 2005). The huge common of them contaminate caterpillars at the larval period of the order "Lepidoptera" (Clem and Passarelli, 2013). They characteristically possess tapered host arrays, often partial to not many interrelated pests (Clem and Passarelli, 2013) even though the most extremely considered affiliate of the family, "*Autographa californica*" numerous nucleopolyhedrovirus (AcMNPV), is capable of infecting at least 30 species from lepidopteran genera (Clem and Passarelli, 2013).

6.3.3   ENTOMOPATHOGENIC FUNGI

Fungi pathogenic on insects, mites, and other fungi are categorized by the capability to dynamically infiltrate the arthropod's body through the cuticle or another orifice. Therefore, they operate by contact and can infect phytophagous insects regardless of feeding habits or age, causing death through the activity of the mycelium or other toxins produced. Relative humidity is the most important factor affecting fungal activity. During the infection period, relative humidity must be close to 100%. For some species, a film of water is also required to limit the use of fungi against epigeous parasites, which can handle high relative humidity rates without exposure to the attacks of cryptogams. The control of terrestrial insects is less problematic. Fungal biocontrol agents possess a distinctive manner of infectivity. In comparison to bacteria and viruses, they do not require to be ingested and can raid their host directly through the cuticle. Fungal biocontrol agents have confirmed effectiveness against a wide array of insect pests like "*Metarhizium*" and "*Beauveria bassiana.*"

6.3.4   NEMATODES

These are uncomplicated, round, colorless, and un-segmented worms devoid of appendages (Stuart et al., 2006). These may be "free-living,""predaceous" or "parasitic" in nature (Stuart et al., 2006) and the majority of the genus cause vital diseases to flora and fauna, and humans, whereas, rest of the genus are helpful in controlling pests, predominantly by sterilizing or else debilitating their "hosts" (Stuart et al., 2006). The recurrently used nematodes in "biological control" are of the genera *Steinernema* (*Neoaplectana*) and *Heterorhabditis* and can be resourcefully used to repress populations of the insect's in a variety of "agro-ecosystems," and in numerous cases, their constructive impacts on crop production (Georgis et al., 2006). Entomopathogenic "nematodes" have been considered to be a capable "biological control" biota (Chitra et al., 2017) and have been on the rampage at length in agricultural units with insignificant impacts on non-target organisms and are rendered as remarkably secure to environs (Chitra et al., 2017). They are coupled with numerous "symbiotic bacteria" that help them kill and digest their host insects. They operate via direct contact, as they are able to infect the host through the cuticle or other orifices, and their harmful effect on insects is related strictly to their symbiosis with bacteria belonging to the genera *Xenorhabdus*. Once these bacteria are released inside the host, they proliferate in the hemolymph of the insect pest and kill them by septicemia (Stuart et al., 2006). Biocontrol of plant diseases has also got a practical application in disease management. The *"Agrobacterium radiobacter"* strain K 84 was registered with the "United States Environmental Protection Agency" (EPA) to manage the "crown gall" in 1979 (Fravel, 2005). After a few decades, *"Trichoderma harzianumi"* ATCC 20476 was registered as the first fungus for the management of plant disease.

6.4   MODE OF ACTION OF BIO-CONTROL AGENTS

6.4.1   ANTIBIOSIS

Antibiosis can be defined as an antagonism interceded by definite or non-definite metabolites of microbial origin, by lytic agents, enzymes, volatile compounds, or other toxic substances (Sharma et al., 2013). It plays an imperative role in biological control, and it displays a condition where the metabolites are discharged by belowground components of plants, soil microorganisms, plant residues, etc. (Sharma et al., 2013). It occurs when the pathogen is inhibited or killed by metabolic products of the antagonists (Junaid et al., 2013). The products include lytic agents, enzymes, volatile compounds, and other toxic substances (Sharma et al., 2013). Bio agents are known to produce three types of antibiotics viz., nonpolar/volatile, polar/ non-volatile, and water-soluble (Junaid et al., 2013). Among all of these, the volatile antibiotics are extra efficient as they can act at the sites away from the site of production (Junaid et al., 2013).

6.4.2   HYPERPARASITISM/MYCOPARASITISM

Hyper-parasitism or mycoparasitism can be defined as the measured and the most direct outward appearance of "antagonism" (Pal et al., 2006).

This phenomenon happens when the opponent attacks the "pathogens" by secreting various types of enzymes (Pal et al., 2006). It is the phenomenon of one fungus being parasitic on another fungus and takes place beneath the control of biocatalysts (Harman, 2000). It has also been accounted that the attachment of "chitinase" and "β-1, 3 glucanase" in the *Trichoderma* arbitrate biological control (Harman, 2000).

6.4.3 COMPETITION

Competition is an important process involved in the mechanism of "bio-control" operation. From the microbial viewpoint, soil and plants are recurrently "nutrient" inadequate environs. Therefore, to settle on the "phytosphere," microbes should successfully contend for the offered "nutrients" (Pal et al., 2006). Microorganism contends for "breathing space," raw materials, and organic "nutrients" to propagate and endures in their innate environment. Bio-control agents struggle for the infrequent but crucial "micronutrients," such as Fe and Mn principally in extremely in well-aerated soil environs (Pal et al., 2006). Competition for micronutrients exists because bio-control agents are more capable of utilizing uptake systems for extracting vital constituents than the "pathogens" (Nelson, 1990). This practice of contest is considered to be an oblique dealing involving the "pathogen" and the "biocontrol agent" (Pal et al., 2006) whereby the "pathogens" are expelled by the diminution of food and place (Loritoet al., 1994). Interaction has been recommended to play a role in the "bio-control" of *Fusarium* and *Pythium* by some strains of fluorescent pseudomonas (Nelson, 1990).

6.5 PRODUCTION OF LYTIC ENZYMES

Production of "lytic enzymes" is an eco-friendly approach to manacle the proliferation of harmful pests. Lysis is the incomplete or entire obliteration of a cell by enzymes. Many microorganisms produce "lytic enzymes" that can "hydrolyze" an extensive range of "polymeric compounds" (including chitin, proteins, cellulose, hemicellulose, and DNA). Appearance and release of enzymes by diverse "microbes" can seldom consequence in the repression of plant "pathogen" actions (Clay et al., 2009; Müller et al., 2010). Many rhizobacteria produce HCN that will help in bio-control of "plant-pathogen" and escalating the produce. Capricious compounds such as NH_3 produced by *"Enterobacter cloacae"* can result in the inhibition of *Pythium ultimum*-induced damping-off of cotton (Howell et al., 1988).

6.6 INDUCED SYSTEMIC RESISTANCE (ISR)

ISR is another potent mode of action as far as biocontrol agents are concerned. Advantageous rhizospheric microbes can perk up plant well-being by leading the complete plant to enhance the defense in opposition to different pathogens and herbivore pests by the process of ISR (Pieterse et al., 2014). The induced resistance is a generic term (Kuc, 1968) for the induced status of resistance in plants triggered by biological or chemical inducers, which protects non exposed plant parts against future attack by pathogenic microbes and herbivorous insects (Kuc, 1968). ISR is a considerable method by which particular "plant growth-promoting bacteria" and fungi in the plant root environment lead the enhanced protection in opposition to a broad assortment of "pathogens" and "herbivore insects" (Mewis et al., 2006; Kim et al., 2008). An extensive diversity of PGPR and PGPF (including *Pseudomonas, Bacillus, Trichoderma,* and mycorrhiza species) incite resistance in plants against herbivores and plant disease (Campos et al., 2014; Pieterse et al., 2014). Generally, ISR bestows an improved stage of safety against a wide continuum of attackers (Walters et al., 2013). Furthermore, it is synchronized by a complex of integrated signaling routes in which "phytohormones" play a crucial role (Pieterse et al., 2012; Mejía et al., 2014).

6.7 CONCLUSION

Biocontrol agents occupy the prime position in sustainable agriculture. Therefore, it is reckoned as an effective eco-friendly approach in attenuating pest swarms through the active participation of natural adversaries without leaving any chemical wastes that can otherwise pose a disastrous threat to other organisms and ecosystems. In order to feed the enormous global population, the incorporation of biocontrol agents in agricultural practices on a broader scale may boost food productivity qualitatively.

KEYWORDS

- *Bacillus thuringensis*
- baculoviruses
- biocontrol
- induced systemic resistance
- parasitoid
- pathogens

REFERENCES

Bailey, A. S., Bertaglia, M., Fraser, I. M., Sharma, A., & Douarin, (2009). Integrated pest management portfolios in UK arable farming: Results of farmer's survey. *Pest Management Science, 65*, 1030–1039.

Campos, M. L., Kang, J. H., & Howe, G. A., (2014). Jasmonate-triggered plant immunity. *J. Chem. Ecol., 40*, 657–675.

Chitra, P., Sujatha, K., & Jeyasankar, K., (2017). Entomopathogenic nematode as a biocontrol agent-recent trends: A review. *Int. J. Adv. Res. Biol. Sci., 4*(1), 9–20.

Clay, N. K., Adio, A. M., Denoux, C., Jander, G., & Ausubel, F. M., (2009). Glucosinolate metabolites required for an *Arabidopsis* innate immune response. *Science, 323*, 95–101.

Clem, R. J., & Passarelli, A. L., (2013). Baculoviruses: Sophisticated Pathogens of Insects. *Plos Pathogens, 9*(11), 1–2.

Contreras-Cornejo, H. A., Macías-Rodríguez, L., Cortés-Penagos, C., & López-Bucio, J., (2009). *Trichoderma virens*, a plant beneficial fungus, enhances biomass production and promotes lateral root growth through an auxin-dependent mechanism in *Arabidopsis. Plant Physiol., 149*, 1579–1592.

Dabbert, S., Haring, A. M., & Zanoli, R., (2004). *Organic Farming Policies and Prospects* (p. 169). Zed Books, New York.

Ehler, L. E., (1998). Conservation biological control: Past, present and future. In: Barbosa, P., (ed.), *Conservation Biological Control* (pp. 1–8). Academic Press, San Diego.

Figueiredo, M. V. B., Seldin, L., De Araujo, F. F., & Mariano, R. L. R., (2010). Plant growth promoting rhizobacteria: Fundamentals and applications. In: Maheshwari, D., (ed.), *Plant Growth and Health Promoting Bacteria* (pp. 2–43). Microbiology Monographs; Springer: Berlin/Heidelberg, Germany.

Fravel, D. R., (2005). Commercialization and implementation of bio control. *Annual Review of Phytopathology, 43*, 337–359.

Harman, G. E., (2000). Myths and dogmas of bio control: Changes in the perceptions derived from research on *Trichoderma harzianum*T-22. *Plant Disease, 84*, 377–393.

Harrison, M. J., (2005). Signaling in the arbuscular mycorrhizal symbiosis. *Annu. Rev. Microbiol., 59*, 19–42.

Howell, C. R., Beier, R. C., & Stipanovi, R. D., (1980). Production of ammonia by *Enterobactercaloacae* and its possible role in the biological control of phythium pre-emergence damping off by the bacterium. *Phytopathology, 78*, 105–1078.

Junaid, J. M., Dar, N. A., & Bhat, T. A., (2013). Commercial biocontrol agents and their mechanism of action in the management of plant pathogens. *International Journal of Modern Plant and Animal Sciences, 1*(2), 39–57.

Kim, J. H., Lee, B. W., Schroeder, F. C., & Jander, G., (2008). Identification of indole glucosinolate breakdown products with antifeedant effects on *Myzuspersicae* (green peach aphid). *Plant J., 54*, 1015–1026.

Kost, T. A., Condreay, J. P., & Jarvis, D. L., (2005). Baculovirus as versatile vectors for protein expression in insect and mammalian cells. *Nature Biotech., 23*, 567–575.

Kuc, J., (1982). Induced immunity to plant disease. *Bioscience, 32*, 854–860.

Lorito, M., Hayes, C. K., Zonia, A., Scala, F., Del, S. G., Woo, S. L., & Harman, G. E., (1994). Potential of genes and gene products from *Trichoderma* sp. and *Gliocladium*sp. for the development of biological pesticides. *Molecular Biotechnology. 2*, 209–217.

McCaffery, A. R., (1998). Resistance to insecticides in heliothine Lepidoptera: A global view. *Philosophical Transactions of the Royal Society of London Series B-Biological Sciences, 353*, 1735–1750.

Mejía, L. C., Herre, E. A., Sparks, J. P., Winter, K., García, M. N., Van, B. S. A., et al., (2014). Pervasive effects of a dominant foliar endophytic fungus on host genetic and phenotypic expression in a tropical tree. *Front. Microbiol., 5*, 479.

Mewis, I., Tokuhisa, J. G., Schultz, J. C., Appel, H. M., Ulrichs, C., & Gershenzon, J., (2006). Gene expression and glucosinolate accumulation in *Arabidopsis thaliana* in response to generalist and specialist herbivores of different feeding guilds and the role of defense signaling pathways. *Phytochemistry, 67*, 2450–2462.

Müller, R., De Vos, M., Sun, J. Y., Sønderby, I. E., Halkier, B. A., Wittstock, U., et al., (2010). Differential effects of indole and aliphatic glucosinolates on lepidopteran herbivores. *J. Chem. Ecol., 36*, 905–913.

Nelson, E. B., (1990). Exudate molecules initiating fungal responses to seed seeds and roots. *Plant and Soil, 129*, 61–73.

Newman, R. M., Thompson, D. C., & Richman, B. D., (1998). Conservation strategies for the biological control of weeds. In: Barbosa, P., (ed.), *Conservation Biological Control* (pp. 371–396). Academic Press, San Diego.ded ecosystems.

Oztemiz, C. S., (2008). *Biological Control in Organic Farming*. Plant Protection Research Institute, Adana (Turkey).

Pal, K. K., McSpadden, B., & Gardener, (2006). Biological control of plant pathogens. *The Plant Health Instructor*, 1–25.

Pangesti, N., Reichelt, M., Van, D. M. J. E., Kapsomenou, E., Gershenzon, J., Van, L. J. J., et al., (2016). Jasmonic acid and ethylene signaling pathways regulate glucosinolate levels in plants during rhizobacteria-induced systemic resistance against a leaf-chewing herbivore. *J. Chem. Ecol., 42*, 1212–1225.

Pieterse, C. M. J., Van, D. D. D., Zamioudis, C., Leon-Reyes, A., & Van, W. S. C. M., (2012). Hormonal modulation of plant immunity. *Annu. Rev. Cell Dev. Biol., 28*, 489–521.

Pieterse, M. J. C., Zamioudis, C., Roeland, L., & Berendsen, R. L., (2014). Induced systemic resistance by beneficial microbes. *Annual Review of Phytopathology, 52*, 347–375.

Pimentel, D., (1963). Introducing parasites and predators to control native pests. Canadian entomologist. In: Delfosse, E., (ed.), *Proceedings of the VI Symposium on Biological Control of Weeds* (Vol. 95, pp. 785–782). August 1984, Vancouver. Agriculture Canada.

Ring, D., (2017). *Bacillus Thuringiensisas a Biopesticide Applied to Plants.* LSU College of Agriculture.

Saleem, M., Law, A. D., & Moe, L. A., (2016). Nicotiana roots recruit rare rhizosphere taxa as major root-inhabiting microbes. *Microb. Ecol., 71*, 469–472.

Sharma, A., Diwevidi, V. D., Singh, S., Pawar, K. K., Jerman, M., Singh, L. B., Singh, S., & Srivastawa, D., (2013). Biological control and its important in agriculture. *International Journal of Biotechnology and Bioengineering Research, 3*(4), 175–180.

Spaink, H. P., (2000). Root nodulation and infection factors produced by rhizobial bacteria. *Annu. Rev. Microbiol., 54*, 257–288.

Stuart, R. J., Barbercheck, M. E., Grewal, P. S., Taylor, R. A. J., & Hoy, C. W., (2006). Population biology of entomopathogenic nematodes: Concepts, issues, and models. *Biological Control, 38*, 80–102.

Vacheron, J., Desbrosses, G., Bouffaud, M. L., Touraine, B., Moënne-Loccoz, Y., Muller, D., et al., (2013). Plant growth-promoting rhizobacteria and root system functioning. *Front. Plant Sci., 4*, 356.

Van, D. M. J. E., De Vos, R. C., Dekkers, E., Pineda, A., Guillod, L., Bouwmeester, K., et al., (2012). Metabolic and transcriptomic changes induced in Arabidopsis by the rhizobacterium *Pseudomonas fluorescens* SS101. *Plant Physiol., 160*, 2173–2188.

Van, L. J. C., (2000). *Criteria for selecting natural enemies to be used in biological control programs. In: Bueno, V. H. P., (ed.), Biological Pest Control: Mass Production and Quality Control* (pp. 1–19). Lavras, Brazil.

Van, L. L. C., (2007). Plant responses to plant growth-promoting rhizobacteria. *Eur. J. Plant Pathol., 119*, 243–254.

Van, P. R., Niemann, G. J., & Schippers, B., (1991). Induced resistance and phytoalexin accumulation in biological control of *Fusarium* wilt of carnation by *Pseudomonas* sp. strain WCS417r. *Phytopathology, 81*, 728–734, 175.

Vardharajula, S., Shaik, Z. A., Vurukonda, S. S. K. P., & Shrivastava, M., (2017). Plant growth promoting endophytes and their interaction with plants to alleviate abiotic stress. *Curr. Biotechnol., 6*, 252–263.

Walters, D. R., Ratsep, J., & Havis, N. D., (2013). Controlling crop diseases using induced resistance: Challenges for the future. *J. Exp. Bot., 64*, 1263–1280.

Zhang, H., Xie, X., Kim, M. S., Kornyeyev, D. A., Holaday, S., & Paré, P. W., (2008). Soil bacteria augment *Arabidopsis* photosynthesis by decreasing glucose sensing and abscisic acid levels in plants. *Plant J., 56*, 264–273.

CHAPTER 7

The Science of Vermicomposting for Sustainable Development

ROHAYA ALI[1] and RUMISA NAZIR[2]

[1] Department of Biochemistry, University of Kashmir, Srinagar – 190006, Jammu and Kashmir, India, E-mail: rohayaali01@gmail.com

[2] Department of Environmental Science, Government College for Women Nawakadal, Srinagar, Jammu and Kashmir, India

ABSTRACT

Composting is a process that involves the breakdown or decomposition of a substrate by living organisms in a progressive manner. In these processes, microbes that are indigenous to substrate; this process is instigated by the disintegration of larger or complex molecules present in the unprocessed substrate to simpler ones by means of microorganisms that are indigenous to substrate. Vermicomposting involves the transformation of organic waste by means of microbes and earthworms to those products that are safe and sound to be used as soil conditioners and bio-fertilizers. The rate and degree of composting depend on temperature, moisture, aeration, and pH.

7.1 INTRODUCTION

Composting is a primitive technology that is practiced at present times, from small compost heaps to grand commercial processes (Fernandes and Sartaj, 1997). In various Biblical and Roman texts, there is mention of composting. Attribute the origin of this process to some particular person or to a single civilization is unjustified. The primitive Akkadian territory of Mesopotamian Civilization utilized manure for agricultural purposes on clay tablets a thousand years before Moses was born. There is also proof that Greeks, Romans, and the ethnic groups of Israel had knowledge about compost. The Talmud also mentions the use of decayed manure grain stalks as fertilizer.

Eminent writers like Walter Raleigh and Francis Bacon, also cited the usage of compost in ancient times. There has also been the discovery of compost usage by the original inhabitants of America and the Europeans living in America. Several New England peasants prepared compost as a recipe of 10 fractions muck to 1 fraction fish, turning their compost piles at regular intervals until the fish collapsed or crumbled down completely. However, the combination of dead fish with compost was not that effectual. Subsequent to this finding, farming practices turned chemical in nature. After that, fertilizers substituted compost in several parts of the world. Albert Howard, a British agriculturist, discovered that the finest compost constituted of three times as much plant material as manure. In this compost, the materials were layered in sandwich style. This method was regarded as the "Indore method." Later an agronomist, Sir Howard, wrote a book titled "An Agriculture Testament." This work renewed the importance of organic methods of agriculture and gave Sir Howard the title of "Father of Organic Farming."

For millions of years, nature has recycled and reshaped its own organic waste through this process. This dynamic process can be seen in action as one walks through the woods. Composting is a feasible option, even in solid waste management in developing countries where there is less availability of resources. Composting is the best way to be "green." However, four R's: Reduce, Recycle, Reuse, and Re-buy should be the center of attention to achieve this goal. Composting has many benefits, like lower equipment and operating costs. Besides, it is an environmentally friendly process. Composting is also associated with a few shortcomings also like slow reaction rate. The main condition for enhanced composting is a better understanding of the basic principles of the process. In the absence of such perceptive, failure of the composting process is practically unavoidable.

7.2 CONCEPT OF VERMICOMPOSTING

In composting, biodegradable solid waste is biologically degraded to a state that is highly stable and easy to store or handle (Ahn et al., 2008). This process is beneficial for the treatment of solid waste, and the product obtained is highly beneficial for agricultural use and in eliminating pathogens and unwanted weeds. Composting is advantageous when the chief factors like nitrogen, oxygen, carbon, dampness, temperature, etc., which influence the process of composting, are appropriately taken care of. Thus, proper management of the composting process is preconditioning for its efficient usage in agriculture (Ahn et al., 2008).

7.3   FAVORABLE CONDITIONS FOR AEROBIC COMPOSTING

Various factors responsible for successful composting are nutrition, temperature, pH, moisture content, etc. These are discussed in subsections.

7.3.1   NUTRITIONAL FACTOR

Nutrients present in waste could be used only to the level that it is obtainable to active microorganism. Nutrients can be assessable in two forms, either physical or chemical. It is available to microorganism chemically, if it is a fraction of a molecule that is susceptible to microbial attack. This process is carried out enzymatically by microorganisms in two ways. In the first way, the microbe may either contain an enzyme indigenously or can synthesize it. Physical accessibility is represented in terms of availability to microorganisms. Availability is a function of the ratio of volume to the surface area of waste material.

7.3.1.1   MACRONUTRIENTS AND MICRONUTRIENTS

The substrate is a source of various nutrients. Nutrients can be categorized into main types, i.e., macronutrients and micronutrients. Carbon, nitrogen, phosphorus, calcium, and potassium are included in macronutrients. However, the requisite quantities of calcium and potassium are much lower than those of carbon, nitrogen, and phosphorus. Since they are needed in trace amounts, so they are often regarded as "essential trace elements." Magnesium, iron, cobalt, manganese, and sulfur are included in micronutrients. A number of trace elements are involved in cellular metabolism. They can become poisonous if present above trace ones.

7.3.1.2   CARBON-TO-NITROGEN RATIO

For the process of composting, various organic resources are used that possess a sufficient amount of nutrients. The chief nutrients needed by the microbes in the process of composting are Carbon, nitrogen, phosphorus, and potassium. The carbon to nitrogen ratio (C/N) is a chief nutrient factor. Too much or even scarce carbon and nitrogen may affect the process. Carbon is a source of energy and growth for microbes. On the other hand, nitrogen is important for the synthesis of proteins. Besides, it also helps in reproduction and development.

7.3.2   PARTICLE SIZE

Dimensions of particles influence the rate of organic matter breakdown. The quantity of the particle's mass exposed to the microbe's attack is determined from the ratio of mass to the surface area. Greater the surface area accessible to the microbe, the less difficult it is for microbes to show their activity. This is because the major task is carried at the boundary between the particle surface and air. Microbes are capable of digesting more, produce a large amount of heat, and reproduce more rapidly with a small amount of substance. Besides, smaller pieces speed up the process of breakdown of organic material. These materials can be torn, cut, beaten, or even perforated to amplify their surface area. Various methods are utilized for these purposes. For example, a lawn trimmer is utilized to chop leaves in a garbage container. Various types of shredders and chippers are accessible in the market for shredding woody materials and foliage. However, safety goggles should be for grating and chopping purposes. Hands must be taken care off while the machine is working. Kitchen leftovers can be cut with the aid of a table knife. Some determined citizens utilize meat grinders to make "garbage soup" from the leftovers of foodstuff and then pour the blend or mixture into their garbage heaps.

7.3.3   ENVIRONMENTAL FACTORS

The most important environmental factors that influence the rate and degree of decomposition in composting are temperature, pH, moisture, nutritious food, and aeration. Deficiency in any of these aspects slows down the composting process. Thus, the scarce factor becomes the rate-limiting factor.

7.3.3.1   TEMPERATURE

The temperature within the compost heap acts as an important factor in the maintenance of microbial growth and activity. Consequently, temperature also affects the rate at which the raw materials decay. Higher temperatures result in a quicker breakdown of organic matter. However, exceptionally high temperatures can slow up the activity of microbes. Composting is most advantageous when the temperature of the composting substance is within the two ranges, i.e., Mesophilic (80–120°F) or Thermophilic (105–150°F). Mesophilic temperatures permit successful composting, but most experts advise maintaining temperatures between 110 and 150°F for composting. The process of composting is influenced at temperatures higher than 65°C. This is on account of the reason that microbes associated with the spore-forming phase, form spores above 65°C. Other microbes either tumble into the resting stage or get killed. An alternate way is to resort to those measures which are premeditated to evade temperatures greater than 60°C.

7.3.3.2 pH

The growth of microbes is also influenced by the alkalinity or acidity of the waste material. The pH values of 6.0–7.5 are preferred by bacteria for decomposition, whereas fungi prefer pH ranging from 5.5 to 8.0. However, gaseous losses like that of ammonia take place if pH exceeds 7.5. There are few materials like wastes from the paper industry, cement factories, and dairy farms, which can enhance pH values. The initial pH levels do not cause hindrance to most of the microbes. Therefore, there is no need of buffering in the initial stages. In fact, it may have an unfavorable result in microbial growth. Nonetheless, the addition of lime proves beneficial in a few cases. It enhances the physical features of the compost, possibly by absorbing moisture.

7.3.3.3 MOISTURE CONTENT

The preferable moisture values of the compost mass range between 45–60% by weight. Low moisture content will hamper the process of composting, as it will divest microbes of water required for their metabolism. This, in turn, will inhibit the activity of the microbes. Extreme desiccation makes piles prone to unprompted combustion (Fernandes and Sartaj, 1997). However, if the compost mass is highly hydrated, it may block the pile's air spaces, creating an anaerobic situation. Many compounds need anaerobic conditions for their breakdown, like halogenated hydrocarbons. There are various methods, which can be utilized to measure anaerobic conditions. Manure and composting test laboratories are there to evaluate moisture content. Moisture can be evaluated by manure and compost-testing laboratory. Besides, moisture meters are also available in the market. Nonetheless, a more realistic and simple approach is the 'squeeze test.' In this test, the mixture is pressed in a gloved hand. If more than a small number of drops of water come out, it means the sample is too much drenched. If it appears to be very dehydrated, more moisture needs to be incorporated. The degree of dampness of compost mass depends on the organic waste's organization. Leaves require less moisture content when compared with wheat or corn stalks. Similarly, food leftovers or grass cuttings need less moisture content. If a compost pile is dried up, it should be hydrated immediately. Few substances like straw, dead leaves, and hay are slowly moistened till they get glistened. Such substances have a propensity to discard or adsorb water only on their exterior surface. It should be noted that if the pile gets saturated with water, it should be turned and restacked.

7.3.3.4 AERATION RATES

Composting can take place under both anaerobic and aerobic situations. Nevertheless, aerobic composting is proficient than anaerobic. Even though the air contains 21% oxygen, aerobic microorganisms can thrive even at O_2 concentrations of 5%. However, the O_2 concentrations of more than 10% are considered as most advantageous in the composting process. The organization of the constituents of the waste decides the degree of aeration. Besides, the techniques used for aeration also govern the amount of aeration required under aerobic conditions. Aeration is essential for successful composting. It can be either active or passive. With an increase in microbial activity, more, and more O_2 utilized. If the O_2 supply is not properly replenished, the process of composting may shift to an anaerobic state. This will, in turn, slow down the composting process. Good aeration enhances aerobic composting. If the heaps are too large, soaked, or have less porosity, aerobic bacteria cannot get adequate O_2, and anaerobic bacteria take charge. However, a stinking smell may be generated in such processes. Few by-products of anaerobic respiration like sulfur compounds are well known for this. The odors' types and intensities depend on the type of feedstock and operating circumstances required in the composting process. Since these odors create a lot of annoyance to the people, therefore it is necessary to control them. Such problems may arise, especially in the regions where the composting process is carried out near a human population. Various methods are used to deal with the stinking smell arising out of composting. One of the most widely used methods is bio-filtration. This is an efficient method to lessen the intensity of odors produced due to the dispensation of organic matter.

7.3.4 NUTRITIOUS FOOD

During composting, microbes break down the complex organic material into simpler forms to obtain energy. This energy is utilized to carry out various life processes and acquire nutrients to continue their populations. Among all the substances needed for microbial decomposition, C and N are the most significant. Carbon acts as a source of energy and growth for microbial cells. Nitrogen is a vital constituent of enzymes, proteins, nucleic acids, coenzymes, etc., that is necessary for cell development and function.

7.3.5 ACTIVE ORGANISMS

Many microbes have been shown to be linked with composts. It is obvious that the microbial community, as a whole, plays a vital role in the decomposition of organic materials. Bacteria (especially mesophiles and thermophiles) and other microbes like fungi are the chief living organisms involved in composting's preliminary active phases. Mostly, minute forms like rotifers, amoeba a few protozoans are foremost to come into sight at the composting site. Ultimately, bigger forms like snails and earthworms (*Eisenia foetida* and *Lumbricus terestris*) become abundant. The compost mass gets reasonably matured once the earthworms come into play. The usage of earthworms in the process of composting has led to the emergence of vermicomposting. Bigger organisms take part in the physical transformation of organic matter into compost. However, they are vigorous in the later stages of composting, like in chopping, chewing, mixing, and digesting of compostable mass. Thus, they break larger particles into smaller pieces, and transform them into a more digestible state for microbes.

7.3.6 RAPID COMPOSTING METHODS

The composting process takes a lot of time, which may range from 100–180 days. Substantial research has, therefore, been carried out to speed up the process of composting, by the introduction of suitable microbes with good proficiency in decomposing of organic matter. Presently, the compost producers use microbial inoculants for the quick decomposition of biodegradable material and repression of foul odor. The government institutes like the Indian Agricultural Research Institute and Indian Institute of Soil Science, have formulated efficient cultures which are being utilized by the compost producers. Some processes through which compost production can be enhanced are described below:

7.3.7   PHOSPHO-SULPHO-NITRO COMPOST

In this method, suitable minerals, fertilizers, and microbial cultures are used to fortify the compost so that the end product contains more nutrients per unit volume or weight. It also makes use of compost accelerating culture and biofertilizers for further nutrient enrichment. This reduces the bulk, which has to be transported.

7.4   EFFECTIVE MICROORGANISM BASED PRODUCTION PROCESS

The idea of effective microorganisms (EM) originated in Japan. EMs consists of mixed cultures of valuable and naturally existing microbes that can be utilized as inoculants to augment microbial diversity. EM consists of selected microbial species, including populations of yeasts and lactic acid (LA) bacteria and few photosynthetic bacteria. All of them are mutually compatible and can live together in liquid culture. Research has revealed that the inoculation of EM cultures to floral bionetwork or to the soil, improves the growth of crops and their production. Besides, it improves soil quality.

7.4.1   MICROBIALLY ENRICHED COMPOST FOR RECYCLING OF MSW

This methodology was developed at the Indian Institute of Soil Science. Compost was prepared by the pit method, and this compost was ready after 2.5 months. The pit used was concrete that prevented the nutrient percolation into the soil.

7.4.2   COIR COMPOST METHOD

Coir pith composting is an aerobic way of composting, which decreases the bulkiness, C: N ratio, lignin, and cellulose contents of coir dust and increases its manurial value. 60% moisture is to be sustained at the time of composting. Aged or matured compost is then utilized within 60 days.

7.4.3 RAPID TECHNOLOGY COMPOSTING

In rapid technology composting, plant substrate is inoculated with cultures of a decomposing cellulose fungus (*Trichoderma harzianum*), for better composting. The medium of growth and development for fungus is sawdust mixed with the leaves of Subabul (*Leucaena leucocephala)*, a leguminous tree. This process takes 21 to 45 days, depending on the plant substrates used. The procedure involves two steps. Firstly, the production of the compost fungus activator is done, followed by the process of composting.

7.4.4 EXCEL TECHNOLOGY

This is the process of aerobic composting, which was devised by M/s Excel Industries, Mumbai. It is widely used in large-scale mechanical composting plants. This procedure recovers over 90% of the initial organic matter like compost. The final product is humus-like material, dark brown in color, and free from stink. It has a huge moisture-holding capacity.

7.5 THE TECHNIQUE OF COMPOSTING

Following factors help in the quick production of compost:

- **Compost Mass Location:** The ideal site for the pile is away from direct sunlight, and speedy winds, which would, which would dry it out, and cool the heap. Besides, extreme moisture content should also be avoided can be avoided by putting the pile on soil with proper drainage.

- **Size and Volume:** The volume of compost mass should hit a balance between being bulky enough to grasp heat and being little enough to let air reach the center. Thus, it must be neither too large nor too small.

- **Pile Construction:** Compost piling can be prepared by adding up material as it becomes available or in even batches.

- **Pest Prevention:** Few food substances should be left out of the heap or added in minute quantities so as to distract pests like rats or other rodents. If the pile is not turned recurrently, fly larvae may start appearing, which may be an issue. Hence, finished compost or some carbon material like straw should be used to cover the pile. This will help in controlling the pest.

- **Health Considerations:** Fungi and molds may lead to allergic reactions. Wearing a dust mask while pile turning or washing off hands after working with compost is enough to avoid an allergic reaction. The use of sanitizers is also recommended.

7.6 COMPOST AND ITS PHYSICAL CHARACTERISTICS

Compost is a humus-like substance that is dark in color and possesses numerous nutrients and minerals, making it highly beneficial for soil microbes. It could be utilized in yards and gardens to generate a healthy and eco-friendly environment. It makes nutrients steadily available to plants and can, therefore, be used in place of chemical fertilizers. Hence, compost is often regarded as "black gold" by the gardeners:

- **Appearance:** Compost is light brown in color and slowly darkens as the process advances. The final product usually has a darkish grey or brown color.

- **Odor:** The smell of the substrate changes within a small number of days after the process initiates. The smell is that of raw garbage, if the substrate is municipal solid waste (MSW). In other cases, the smell can be collectively referred to as a "faint cuisine" smell. On the other hand, in the case of anaerobic conditions, the prominent smell is that of the putrefying process.

- **Particle Size:** As the decomposition occurs, it makes the substrate brittle and granular, if the substrate is amorphous. As a result, the particle size decreases and becomes smaller.

- ➤ **Finished Compost Qualities:** Quality of compost may vary depending on the conditions in which composting takes place. Quality depends on the state of biological activity, the composting process used, and above all, the intended use of compost. Following are the qualities of good compost:

  - Good amount of organic matter;
  - 15:1 C/N ratio;
  - Neutral pH 6–8;

- Less amount of soluble salts;
- Absence of phytotoxic compounds;
- Appreciable seed germination;
- Weed-free conditions.

7.7 SUBSTANCES THAT COULD BE COMPOSTED

7.7.1 GREEN WASTE OR ORGANIC SOLID WASTE

Solid waste can be converted into compost and is highly beneficial for agricultural purposes. Composting converts decomposable organic matter into a valuable stable product. It revitalizes the soil vitality by means of overcoming phosphorus depletion. A co-composting process is utilized that combines solid waste with dewatered bio-solids. Industrial composting techniques are being utilized in solid waste management systems, which are alternative to landfills. Besides, other advanced systems are also used for industrial composting. In a landfill, untreated waste breaks down anaerobically, and fugitive greenhouse gas methane is produced. Therefore, treating biodegradable waste before it goes into landfill decreases global warming.

7.7.2 ANIMAL BEDDING AND MANURE

The fundamental composting constituents on farms are cattle dung and straw or sawdust bedding. Besides, newspapers and chopped cardboards are also used as bedding materials, which also act as an ingredient for compost. The quantity of manure formed from livestock waste depends on climatic conditions, cleaning patterns, land accessibility, etc. Different type of manures possesses unique physical, chemical, and biological features. Manures of horse or cattle, when combined with bedding, form good compost. When the bedding is mixed with cattle and horse manure, it offers good qualities for compost formation. Besides, poultry manure (PM), mixed with carbonaceous materials like straw or sawdust, results in efficient composting.

7.7.3 HUMAN WASTE AND SEWAGE SLUDGE

Human waste being rich in nitrogen can be used in the composting process. This can be done either directly or in composting toilets. Apart from this, it can be used when mixed with water or in the form of sewage sludge once it has been treated in a sewage treatment plant (Epstein et al., 1976).

7.7.4 URINE

Humans excrete water greater amount of soluble nutrients like phosphorus, potassium, nitrogen, etc., in urine compared to that in feces. The urine can be utilized directly as fertilizer or mixed in manure. The addition of urine elevates the temperature of compost, which enhances the ability of pathogens to destroy unwanted seeds. Besides, human urine does not attract houseflies or blow files, which otherwise may spread disease. Human urine does not contain pathogens, such as parasitic worm eggs (Epstein et al., 1976).

7.7.5 HUMANURE

Joseph Jenkins (1994) used the term humanure for the first time in his book. This book supports the use of organic soil amendment (Brandjes et al., 1989). Humanure is recycled through composting for agricultural purposes.

7.8 MARKETS FOR COMPOST

Composting is one of the ways of organic waste management. In farms, composting of manures is a good way of manure management. Generally, the end product is a precious soil resource (Robert and Davila, 2000). Materials obtained after composting could substitute substances such as topsoil or peat as mulches, seed starters, soil amendments, container mix, turf, fertilizers, etc., in farms, industrial greenhouses, farms, land remediation, landscaping, bioremediation, etc. (Figure 7.1).

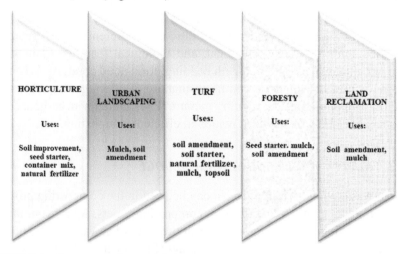

FIGURE 7.1 Compost markets and their use.

7.9 USES

Compost can be added to soil and other materials like peat and coir, thereby providing humus or nutrients to them. This acts as a highly nutritious medium for plants. It also acts as an absorbent, which prevents dryness by absorbing water and soluble minerals. It is combined with mud, gravel, clay, or bark pieces, etc., to form loam. Compost, when added directly to the soil or in a growing medium, can enhance soil fertility. This improves the level of organic matter in the soil.

Usually, direct seeding into compost is not suggested because its drying speed and potential occurrence of phytotoxins that inhibit germination. It is very common to observe that up to 30% compost blends are utilized for transplanting seedlings at the cotyledon stage or later.

7.10 CO-COMPOSTING

Composting in which more than single organic material is utilized is referred to as co-composting. In a few co-composting methods, trial and error techniques are used to determine the quantity or proportion of each material present in compost (DEH, 1975). In order to achieve the finest ingredients for composting in an optimum time interval and free from noxious smell, mixing practices based on chemical or physical parameters of the composting substances should be followed (Lopez et al., 1996). The look and feel method is the only alternative when the material characteristics are unknown.

The dampness of compost may be critical. Excessive moisture may lead to an anaerobic situation. Besides, noxious smell and slowing down decomposition may also occur due to dampness. Therefore, for better co-composting, moisture content should be optimized (DEH, 1975).

7.11 VERMICOMPOSTING

Vermicomposting is an easy biotechnological method of decomposition in which selected species of earthworms are utilized to speed up the progression of waste recycling and generate an enhanced final product. Vermicomposting varies from normal composting in a number of ways. Vermicomposting is a mesophilic method in which microbes and earthworms (active at 10–30°C) are utilized. Besides, the vermicomposting is process is quicker than simple composting. In this method, the substrate passes through the gut of the earthworm, and the end product, i.e., earthworm castings, also called worm manure, is formed (DEH, 1975). Mainly brandling worms and red worms like *Eisenia fetida* and *Lumbricus rubellus* are used for vermicomposting (Will, 1998). They are generally found in aged manure piles with alternating red and buff-colored stripes. The nutrient content of vermicompost is shown in Figure 7.2.

Garden earthworms prefer ordinary soil, but at times they feed on the bottom of a compost heap. Around 500,000 earthworms are found in an acre of land, which can recover as much as five tons of soil per year. However, brandling worms and red worms prefer the compost environment. Recycled organic matter passes through the gut of these earthworms and is excreted as worm casting (worm manure). Worm manure is rich in nutrients and has a fine texture (Senn, 1974; Park et al., 1992).

Earthworms are regarded to be nature's plowman, and it acts as a gift to farmers as it generates high-quality humus. Such humus fulfills the nutritional requirement of crops in the best possible manner. Vermicompost offers numerous benefits to industries, farmers, and also the entire economy. These are listed below:

1. **To Industries:**
    • Less expenditure on pollution abatement.

2. **To Environment:**
    • Pollution-free and hence eco-friendly;
    • Enhancement in soil fertility.

3. **To Farmers:**
    • Least dependence on external inputs leading to less expenditure;
    • Augmented soil production through enhanced soil quality;
    • Enhanced quality and quantity of crops;
    • Generation of an additional source of income for landless people.

4. **To General Economy:**
    • Advancement to pastoral economy;
    • Less investments in purchased inputs;
    • Low wasteland generation.

Earthworms consume a range of organic materials and reduce the final volume up to 60%. A single earthworm weighing around 0.5 g can consume waste equal to its body weight and produce cast equal to half of the waste in a single day. These castings contain a reasonable amount of macro and micronutrients when compared to garden compost. Besides, the content of moisture in these casting varies between 30–60% and has a neutral pH.

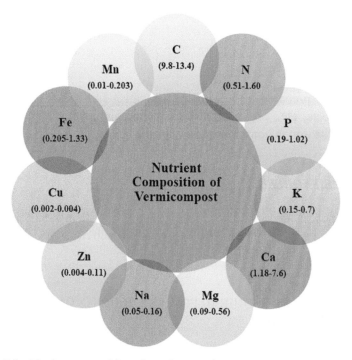

FIGURE 7.2   Nutrient composition of vermicomposting.

7.12   METHODS OF VERMICOMPOSTING

7.12.1   PITS USED FOR COMPOSTING

Pits prepared for the purpose of vermicomposting should be around 1 min in depth and 1.5 m in width. The extension in length varies as per the requisite.

7.12.2   HEAPING ABOVE THE GROUND

Waste materials are laid down on a polythene cover and then put on the ground. It is followed by covering of heap with cattle dung (Sunitha et al., 1997). Bearing in mind the biodegradation of wastes as the principle, the heap method of vermicomposting is considered superior than the pit method. It has also been observed by various agronomists that the population of

Eudrilus eugenae earthworms is always higher in the heap method compared to the pit method. Besides, biomass generation is also more in the heap method. Thus, in the heap method, the overall vermicompost production is greater compared to the pit method.

7.12.3   TANKS

Tanks that are prepared from different substances like rocks, gravel, bricks, asbestos sheets, etc., are evaluated for vermicomposting. Tanks could be made with the dimensions appropriate for the operation.

7.12.4   REINFORCED RINGS

Vermicompost is also made above the ground by utilizing reinforced rings (ICRISAT and APRLP, 2003). The dimension of such rings ring should be 30 cm in height and around 90 cm in width.

7.13   RESOURCES REQUIRED FOR THE PROCESS OF VERMICOMPOSTING

A variety of agricultural remains like sorghum or rice straw, dry or desiccated leaves of plants and trees, vegetable wastes, sugarcane trash, pigeon peas talks, groundnut peelings, soybean waste, weeds before flowering, fiber from coconut trees, etc., can be transformed into valuable vermicompost. Apart from this, animal bedding and manures, dairy or poultry products, food industries waste material, biogas sludge, urban solid wastes, and residues from sugarcane industries also furnish a reasonable amount of raw material for vermicomposting (Alberta, 1999).

7.14   CONCLUSION

Composting is an environmentally friendly process that involves recycling raw organic matter into precious soil amendments associated with a number of benefits. The process of composting is carried by microbes naturally occurring in soil. Under normal circumstances, earthworms, nematodes, and soil insects like bugs, beetles, ants, springtails, mites, etc., perform the preliminary mechanical disintegration of organic matter into simpler molecules. In order to accelerate the composting process, suitable microorganisms with good efficiency in the rate of organic matter decomposition are introduced. The compost producers are presently using microbial inoculants to decompose biodegradable material and suppress foul odor. Compost, the end product of the process is easier to manage than manure and other raw organic substances. Besides, it is easy to store and is odor-free.

KEYWORDS

- **composting**
- **manure**
- **night soil**
- **sludge**
- **sustainable development**
- **vermicomposting**

REFERENCES

Ahn, H. K., Richard, T. L., & Glanville, T. D., (2008). Laboratory determination of compost physical parameters for modeling of airflow characteristics. *Waste Management, 28,* 660–670.

Alberta, (1999). Environment. *Leaf and Yard Waste Composting Manual-Revised Edition.* Alberta Environment, Olds College: Edmonton, Canada.

Brandjes, P., Van, D. P., & Van, D. V. A., (1989). *Green Manuring and Other Forms of Soil Improvement in the Tropics.* Agromisa Wageningen The Netherlands.

Department of Environmental Health (DEH), (1975). Sanitary effects of urban garbage and night soil composting. *Chinese Medical Journal, 1*(16), 407–412.

Epstein, E., Willson, G. B., Burge, W. D., Mullen, D. C., & Enkiri, N. K., (1976). A forced aeration system for composting wastewater sludge. *Journal Water Pollution Control Federation, 48*(4), 688.

Fernandes, L., & Sartaj, M., (1997). Comparative study of static pile composting using natural, forced and passive aeration methods. *Compost Science and Utilization, 5*(4), 65–77.

Lopez, R. J., & Baptista, M., (1996). A preliminary comparative study of three manure composting systems and their influence on process parameters and methane emissions. *Compost Science and Utilization, 4*(3), 71–82.

Park, K. J., Choi, M. H., & Hong, J. H., (1992). Control of composting odor using biofiltration. *Compost Science and Utilization, 10*(4), 356–362.

Robert, B., & Davila, J., (2000). *The Peri-Urban Interface: A Tale of Two Cities*. University of Wales, Bangor development planning unit (DPU), DFIDnatural resources systems programme: Hemel Hempstead, Herts.

Senn, C. L., (1974). Role of composting in waste utilization. *Compost Science, 15*(4), 24–28.

Sunitha, N. D., Giraddi, R. S., Kulkarni, K. A., & Lingappa, S., (1997). Evaluation methods of vermicomposting under open field conditions. *Karnataka Journal of Agricultural Sciences, 10*(4), 987–990.

Will, B. J., (1998). *On farm Composting Evaluation of Manure Blends and Handling Methods for Quality Composts*. Woods End Research Laboratory, Report to USDATech Center, Chester PA.

CHAPTER 8

Consolidation of Green Chemistry into Biorefineries: A Pavement for Green and Sustainable Products

NOWSHEEBA RASHID,[1] AMIR HUSSAIN DAR,[2] and IFRA ASHRAF[3]

[1] Amity Institute of Food Technology, Amity University Noida, Uttar Pradesh, India

[2] Department of Food Technology, Islamic University of Science and Technology, Awantipora, Pulwama, Jammu and Kashmir, India

[3] College of Agricultural Engineering and Technology, Sher-e-Kashmir University of Agricultural Sciences and Technology of Kashmir Shalimar Campus, Srinagar, Jammu and Kashmir, India

ABSTRACT

The concept "green bio-refinery" is developed to make use of green biomass as basic material for the fabrication of bio-based goods such as fibers, lactic acids (LAs), proteins, and power (by means of biogas). Upcoming supply chains for actually sustainable and green chemical entities can be established by the assimilation of green chemistry into bio-refineries and also through the implementation of least environmental impact technologies. The first and foremost step in these upcoming bio-refineries ought to be the benevolent extrication of surface chemicals; at this point, the utilization of greener solvents like supercritical CO_2 (carbon dioxide) and bio-ethanol, ought to be well-thought-out. The filtrates will frequently be loaded with lignocellulosic, and the efficient partition of the cellulose is a basic confront which might, in the coming time, be supported by greener solvents, like ionic liquids. In order to persuade the huge as well as the varied industrial requirement for aromatics, it is necessary to develop novel ways to procure simple aromatic building blocks out of lignin, which is considered nature's most important source of aromatics. Fermentation is the other process used nowadays to convert green grassland into an extensive range of bio-platform chemicals besides ethanol. Their green chemical translation to superior chemicals is as imperative as their competent production. Hence, clean technologies like catalysis, especially bio-catalysis as well as heterogeneous catalysis also the employment of benevolent solvents, and energy competent reactors are important. Another process called the thermo-chemical processes will also play a vital role in potential bio-refineries for the translation of biomass; thus, once more green chemistry methods should be utilized to go to superior value downstream chemicals.

8.1   INTRODUCTION

In the present epoch, the most significant contributions to rural economic development, energy security, and also the environmental quality are attributed to biomaterials, renewable fuels, chemicals, and power derivatives of plant biomass. Specifically, fossil energy reliance is able to be bridged through hastening the growth of renewable substitutes to immobile power and carrying fuel, and the US proposes to relocate about 30% of the country's gasoline expenditure and about 10% of overall industrial as well as electric initiator power requirement near to 2030 (AEI, 2009). The underutilized basin of lingo-cellulosic biomass is majorly agricultural residues. Because of this fact, these agricultural residues acquire an immense prospective as feedstocks for the manufacturing of renewable chemical products and bio-based fuels. Besides, they could eventually dislodge a nontrivial division of contemporary utilization of fossil fuel. On the other hand, the contest connected to feedstock logistics (Hess et al., 2009; Sokhansanj et al., 2009) and translation technology (Kazi et al., 2010; Eggeman and Elander, 2005; Klein-Marcuschamer and Blanch, 2013) are chief fiscal hurdles hampering the commercialization of lignocellulose-based bio-refinery. As already mentioned, one of the major impediments that prevent the utilization of these feedstocks for bio-energy creation is the collection of feedstock as well as its transportation (Hess et al., 2009; Sokhansanj et al., 2009). As aforementioned, because of the dispersed nature of green residues, yielding such an enormous quantity of biomass and then its distribution to national bio-refineries forms a logistical confront, and competent as well as reasonable deliverance systems have yet to be recognized. In addition, there are millions of animal maneuvers that are comparatively consistently disseminated in the US (USDA, 1997).

Another most prominent obstacle to lignocellulose-based biorefining is the conversion technology. Present conversion procedures engross chemical pretreatment approach, enzymatic hydrolysis via various cellulolytic enzymes followed by microbial decomposition to certain metabolites, commonly ethanol. The involved tools acquire certain lacunas like the accretion of fermentation inhibitors, the failure of intermediary monomeric sugars for finalized energy or chemical making, and also the requirement for a huge quantity of water and energy. Because of these reasons, extra benevolent measures are required to improve these inconsiderate and energy-intensive chemical conversion procedures (Liao et al., 2014). Thus, the biggest confront to the sustainable market is to fabricate such goods for people that are green and sustainable supply chains. The need of an hour is to understand the fact that the resources cannot be used sustainably more rapidly than they are created, and the waste cannot be sustainably generated more rapidly than the soil can formulate it again to functional and valuable resources. During the commencement of every maintainable supply chain short cycle, renewable resources are required. As a consequence, here is an ever-growing yearning commencing via industry to hunt for justifiable and substitute resources for key service chemicals and otherwise appropriate counterparts. Figure 8.1 (Clark et al., 2009) is a summarization of an idealized bio-refinery, the center of attention being commodity chemical products.

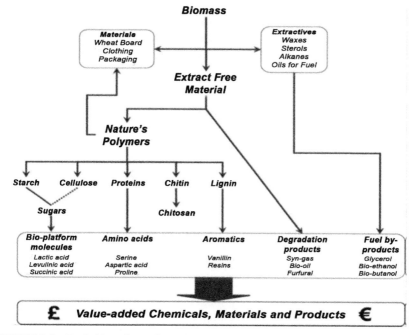

FIGURE 8.1   Basic bio-refinery flow diagram.

Source: Reprinted with permission from Clark et al., 2009. © John Wiley and Sons.

The merely maintainable substitute for fossil fuels as a resource of carbon intended for chemical and material requirements is biomass. As is the situation for fossil resources, they are obtainable for cycle time calculated in hundreds of millions of years; however, the reverse case is with biomass cycle time is calculated in years only. Therefore, a study among the part of restoring chief petro-originated goods and chemicals to biomass originated alternatives is necessary (Klass, 1998). Through the passing time, 'cheap' oil is approaching to its finishing stage, so its need of an hour to go back to an industrial civilization functional on plant-derived feedstocks (National Research Council, 2000). The novel conception of a bio-refinery is the vital step to unlock biomass as a feedstock (raw material) intended for the chemical industry. Upcoming bio-refineries will integrate the fabrication of energy, value-added chemicals, and fuels by means of the dispensation of biomass, into a distinct place (Lyko et al., 2009). One of the major prerequisites of a bio-refinery is that it is capable to deal with varying feedstock compositions as biomass varies significantly in composition among various species and also from season to season; contrasting to the comparatively generic and composition reliable oil. Harvesting of biomass is not promising right through the entire year as it encounters seasonal changes, unlike petroleum. Hence, a shift from crude oil to biomass possibly will necessitate an alteration in the competence of chemical industries, through an inevitability to produce the materials as well as chemicals in a regular manner. On the other hand, biomass possibly will have to be alleviated proceeding to enduring storage to regulate guaranteed uninterrupted year-round function of the bio-refinery (Clark and Deswarte, 2008). A good example is of Austrian green bio-refinery there this problem is tackled by procuring not just simply pulverized grasses but fodder as well, which could be produced in budding period and later preserved in a silos (Koschuh et al., 2005; Thang and Novalin, 2008).

In the U.S., there are about 450,000 animal feeding operations (AFOs), generating more or less 1.3 billion damp tons (i.e., 335 million desiccated tons) of animal squander annually (USDA, 2009; Golan, 2013). Because of smell crisis, potential surface and groundwater pollution, and greenhouse gas emission, animal wastes are of meticulous environmental distress. The latest inclination in animal waste administration and supervision is the transformed attention in utilizing anaerobic digestion (AD) technology for carbon sequestration and energy creation (Gunaseelan, 1997; Liu et al., 2013). Despite the fact that AD is an efficient technique for generating methane energy and plummeting volatile organics, but at the same time, it is lacking the ability to confiscate all carbons and take away nutrients out of animal wastes. Subsequent to digestion, solid digestate still contains a high amount of carbon residues (Teater et al., 2011;

Yue et al., 2010), and liquid digestate hold a considerable quantity of total solids, nitrogen, and phosphorus (Qureshi et al., 2014; Liu et al., 2015).

In order to treat liquid digestate like chemical coagulation and flocculation (Tyagi et al., 2010), active carbon adsorption (Rodrıguez et al., 2004), ozone treatment (Battimelli et al., 2003), and UV treatment (Apollo et al., 2013) various studies have been carried out. In spite of the excellent management performance of these processes, their energy intake, and also their chemical convention makes them least attractive to be commercially put into practice. In the meantime, electrocoagulation (EC) in recent times has been deliberated to take care of high-strength wastewaters, i.e., wastewaters with soaring solids concentrations well as high chemical oxygen demand (Liu et al., 2015). Owing to its elevated elimination competence as well as chemical-free nature, EC technology possesses a diminutive withholding time and circumvents secondary pollution problems (Mollah et al., 2001).

The prospect to improve logistical feedstock tribulations and advance conversion effectiveness, system combination approaches by means of a parallel engineered set of processes may be more useful. As eminent from the title, this chapter also outlines consolidated farm-based bio-refinery concept that unites algae cultivation, AD, and bio-ethanol creation by means of ligno-cellulosic feedstock, i.e., animal dung and corn stover. Thus, manufacturing employment of synergies among course streams produce numerous fuel and chemical products together with algal biomass, methane, and ethanol (Liao et al., 2014). It engrosses the improved carbon employment efficiencies, at the same time potentially civilizing the finances of the net processes. A variety of procedures concerning mass and energy flow in the consolidated farm-based bio-refinery are presented in Figure 8.2.

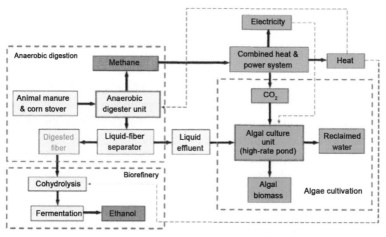

FIGURE 8.2 Schematic diagram of the projected consolidated farm-based bio-refinery (IFBBR) black arrows signify mass flow; dotted blue lines signify energy flow.

Source: Adapted with permission from Liao et al., 2014. © Elsevier.

Each and every stage of abstract bio-refinery discusses how a variety of procedures and methods utilized could be executed amid negligible environmental impact by means of the standard practices of green chemistry.

8.2 INTEGRATED FARM-BASED BIOREFINERY (IFBBR)

The IFBBR take account of the unit operations of open-pond algal culture, bio-ethanol manufacture, AD, as well as shared heat and power element (Figure 8.2). In this particular consolidated process, corn stover and dairy manure are the supply stocks. The corn stover is produced and stocked up instantly at the dairy farm by means of the currently owed storage room for bedding materials and animal feed. The dairy manure is formed every day by the animal process. The dairy manure and corn stover are assorted with each other and then supplied into the digester. For the generation of methane gas, a small proportion of organic matter in the compost and corn stover is inspired by anaerobic absorption. Methane is converted into electricity and heat by the combination of heat and power (CHP) systems. For sequestration of CO_2 the exhaust outlet gas is then transported to an algal cultivation system. This is followed by the cultivation of algae by sending the liquid digestate, which is a rich source of nutrients like N and P (Chen et al., 2012), to a shaft open-pond structure. The resultant algal biomass loaded with CHO and protein can be utilized as a nutrient appendage for either manure fabrication or fermentation of bio-ethanol (Chen et al., 2012; Michalak and Chojnacka, 2013). The domesticated water out of the open pond is cast back as a water resource for the main procedure as well as animal procedures. In addition, the unsolvable solid fraction out of AD, i.e., AD fiber is comprising incompletely digested plant cell wall material thus has an elevated cellulose-to-hemicellulose ratio as compared to the original plant material, therefore serves as the best feedstock for the manufacture of bio-ethanol.

Numerous studies have signified that, as a bioethanol feedstock, AD fiber is not merely of better quality to the novel manure, nevertheless it in addition analogous to other energy crops and remainders like corn stover and switch-grass (Teater et al., 2011; Yue et al., 2010). The size of the fiber in the manure and also in corn stover is considerably abridged by AD, which will appreciably make simpler the management of feedstock in the manufacturing process of bio-ethanol. Because of the elevated alkalinity of anaerobically digested fiber (Table 8.1), treatment with mild NaOH can be done in order to pretreat the respective fiber. Subsequent to pretreatment, batch enzymatic hydrolysis with the enzyme cellulases is able to be utilized for transforming pretreated AD fiber to mono-sugars such as C-5 and C-6 sugars, chased by solid-liquid partition. The solid leftovers obtained after separation are a rich in lignin that can be utilized for ignition or additional purposes like antioxidant production, carbon fiber, and board binder (Dong et al., 2011; Kumar et al., 2009; Hüttermann et al., 2001). A non-recombinant, technologically strong yeast strain, commonly *Saccharomyces cerevisiae*, can be utilized for the fermentation of the C-6 sugar hydrolyzate. Recovery of the Ethanol from the fermentation broth is done by the process of distillation. After the process is over, the left out water from the algae pond as well as distillation is reversed again to the process. The waste matter out of distillation contains C-5 sugars and acetate, which is unconfined all through the pretreatment process is utilized by the AD for the enhancement of biogas production (MacLellan et al., 2013).

TABLE 8.1 Chief Components of AD Fiber, Dairy Manure, and Corn Stover

	AD Fiber	Corn Stover	Raw Dairy Manure
Dry matter (%)	31.0	92.1	10.7
Ash (% dry basis)	3.4	5.2	6.60
Rudimentary protein (% dry basis)	6.0	2.5	16.44
Hemicellulose (% dry basis)	17.4	29.9	18.8
N (% dry basis)	1.22	0.61	2.63
Lignin (% dry basis)	18.5	8.9	14.2
C (% dry basis)	44.60	43.65	45.49
Cellulose (% dry basis)	41.5	39.7	26.6
C:N ratio	–	71.55	17.30
pH	7.8	–	6.87
Alkalinity (mgCaCO$_3$/L)	400	90	1.370

Source: Yue et al. (2013); Liao et al. (2014).

8.3 INTEGRATED SCHEME FOR GREEN BIOMASS EXPLOITATION

The exploitation of green biomass, i.e., grasslands meant for the generation of bio-based produces involving fibers, lactic acids (LAs), proteins, and energy in the form of biogas, is referred to as Green Bio-refinery. The grassland utilization possibly can pose captivating negative impacts like management of cultural terrains, as well as the development of the "stay option" of agrarians. The stuff grounded on renewable reserves are assumed to befit an imperative alternative for the products based on petrol in the subsequent two decades. Generally, there exists a big incongruity among the projected yearly biomass production and the essentially exploitable biomass for the inedible segment worldwide (Kromus et al., 2004).

The exploitation of most of the renewable resources except timber is constrained owing to their dependence on intense farming in addition to relatively restrained expanse for their growing, particularly in Europe. Instances are sugar beet and corn for ethanol production or different bio-based substances like rape to produce biodiesel or hemp for fiber exploitation. Conversely, grassland meadow is mostly accessible by offering elevated vintages in Europe. Furthermore, grassland can be tended in a justified way.

Thus, a Green Bio-refinery is an incorporated conception of exploitation of green biomass as a copious and expedient raw material for the production of industrial commodities. This concept has progressed immensely and is presently in a higher phase of expansion in various European countries, principally in Switzerland, Denmark, Germany, The Netherlands, and Austria (Benjamin and Van Weenen, 2000; Xiu et al., 2013; Kromus et al., 2004). The central scheme of this idea is to operate on the green biomass like alfalfa, grass, and many other sources, and engender a gamut of treasured products which may even lead to further lines of production. Not only the bio-based products are produced by this technology, but also the same can furnish us with the energy source in the form of biogas.

Therefore these green bio-refineries put forward not only a very elevated monetary prospective, nevertheless possibly supports sustainable advancement efforts, particularly in rural areas. Green bio-refineries might, therefore, considerably add to sustainable growth and expansion of the bio-industry segment.

8.4   FUNDAMENTAL TECHNOLOGY AND YIELD OUT OF A GREEN BIO-REFINERY

Since 1999 the technological aspects of the Austrian green bio-refineries technological aspects are urbanized in an extensive interdisciplinary involvement of farmers, engineers, scientists, and the industry since 1999 (Kromus et al., 2004). The primary step integrated the recognition of motivating foodstuffs which can be subjugated from the biomass grassland in a practicable manner. The succeeding step, which was initiated in 2000, integrated the in place field fabrication of the raw material as well as the initiation of down streaming and fractionation assessments. The consequences are supposed to lead to the setting up of a fundamental pilot plant in the year 2004. Since grassland biomass in no way offers an explicit main constituent, approximating corn (starch) and sugar beet (sugar) or it is convincing to set up a multi invention arrangement (Kormus, 2002). The majority of capable products which have been acknowledged are (Figure 8.3):

- **Fibers:** As padding substance.
- **Proteins:** As nosh or food (cheap), hydrolyzed as amino acids (AA) for utilization in cosmetics.
- **Lactic Acid:** For neutralization/buffering.
- Specialty products such as chlorophyll, xanthophylls, carotenoids, etc.
- **Energy:** Remains will be transformed into biogas.
- For setting up such a complex system, the technology has been separated into six modules:
- Supply chain of basic raw material like cultivation, harvest, ensiling, and storage;
- Pressing into juice and press cake;
- Down-streaming of proteins resp. AA;
- Down-streaming of LA;
- Manufacture of grass-fiber products; and
- Biogas creation by fermentation of bioorganic residues (Kormus, 2002).

On the other hand, present equipment, and machinery growth convolutes incorporated progression units to coalesce elements 3 and 4. As a result, every grassland yield is dumped in clamp-silos subsequent to inoculation with particular starter bacteria, which are believed to augment LA fermentation in the storage towers (Kromus et al., 2004). Subsequent to almost two to three weeks of fermentation, an itinerant pulverization and pressing system will dub at the farmhouse to separate the resultant raw material into a fluid and solid fractions. Elements of the stream possibly will experience their own stream processing or being used in the biogas plant for the creation of energy unswervingly on the farm. The fodder fluid, whether clarified or/and concentrated, could be elated to a national industrial location for advance downstream processing, correspondingly product advancement.

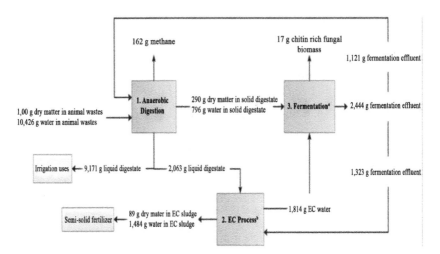

FIGURE 8.3 Self-sustaining bio-refinery mass balance on the whole mass balance study was based on 1000 g dry animal wastes. (a) The EC process utilized the combination of fermentation effluent and liquid digestate to produce the EC water for the fermentation utilization, (b) the mass balance for fungal fermentation was deliberated based on 50 mL flask data.

Source: Reprinted with permission from Liu et al., 2016. Open access. http://creativecommons.org/licenses/by/4.0/

8.5 OUTLOOK OF IFBBR

IFBBR can contribute significantly to next-generation biofuel making and present an amicable solution for both the bio-refining industry as well as agricultural operations. Farmland surrounded by cattle and dairy producers is the best place to establish a centralized regional bio-ethanol refinery. About 8.5 tons of dehydrated manure is produced every day via a medium-sized dairy farm containing 1,000 cows. It is possible to produce 1.3 tons of dry cellulose per day by each farm if AD is utilized to treat the farm-generated manure assorted with corn stover (weight ratio 4:1). Hence it is enumerated that about 439 medium-sized dairy and cattle farms may perhaps create the cellulose for the production of 20 million gallons of ethanol. The execution of AD (Table 8.2) on a nationalized range with 1.2 million cattle procurers would capitulate around 18.1 million dehydrated tons of cellulose yearly as a cellulosic ethanol feedstock in the US on its own. The continual action of these biomass procurers as matched up with regular grain-based feedstock and the comparatively huge room for rummage dumping on dairy and cattle farms may well offer a neighboring delivery system for biomass allocation, extensively plummeting the shipping and storage expenses for lignocellulosic ethanol creation.

The substitution of fossil fuel with bio-ethanol, algae, and methane from the incorporated system may perhaps then direct to confiscate an overall of 13 million tons of carbon per annum in the US. For that reason, the incorporation of algae cultivation, lignocellulose bio-refining, and AD can construct an amicable way out for cattle operations, ethanol production, and the surroundings. The sustainability of animal making systems could be enhanced by plummeting greenhouse gas discharge, probable surface and groundwater contamination from fertilizer, and poisonous whiffs, although at the same instance creating power, algal biomass, and AD fiber that might wholly augment farm and industry income. For financial evaluation, mass and energy balances were taken as the basis, and a biogas plant has been well-thought-out as standard (Koschuh et al., 2003; Kromus et al., 2003).

TABLE 8.2 Energy Stability of Healthy Bio-Refinery

Energy Balance	EC Process	AD	Fungal Fermentation[b]
Energy put in (MJ/kg dry feedstock)	-1.47^d	-0.16^c	-3.63^e
Energy yielded (MJ/kg dry raw material)	0	6.95^f	0
Net energy (MJ/kg dry feedstock)	-1.47	6.79	-3.63
Total net energy (MJ/kg dry feedstock)	–	1.69	–

Note: Every single input is negative, and every output is constructive.

[a]*Records were deliberated and accustomed based on 1 kg dry animal wastes.*

[b]*The fungal fermentation comprises unit operations such as enzymatic hydrolysis, pretreatment, and fungal fermentation.*

[c]*The power contribution for the AD unit consists of both heat and electricity.*

[d]*The energy contribution for the EC unit is 446.65 kJ/L liquid digestate.*

[e]*The energy put in case of pretreatment, enzymatic hydrolysis, fungal fermentation, and post-processing is 1.25 MJ/L fermentation broth (unpublished data).*

[f]*The energy yielded out of the AD is the methane energy. A low heat value of methane of 50 kJ/g methane was utilized for the estimation.*

Source: Reprinted with permission from Liu et al., 2016. Open access. http://creativecommons.org/licenses/by/4.0/

8.6 CONCLUSION

It is essential at the time of developing chemistry for upcoming bio-refineries that the processes and procedures utilized to reduce the adverse effect to the surroundings to facilitate sustainability broaden downstream, along with the goods being actually green and ought to strengthen. Transformation of biomass by means of thermo-chemical conversion is a pathway to a family of chemicals out of biomass that none supplementary technique at present proposes, and is above all priceless in the making of sophisticated fuels. Reusable, abundant, diverse, and geographically dissolved source of carbon is provided to us by nature, a diminutive proportion of which can endow us with all the chemicals as well as resources society necessitates. The confront for the researchers is to understand that prospective, sustainably, and efficiently, so that those indefensible practices, as well as products that make threats to our very survival, can be restored quickly.

KEYWORDS

- **aromatics**
- **catalysis**
- **fermentation**
- **heterogeneous**
- **ionic liquids**
- **lignin**
- **lignocellulosics**
- **solvents**

REFERENCES

Agricultural Statistics Board, USDA, (2009). *National Agricultural Statistics Service: Farms, Land in Farms, and Livestock Operations.*

Apollo, S., Onyango, M. S., & Ochieng, A., (2013). An integrated anaerobic digestion and UV photocatalytic treatment of distillery wastewater. *Journal of Hazardous Materials, 261,* 435–442.

Battimelli, A., Millet, C., Delgenes, J. P., & Moletta, R., (2003). Anaerobic digestion of waste activated sludge combined with ozone post-treatment and recycling. *Water Science and Technology, 48*(4), 61–68.

Benjamin, Y., & Van, W. H., (2000). *Crops for Sustainable Enterprise.* European foundation for the improvement of living and working conditions. Dublin Irlanda.

Bush, G. W., (2006). *The White House National Economic Council.* Advanced Energy Initiative (AEI).

Chen, R., Li, R., Deitz, L., Liu, Y., Stevenson, R. J., & Liao, W., (2012). Freshwater algal cultivation with animal waste for nutrient removal and biomass production. *Biomass and Bioenergy, 39,* 128–138.

Chen, R., Yue, Z., Deitz, L., Liu, Y., Mulbry, W., & Liao, W., (2012). Use of an algal hydrolysate to improve enzymatic hydrolysis of lignocellulose. *Bioresource Technology*, *108*, 149–154.

Clark, H. J., Deswarte, E. I. F., & Farmer, J. T., (2009). The integration of green chemistry into future biorefineries. *Biofuels, Bioproducts and Biorefining*, *3*(1), 72–90.

Clark, J. H., & Deswarte, F. E., (2008). The biorefinery concept: An integrated approach. In: *Introduction to Chemicals from Biomass* (pp. 1–18). Wiley, Chichester, United Kingdom.

Dong, X., Dong, M., Lu, Y., Turley, A., Jin, T., & Wu, C., (2011). Antimicrobial and antioxidant activities of lignin from residue of corn stover to ethanol production. *Industrial Crops and Products*, *34*(3), 1629–1634.

Eggeman, T., & Elander, R. T., (2005). Process and economic analysis of pretreatment technologies. *Bioresource Technology*, *96*(18), 2019–2025.

Golan, E., (2013). *The US Food Waste Challenge*. Washington, D.C.: Department of Agriculture.

Gunaseelan, V. N., (1997). Anaerobic digestion of biomass for methane production: A review. *Biomass and Bioenergy*, *13*(1, 2), 83–114.

Hess, J. R., Kenney, K. L., Wright, C. T., Perlack, R., & Turhollow, A., (2009). Corn stover availability for biomass conversion: Situation analysis. *Cellulose*, *16*(4), 599–619.

Hüttermann, A., Mai, C., & Kharazipour, A., (2001). Modification of lignin for the production of new compounded materials. *Applied Microbiology and Biotechnology*, *55*(4), 387–394.

Kazi, F. K., Fortman, J., Anex, R., Kothandaraman, G., Hsu, D., Aden, A., & Dutta, A., (2010). *Techno-Economic Analysis of Biochemical Scenarios for Production of Cellulosic Ethanol* (No. NREL/TP-6A2-46588). National Renewable Energy Lab (NREL), Golden, CO (United States).

Klass, D. L., (1998). *Biomass for Renewable Energy, Fuels, and Chemicals*. Elsevier.

Klein-Marcuschamer, D., & Blanch, H. W., (2013). Survival of the fittest: An economic perspective on the production of novel biofuels. *AIChE J., 59*(12), 4454–4460.

Koschuh, W., et al., (2003). *Green Biorefinery-Extraction of Grass Juice Proteins, Factory of the Future Project of the Ministry of Transport*, Innovation and Technology (804136), final report. Kornberg Institute, Feldbach.

Koschuh, W., Thang, V. H., Krasteva, S., Novalin, S., & Kulbe, K. D., (2005). Flux and retention behavior of nanofiltration and fine ultrafiltration membranes in filtrating juice from a green biorefinery: A membrane screening. *Journal of Membrane Science, 261*(1/2), 121–128.

Kromus, S., (2002). *The Green Biorefinery Austria-Development an Integrated System for the Use of Grassland Biomass*. PhD thesis, Graz University of Technology, Graz.

Kromus, S., et al., (2003). *Green Biorefinery-Recovery of Lactic Acid from Silage Juice, Factory of the Future Project of Ministry of Transport, Innovation and Technology (804141), Final Report*. Kornberg Institute, Feldbach.

Kromus, S., Wachter, B., Koschuh, W., Mandl, M., Krotscheck, C., & Narodoslawsky, M., (2004). The green biorefinery Austria-development of an integrated system for green biomass utilization. *Chemical and Biochemical Engineering Quarterly, 18*(1), 8–12.

Kumar, S., Mohanty, A. K., Erickson, L., & Misra, M., (2009). Lignin and its applications with polymers. *Journal of Biobased Materials and Bioenergy, 3*(1), 1–24.

Liao, W., Liu, Y., & Hodge, D., (2014). Integrated farm-based biorefinery. In: *Biorefineries* (pp. 255–270). Elsevier.

Liu, Z., Liao, W., & Liu, Y., (2016). A sustainable biorefinery to convert agricultural residues into value-added chemicals. *Biotechnology for Biofuels*, *9*(1), 197.

Liu, Z., Ruan, Z., Xiao, Y., Yi, Y., Tang, Y. J., Liao, W., & Liu, Y., (2013). Integration of sewage sludge digestion with advanced biofuel synthesis. *Bioresource Technology*, *132*, 166–170.

Liu, Z., Stromberg, D., Liu, X., Liao, W., & Liu, Y., (2015). A new multiple-stage electrocoagulation process on anaerobic digestion effluent to simultaneously reclaim water and clean up biogas. *Journal of Hazardous Materials*, *285*, 483–490.

Lyko, H., Deerberg, G., & Weidner, E., (2009). Coupled production in biorefineries—combined use of biomass as a source of energy, fuels and materials. *Journal of Biotechnology*, *142*(1), 78–86.

MacLellan, J., Chen, R., Kraemer, R., Zhong, Y., Liu, Y., & Liao, W., (2013). Anaerobic treatment of lignocellulosic material to co-produce methane and digested fiber for ethanol biorefining. *Bioresource Technology*, *130*, 418–423.

Michalak, I., & Chojnacka, K., (2013). Algal compost-toward sustainable fertilization. *Reviews in Inorganic Chemistry*, *33*(4), 161–172.

Mollah, M. Y. A., Schennach, R., Parga, J. R., & Cocke, D. L., (2001). Electrocoagulation (EC)—science and applications. *Journal of Hazardous Materials*, *84*(1), 29–41.

National Research Council, (2000). *Biobased Industrial Products: Research and Commercialization Priorities*. National Academies Press.

Qureshi, N., Hodge, D., & Vertes, A., (2014). *Biorefineries: Integrated Biochemical Processes for Liquid Biofuels*. Newnes.

Rodrıguez, J., Castrillon, L., Maranon, E., Sastre, H., & Fernandez, E., (2004). Removal of non-biodegradable organic matter from landfill leachates by adsorption. *Water Research*, *38*(14/15), 3297–3303.

Sokhansanj, S., Mani, S., Turhollow, A., Kumar, A., Bransby, D., Lynd, L., & Laser, M., (2009). Large-scale production, harvest and logistics of switchgrass (*Panicum virgatum* L.): Current technology and envisioning a mature technology. *Biofuels, Bioproducts and Biorefining*, *3*(2), 124–141.

Teater, C., Yue, Z., MacLellan, J., Liu, Y., & Liao, W., (2011). Assessing solid digestate from anaerobic digestion as feedstock for ethanol production. *Bioresource Technology*, *102*(2), 1856–1862.

Thang, V. H., & Novalin, S., (2008). Green biorefinery: Separation of lactic acid from grass silage juice by chromatography using neutral polymeric resin. *Bioresource Technology*, *99*(10), 4368–4379.

Tyagi, V. K., Khan, A. A., Kazmi, A. A., & Chopra, A. K., (2010). Enhancement of coagulation-flocculation process using anionic polymer for the post-treatment of UASB reactor effluent. *Separation Science and Technology*, *45*(5), 626–634.

USDA, (1997). ERS. *Confined Animal and Manure Nutrient Data System*.

Xiu, S., Shahbazi, A., & Wang, L., (2013). *Utilization of Alfalfa Through Integrated Green Biorefinery: Effects of Fractionation Processes on Green Juice Extraction* (Vol. 24, p. 1). In. Kansas City, Missouri. American Society of Agricultural and Biological Engineers.

Yue, Z. B., Chen, R., Yang, F., MacLellan, J., Marsh, T., Liu, Y., et al., (2013). Effects of dairy manure and corn stover co-digestion on anaerobic microbes and corresponding digestion performance. *Bioresour Technol., 128*, 65–71.

Yue, Z., Teater, C., Liu, Y., MacLellan, J., & Liao, W., (2010). A sustainable pathway of cellulosic ethanol production integrating anaerobic digestion with biorefining. *Biotechnology and Bioengineering*, *105*(6), 1031–1039.

CHAPTER 9

Microbial Interventions and Biochemistry Pathways for Degradation of Agricultural Waste

MONICA BUTNARIU,[1] RAMONA STEF,[1] and ALINA BUTU[2]

[1] Banat's University of Agricultural Sciences and Veterinary Medicine, "King Michael I of Romania" from Timisoara – 300645, Calea Aradului 119, Timis, Romania, E-mail: monicabutnariu@yahoo.com (M. Butnariu)

[2] National Institute of Research and Development for Biological Sciences, Splaiul Independentei, 296, Bucharest – 060031, Romania

ABSTRACT

Biochemical cycles and microorganisms (primarily heterotrophic microorganisms) are responsible for the maintenance of the biosphere. They exploit the chemical reactions favorably to obtain carbon and energy from biomass. As a result of the microbial degradation processes, the essential nutrients present in the biomass of a generation of organisms are available for the next generation. The impact of human activities on the quality of environmental factors has intensified over the last decades due to population growth and extensive exploitation of natural resources, including soils. The following processes can be mentioned as the main sources of impact on the quality of environmental factors: atmosphere emissions-mainly from industry and traffic; agricultural practices, in particular, the use of organic or mineral fertilizers and pesticides; waste deposited on the soil surface. The rate of production and dispersion of pollutants from agricultural waste has now surpassed biodegradation's natural processes. In the search for technological remedies for pollution with pollutants from agricultural waste, physical, chemical, and biochemical processes can be essential, but microbiological processes also provide important insights. The purpose of this review is to support the knowledge of biochemical cycles involving microorganisms and the need to include biological parameters in environmental impact assessment studies and environmental quality monitoring strategies that are currently based only on the determination of physicochemical parameters. In natural ecosystems, there are fluctuations in the nutrient and energy concentrations, against which microorganisms constantly modulate their rate of energy-producing reactions (through the genetic regulation of enzyme synthesis involved, induction/repression) to match energy-consuming reactions and ensure the survival of cells.

9.1 INTRODUCTION

The soil pollutants originating from agricultural waste are subjected to biogeochemical transformation processes. These processes affect the structure of the pollutant by a chemical process that takes place in a geological environment and can be achieved by a biological organism. Biodegradation is defined as the biologically catalyzed reaction that has the effect of reducing the complexity of a chemical compound. A biodegradable compound can be transformed under the influence of microorganisms (MO) into another compound with a simpler structure but not necessarily less toxic than the compound of origin. A compound may be recalcitrant if it cannot be biodegraded in any form. A compound is persistent when it is biodegradable, but only under certain conditions that favor biodegradation (Abraham and Gajendiran, 2019). Mineralization means the complete conversion of an organic compound into the final degradation products: CO_2 and H_2O. It is called primary biodegradation, the single transformation of a compound; partial biodegradation is a more advanced transformation than primary biodegradation; however, without achieving mineralization (Huang et al., 2019). Considering the pollutants from the organic waste of organic nature, the microbial transformation is due to the fact that MOs can use these compounds for growth and reproduction. Organic pollutants have a double role: carbon source, the foundation stone of any cellular construction, and an electron supplier, which MO can extract to get energy. MO acquires energy by catalyzing redox reactions with energy release.

The pollutant, as an electron donor, will be oxidized while an electron acceptor will be reduced. The electron donor and acceptor are essential for cell growth, as they are commonly known as the primary substrate (Ghaffar et al., 2018).

Most MO use molecular oxygen (O_2) as an electron acceptor. In this case, we can talk about aerobic degradation, where O_2 is used to oxidize a part of the carbon from the pollutant to carbon dioxide (CO_2), the rest of the carbon being used to produce new cell mass. In this process, the O_2 is reduced, forming water. Thus, the main products of aerobic degradation are CO_2, H_2O, and an increased MO population. There is MO that uses other electron acceptors, surviving in the absence of O_2. These MOs perform the anaerobic degradation of pollutants from agricultural waste, using electron donor ions (NO^{3-}), sulfate (SO_4^{2-}), iron (Fe^{3+}), manganese (Mn^{4+}), or even CO_2 (Sayen et al., 2019).

In addition to the newly formed cell mass, several other anaerobic degradation products will form, such as molecular nitrogen (N_2), hydrogen sulfide (H_2S), metals in reduced form ($Fe^{2+,}$ Mn^{2+}), or methane (CH_4), depending on the electron acceptor used. Some MOs can also use as electron donor inorganic substances: ammonium ions (NH_4^+), nitrite (NO_2^-), reduced metals (Fe^{2+}, Mn^{2+}), H_2S. When these inorganic compounds are oxidized (NO^{2-}, NO^{3-}, Fe^{3+}, Mn^{4+}, SO_4^{2-}), energy is produced for cellular growth, and electrons are taken up by an electron acceptor (usually O_2). In most cases, these MOs use atmospheric carbon dioxide as a source of carbon, thus achieving CO_2 fixation (El-Naggar et al., 2019).

A particular form of metabolism, which can play an important role in O_2-free environments, is fermentation. This process does not require an external electron acceptor, the pollutant playing both donor and electron acceptor functions. Through a sequence of electrons' internal transfers, catalyzed by MO, the pollutant is transformed into harmless fermentation products: acetates, propionates, ethanol, hydrogen, and carbon dioxide.

Other bacteria can biodegrade these fermentation products to the final products: Sometimes, the MO can produce the transformation of pollutants from agricultural waste, even if this process brings little benefit to the cell. Such biotransformation is generally known as a secondary use. A particular case of secondary use is co-metabolism.

During this process, the transformation of the pollutant is the result of an accidental reaction catalyzed by enzymes involved in normal cell metabolism. For example, in the methane oxidation process, some bacteria can degrade chlorinated solvents, which they are usually incapable of destroying. In a methane oxidation process, MO produces enzymes that destroy the chlorinated solvent, even if it does not support microbial development. Methane is the primary electron donor, being the primary source of the MO feed, while the chlorinated solvent is a secondary substrate (Ganser et al., 2019).

Another form of microbial metabolism is reducing dehalogenation. This is important for the destruction of organic halogenated pollutants from agricultural waste, in particular chlorinated solvents. During this process, MO catalyzes, a reaction by which the halogen atom in the pollutant molecule is replaced with a hydrogen atom. For the reaction to occur, it is necessary to have an electron donor other than the pollutant to provide the pair of electrons necessary for the reduction. Potential donors are hydrogen and low molecular weight organic compounds (lactate, acetate, methanol, glucose).

Regardless of the mechanism by which MO produce degradation of pollutants from agricultural waste, their cellular composition is relatively constant: 50% C; 14% N; 3% P; 2% K; 1% S; 0.5% Ca, Mg, Cl; 0.2% Fe. If any of these elements is deficient in relation to carbon in the organic pollutant, the nutrition competition between MO can limit the overall cell mass growth and slow down the removal of the pollutant (Petousi et al., 2019).

9.2   AEROBIC DEGRADATION OF ORGANIC COMPOUNDS

The vast majority of natural or anthropic organic compounds degrade under aerobic conditions with O_2 as an electron acceptor. As long as oxygen is available, it is the favorite electron acceptor of microbial degradation processes occurring in nature. The most important classes of organic pollutants are the components of crude oil and halogenated petrochemicals. These compounds degrade rapidly and completely under aerobic conditions.

To study aerobic MO's ability to degrade such compounds, several studies reported the use of model molecules such as aliphatic, aromatic, and halogenated derivatives. MO capable of degrading pollutants from organic agricultural waste from contaminated areas is chemo-organotrophic bacteria, which have the ability to use a large number of natural and synthesis compounds as a source of carbon and as electron donors. Although many bacteria can metabolize pollutants from organic agricultural waste, a single species does not possess the enzymatic ability to degrade all or most of the pollutants from agricultural waste existing in the soil. Bacteria carry out biodegradation of pollutants from agricultural waste through both growth and co-metabolism (Li et al., 2018).

Mixed microbial communities have the most potent biodegradability potential, as the genetic information of several organisms is required for the degradation of complex mixtures of pollutants from agricultural waste.

In the first stage, the molecule's molecular mechanism is degraded into a reaction catalyzed by a monooxygenase or dioxygenase, depending on the nature of the substrate and the enzyme possessed by the microorganisms. Long-chain (C10-C24) n-alkanes are rapidly degraded to fatty acids that are metabolized. Short-chain (C5-C9), n-Alkanes are toxic to most MO, but they evaporate rapidly from oil-contaminated soils. Cycloalkanes are relatively resistant to biodegradation.

The absence of a methyl group slows oxygen's primary attack, while a side chain facilitates biodegradation. While there is only a few MO that is able to use cycloalkanes as the sole source of carbon, most often, their biodegradation takes place by co-metabolism. In the case of aromatic compounds, whether hydrocarbons (benzene, toluene, ethylbenzene, xylenes, naphthalene), phenols, and chlorophenols, amino acids (AA), quinones, and hydroquinones, they can be converted enzymatically to intermediate natural biodegradation products: 1,2-dihydroxybenzene (catechol) 4-dihydroxybenzoic acid (protocatechuic). They are subsequently decomposed, in several steps, either to acetyl-CoA and succinate, or to acetaldehyde and pyruvate (Armstrong et al., 2018).

Benzene and its derivatives generally have a higher thermodynamic stability than aliphatic compounds. Oxidation of benzene takes place via a dioxin-catalyzed hydroxylation. Diol formed is then converted into catechol. Hydroxylation followed by dehydrogenation also occurs in the case of biodegradation of other aromatic hydrocarbons, such as 3 and 4 cycle polycyclic aromatic hydrocarbons (PAHs). The existence of a substituent in the benzene ring allows either the side chain attack or the oxidation of the aromatic ring by alternative mechanisms. Methanotrophic bacteria can use methane or other C1 compounds as carbon and energy source, oxidizing methane to CO_2 via methanol and formaldehyde.

The first stage of oxidation of CH_4 is catalyzed by methane monooxygenase (MMO), a non-specific enzyme able to oxidize several other complex compounds: alkanes, arenas, trichloroethylene (TCE). It was later found that many other groups of aerobic bacteria that are capable of oxidizing propane, ethene, toluene, phenol, ammonia are able to co-metabolize halogenated organic compounds. The practical use of a co-metabolic aerobic biodegradation system of halogenated derivatives is still difficult: the co-metabolite must always be present to support the reactions, but excess methane and high oxygen concentrations inhibit the oxidation of chlorinated compounds (Jin et al., 2017).

Moreover, studies report that the TCE degradation products are toxic to certain methanotrophs, and perchloroethylene (PCE) inhibits TCE degradation. Another limitation is the fact that methanotrophs cannot degrade PCE or polychlorinated compounds with a greater number of chlorine atoms in their molecules. Another category of organisms capable of degrading pollutants from agricultural waste are fungi. They live in various environments (freshwater, seawater, soil, plant, and animal waste, living organisms, etc.).

Molds and yeasts can be considered as microfungi. It is well known the ability of yeasts to degrade aliphatic hydrocarbons to existing in oil and oil products. C24-C10 n-alkanes are the most easily degraded in the presence of yeasts such as *Candida lipolytic, C. tropicalis, Rhodoturula rubra,* and *Aureobasidion (Trichosporon)* or molds *Cunninghamella blakesleeana, Aspergillus niger,* and *Penicillium frequentans.* C5-C9 is toxic to fungi and bacteria. Since the higher alkanes are practically insoluble in water, the fungus produces the biosurfactants, which are dispersing the substrate into an emulsion, leading to an increase in interfacial area growth and to increased bioavailability of hydrocarbons (Amodeo et al., 2018). In the microfungi, the alkanes are enzymatically oxidized in the presence of monooxygenases to the corresponding primary alcohols:

$$R–CH_2–CH_3 + O_2 + NAD (P)H_2 \rightarrow R–CH_2–CH_2–OH + NAD(P) + H_2O$$

Subsequently, they are oxidized to aldehydes and then to fatty acids via pyridine nucleotide-linked dehydrogenases. The fatty acids produced are always metabolized by β-oxidation and finally to CO_2. Unlike bacteria, fungi cannot use isoalcans or cycloalkanes as a unique source of carbon. Although some yeasts and molds may use aromatic compounds as a growth substrate (Table 9.1), it is more important for them to degrade by co-metabolism.

Hydroxylation enzymes and aromatic ring breakage enzymes from microfungi are relatively non-specific, and they also commonly convert related compounds, including halogenated and nitroderivatives. Microfungi can also co-metabolize many aromatic pollutants, including PAHs, biphenyls, dibenzofurans, nitroaromatics, pesticides, plasticizers (Kim et al., 2016).

Typical transformations are glycosylation, hydroxylation, and aromatic ring breakage, methoxylation, reduction of nitro groups to amino groups. It is important to note that macrofungi (mushrooms) can contribute to the aerobic degradation of pollutants from organic agricultural waste.

TABLE 9.1 Yeast and Mold Species Using Aromatic Compounds as a Growth Substrate

Species	Growth Substrate
Yeasts	
Aureobasidium pullulans	Phenol, o-crezol, p-cresol, benzoic acid
Candida maltosa	Phenol, pyrocatechin, benzoic acid
Exophiala jeanselmei	Phenol, styrene, benzoic acid, acetophenone
Rhodotorula glutinis	Phenol, m-crezol, benzoic acid
Trichosporon cutaneum	Phenol, p-crezol, benzoic acid, salicilic acid
Molds	
Aspergillus niger	2,4-dichloro-phenoxyacetic acid, benzoic acid, salicylic acid, monoclorobenzoic acids
Aspergillus fumigatus	Phenol, p-cresol, 4-ethylphenol, phenylacetic acid
Fusarium flocciferum	Phenol, resorcinol Penicillium frequentans phenol, p-cresol, resorcinol, floroglucin, anisole, benzyl alcohol, benzoic acid, salicylic acid, gallic acid, phenylacetic acid, 1-phenylethanol acetophenone
Penicillium simplicissimum	Phenol, floroglucine, monofluorophenols

Some of these, the most representative being *Trametes versicolor, Phanerochaete chrysosporium, Pleurotus ostreatus, Nematoloma frowardii, Agaricus bisporus, Agrocybe praecox, Stropharia coronilla*, have developed an efficient enzymatic degradation and mineralization system of lignin. Lignin degradation is achieved by the synergistic action of some oxidoreductases, ligninolytic enzymes. These enzymes act through a high reactivity depoly-merization mechanism by free radicals, a mechanism that is also suitable for the degradation of organic pollutants from recalcitrant and toxic agricultural waste, such as polychlorinated dibenzodioxins, dibenzofurans, chlorinated aromatic compounds, nitroaromatic compounds (explosives) or CGA-class aromatics from the HAP class. Although, recent research demonstrated that at laboratory level, the use of macrofungi for the biodegradation of certain pollutants (PAHs, TNT, polychlorophenols) from agricultural waste is a cheaper remediation approach. However, macrophages such as *Stropharia rugosoannulata*, which develop in the upper soil and humus layers, have been successfully tested for decontamination of soils polluted with TNT or HAP (Kandasamy and Dananjeyan, 2015).

9.3 ANAEROBIC DEGRADATION OF ORGANIC COMPOUNDS

Anaerobic degradation processes have always been considered less effective than aerobic degradation, especially from a kinetic point of view. Anaerobic bacteria produce much less energy by substrate conversion than aerobic bacteria, and therefore produce much less biomass. One mole of glucose degraded aerobically to 6 moles of CO_2 produces 2870 kJ, while the same mole of glucose degraded anaerobically to 3 moles of CO_2 and 3 moles of CH_4 produces only 390 kJ of energy that will be divided into at least three different groups of bacteria that perform metabolisation. The degradation of organic matter in the absence of oxygen can be coupled with the reduction of alternative electron acceptors in an order that depends on the redox potential values of the respective systems. Reduction of electron acceptors by organic matter (global redox potential for global transformation: glucose $\rightarrow$ $6CO_2$ is -0.434 mV) will lead to different amounts of energy, thus influencing the biochemistry of anaerobic biodegradation (Valenti et al., 2018).

With all the disadvantages of anaerobic degradation, the presence of oxygen is not always advantageous in degradation processes. Oxigenases introduce hydroxyl groups into aromatic nuclei; Further, the oxygen can lead to the formation of phenolic radicals that initiate uncontrolled polymerization and polycondensation to compounds similar to humic derivatives in the soil, very difficult to degrade, either aerobically or anaerobically. Saturated aliphatic hydrocarbons are relatively stable under anaerobic conditions. Reducing sulfate and nitrifying bacteria can assimilate long-chain (C12-C20) or medium (C6-C16) hydrocarbons, but the process is very slow. For the anaerobic degradation of mononuclear aromatic compounds, at least three pathways are known: by benzoyl-CoA (most important), resorcinol, and fluoroglucin. In all three cases a 1,3-dioxo compound is formed in a reducing step, which subsequently allows nucleophilic attack to one of the keto ring atoms of the nucleus, leading to its rupture. Depending on the aromatic substrate, eventually will obtain either a pemellar (C7-dicarboxylic) residue bound to coenzyme A or a partially oxidized caproic (C6-monocarboxylic) residue which passes into the acetate by β oxidation (Ning et al., 2018).

Aromatic compounds that do not have a carboxyl group (phenol, aniline, etc.), are first carboxylated to a p-hydroxy or p-aminobenzoic residue which is subsequently activated with coenzyme A. If there is little data on the anaerobic degradation of benzene and the process activation biochemistry is not yet fully elucidated, the mechanism of anaerobic degradation of toluene, ethylbenzene, crezoles, by nitrate-reducing and sulfate-reducing bacteria is known.

The halogenated derivatives can be dehalogenated by three basic mechanisms: oxidative, hydrolytic, and reducing. The most common mechanism of anaerobic bacteria is the reduction of halogen, first highlighted in 1982. As a rule, reductive declination is preferred for highly halogenated compounds. The redox potential of the process is significant (+250 + 580 mV), and the nucleophilic attack on the halogenated carbon is more likely than an oxidative reaction (Li et al., 2018).

9.3.1 MIXED DEGRADATION OF POLLUTANTS FROM PERSISTENT ORGANIC AGRICULTURAL WASTE

Organic pollutants from agricultural waste are even more recalcitrant as the degree of halogenation is higher. Substitutions at the aromatic ring of halogens, nitro or sulfonic groups lead to increased system resistance to electrophile attack by oxygenase of aerobic bacteria.

Compounds resistant to these attacks include polychlorobiphenyls (PCBs), chlorinated dioxins, some pesticides (DDT, lindane). In the case of these pollutants from agricultural waste, degradation is done by cooperation between aerobic and anaerobic bacteria.

In a first step, reductive dehalogenation takes place under the anaerobic conditions of the primary pollutant. The process reduces the degree of halogenation and makes subsequent mineralization possible under aerobic conditions. Thus, for example, PCB degradation occurs (Hultberg and Bodin, 2018).

The microbial metabolism represents the total biochemical reactions involved in the biological activities of the MO, by which they take the energy and biogenic chemical elements (as such or in the form of combinations) from environment and use them in biosynthesis reactions, in energy biodegradation reactions, as well as for growth and other physiological activities (transmembrane transport, mobility, bioluminescence, etc.).

Substances are taken from the environment through passive or active (energy-consuming) transport processes and, by their nature, are transformed into cellular constituents that can be secreted by the metabolism), energy. These metabolic reactions occur in the MO in general and in the bacteria in particular, respecting the fundamental principle in biology, namely the principle of economy and optimality or maximum efficiency, meaning that the reactions result from minimal energy consumption and its maximum use for biosynthesis, thus resulting the largest number of cells in the time unit.

Rapid multiplication and the existence of a large number of MO is the fundamental condition for survival in nature, being the main mechanism of competition with other associated organisms, as well as resistance to unfavorable environmental conditions. Metabolism is a self-regulated cyclic process due to special chemical reactions regulating the rhythm of producing reactions of various metabolic pathways, with a role in maintaining cell stability (Geed et al., 2018).

Metabolic pathways are sequences of metabolic reactions in several steps, each step being catalyzed by a specific enzyme. Within a metabolic pathway, the metabolic substrate is transformed into intermediate products and furthermore, those are transformed in the final product. An individual metabolic pathway can be manifested in several ways: linear, cyclic, or branched. Microbial metabolism is achieved by two main metabolic pathways:

- **Catabolic Reactions:** i.e., biodegradation, with the release of energy, i.e., exergonic reactions; and

- **Anabolic Reactions:** i.e., biosynthesis, made with energy consumption, i.e., endergonic reactions (Song et al., 2018).

From a functional point of view, the two types of metabolic pathways are interconnected because energy and some of the products resulting from catabolism reactions are used as energy and intermediates in anabolic reactions. Therefore, central metabolic pathways that release energy can also provide precursors for other metabolic pathways, these pathways being called amphibolic (auxiliary) pathways.

Anaplerotic routes are all auxiliary pathways that occur when the development of a major metabolic pathway is blocked by the use of intermediary products in other metabolic pathways. Anaplerotic pathways are of particular importance as ways of replenishment with intermediary products, resulting in another major metabolic pathway, avoiding its blocking. The operation and interaction of the four types of pathways are perfectly coordinated in the cell so that it works with optimal efficiency (Dourou et al., 2018).

9.3.2   CATABOLIC PATHWAYS (CATABOLISM OR ENERGY METABOLISM)

Catabolic pathways, i.e., catabolism is a succession of biochemical reactions involved in the degradation of nutrients and the release of energy necessary for the functioning of other metabolic pathways and other physiological activities of the cell. Catabolic pathways occur in three successive phases:

- **Phase 1:** Macromolecules are enzymatically decomposed into basic units: proteins → AA, lipids → fatty acids and glycerol, carbohydrates → monoglucides; this phase occurs frequently at the exterior of the bacterial cell, being performed by exoenzymes. From these reactions, ~ 1% of the total energy of the macromolecule, inaccessible to the cell, is released in the form of heat.

- **Phase 2:** The molecules produced in the previous phase are incompletely degraded, releasing 1/3 of total energy (E), producing, besides CO_2 + H_2O, a small number (12) of essential metabolic products called intermediates metabolic of the central metabolic pathways. These compounds are the same in all organisms, a proof of metabolic unity in the living world. For example, AA are used in different ways and their catabolism leads to the formation of acetyl-CoA or intermediates of the ring tricarboxylic acid, i.e., the Krebs cycle.

- **Phase 3:** Deviate differently, depending on the respiratory type of MO considered: aerobic MO, which can degrade the entire substrate to CO_2 + H_2O, the major course, with energy release is the Krebs cycle coupled with oxidative phosphorylation and releasing a large amount of energy stored in ATP. The anaerobic MO follow the fermentation pathway (alcohol, lactic, butyric, propionic, etc.), whose degradation products serve as electron donors or acceptors of H^+ in coupled oxidation-reduction reactions, releasing small amount of energy (ŋ small). Degradation occurs through a series of reactions in which an e^-/H^+ donor substance, D, is oxidized and another substance, A, as an e^-/H^+ acceptor is reduced. The process of degradation of nutrients through biological oxidation reactions is called cellular respiration (Lu et al., 2016).

9.3.3 ANABOLIC PATHWAYS (ANABOLISM OR METABOLISM OF ASSIMILATION OR BIOSYNTHESIS)

Anabolic pathways are pathways whose evolution is in the reverse direction of catabolic pathways. It represents the totality of biochemical reactions by which the MO synthesizes their own cellular constituents from simple molecules. Central metabolic pathway intermediates may also be used. Two categories of macromolecules are synthesized: storage (deposit) consisting of monomers of the same type (glycogen, starch, etc.); genetically engineered macromolecules, essential for biological systems (proteins, nucleic acids).

The synthesis of macromolecules is very effective and is under the action of genetic information encoded in DNA. The bacterial cell first synthesizes monomers (AA, nitrogenous bases) which they subsequently arrange in a specific order, genetically dictated, which determines the primary structure of the respective macromolecules through the cellular diataxis process.

Periodically, there are crashes such as mutations that can lead to the formation of non-functional molecules, but at the time of diataxis they are recognized and their binding is avoided. The synthesis of macromolecules in small molecules provides great synergy efficiency. Biological specificity consists in the different arrangements of a limited number of monomers (structural units: 20 AA, 5 nitrogen bases) to form an impressive number of biological macromolecules with different structures and functions (Dodd et al., 2018).

9.4 PARTICULARITIES SPECIFIC TO MICROBIAL METABOLISM

Previously, on the basis of small dimensions and relative structural simplicity, it was thought that microbial metabolism was rudimentary. Modern biochemical research has demonstrated the similarity of central metabolic pathways to all life forms using common metabolic pathways. Most major metabolic pathways were first discovered at the MO and later were extrapolated to the upper organs. However, bacteria exhibit unique metabolic pathways in the living world: biological fixation of atmospheric N_2, anaerobic respiration, synthesis of certain antibiotics, anoxigenic photosynthesis (Lyu et al., 2018). Though similar to the metabolism of higher organisms, bacterial metabolism has some general features:

The nature and variety of nutrients used, which differentiates MO in general and bacteria in particular, is their ability to use a wide range of substances ranging from simple inorganic to complex organic substances, including several known as inhibitors of growth. Ex: acids (formic, oxalic, sulfuric), lignin, chitin, cellulose, antibiotics, phenols, asphalt, petroleum, paraffins, plastics, chemical synthesis in general. Therefore, they are able to use chemical synthesis substances or so-called xenobiotic substances. Thus, MO is considered to be the most commonly known omnivorous organisms. This particularity explains that although large quantities of dead organic matter, excretion products, as well as wastes of human activity have been deposited in nature, they have not accumulated, but after their decomposition by MO, they were reintroduced into the biogenic elements. It has been shown that biodegradable organic compounds can be degraded mainly by MOs in associations such as polyspecific biofilms adhering to surfaces (including aquatic sediments), whose metabolic activity is more diverse and more effective than planktonic cells (Lu et al., 2018).

There are individual differences in bacteria, some bacterial species can use a lot of nutrients (e.g., *Pseudomonas fluorescens*) and others specialize in using only a particular substrate; there are also physiological groups of bacteria:

- Cellulosics use as C source only cellulose, atmospheric N_2 fixative bacteria;

- Methylotrophs use only C1 compounds.

The plasticity of bacterial metabolism refers to the ability of bacteria to use alternative sources of nutrients. The bacteria use preferentially certain sources of carbon and nitrogen, but in their absence, they use alternative substrates, the synthesis of the necessary enzymes being induced by the presence of these substrates.

Plasticity gives MO the ability to adapt to the type and amount of nutrients, going on the principle of maximum economy and based on the existence of very complex enzyme equipment. For example, *E. coli* preferentially uses glucoses and AA if they exist in the environment. In the case where the average AA exist simultaneously and NH_4^+, then it uses AA as the source of N and NH_4^+, thereafter.

The diversity of enzymatic mechanisms and resulting products, bacteria, MO generally do not present a metabolic pathway for a product, but have multiple alternative pathways to adapt to varying environmental conditions; metabolic pathways or shunts occur, each pathway leading to the production of other compounds (Lyu et al., 2018). For example, glucose degradation is done in several ways: EMP, HMP, ED, and FC, depending on the environmental conditions:

- The Embden-Meyerhoff-Parnas (EMP) pathway = the glycolysis cycle;

- The hexosomonophosphate (HMP) pathway;

- The Entner-Doudoroff pathway;

- The phosphocellosis pathway.

The first two are also present in the upper organisms, and the last two are only for bacteria. In all of these ways the pyruvate occupies the position of a key intermediate because it is located at the intersection points of the metabolic pathways.

The EMP pathway, the major pathway of glucose degradation in most MO as well as the eucaryotic vegetal and animal organisms. It is an anaerobic pathway, present not only in purely anaerobic bacteria, but also in aerobic organisms, in the absence of partial molecular oxygen (this is how is explained the use of the term glycolysis instead of fermentation). It comprises a sequence of 10 enzymatic reactions by which a glucose molecule is degraded to 2 pyruvate molecules. Unlike animal cells in which glucose is converted to LA, this mode of evolution is present only in homofermentative lactic bacteria, whereas other MO use the pathway to pyruvate, then acetaldehyde and finally ethanol are formed. This pathway does not explain the synthesis and use of pentoses as a source of energy, as well as for the synthesis of nucleic acids. From the point of view of energy efficiency is the major pathway of ATP synthesis, present in aerobic and anaerobic MO in complex environments.

The HMP pathway, the pathway of hexosophosphates or the pentoso-phosphate pathway, is an aerobic pathway of Glu-degradation and bypassing EMP. Less efficient (1/2 of the amount of ATP resulting from the glycolysis pathway) is a pathway used for the synthesis of nucleic acid precursors (pentosophosphates required for nucleotide synthesis) and for the production of $NADPH_2$ as reducing power used in other metabolic pathways.

The ED pathway is present only in strictly aerobic bacteria and some parasitic worms, which are partly linked to the HMP shunt, but it can also function independently. It was described in *Pseudomonas sp.* As an independent EMP pathway resulting in ATP + $NADPH_2$. It will result intermediates of degradation of carbohydrates that can be also used in EMP and HMP pathway. It provides precursors for DNA biosynthesis, RNA, vitamins, aromatic acids. These three pathways can work alternately or in combinations.

FC pathway, described in *Leuconostoc mesenteroides* (heterofermentative lactic bacteria). 1ATP is results; the first three reactions are common with HMP, being a variant of this pathway. The enzyme involved is phosphocetholase, which cleaves acetyl phosphate from C5 or C6 compounds (Dourou et al., 2018).

For most MO, there is a, b, c, d, alternately, as a function of necessities. The pathways are interconnected and have a number of common enzymes, stages, or intermediate products.

Some MO, are preferentially using certain pathways, while others, depending on cultivation conditions will use alternative routes of degradation.

The intensity of bacterial metabolism is exceptionally high in relation to that of homologous activities of higher organisms. This property derives from a structural property, that is, from the large surface/volume ratio (S/V increase). The large area of contact with the environment and the absorption of nutrients determines the high intensity of metabolic reactions (biosynthesis and biodegradation) and implicitly the high rate of multiplication, this being the very survival strategy of MO in nature, for compensating for the losses due to the variation of abiotic factors, as well as the antagonistic relations with other species.

Other incriminated causes: the great variety of reactions they can make; the small ratio of the amount of genetic material/cytoplasm; the very high enzymatic activity of bacterial enzyme systems compared to those from plant or animal tissues. Intensity is manifested both in biodegradation reactions of nutrients and in biosynthesis reactions (Ugwuanyi et al., 2005).

The enormous capacity for biosynthesis, in particular for proteins, explains the high capacity for growth and reproduction of the MO with practical applications in biotechnology: yeast and bacterial single-cell production produced by single-celled organisms SCP (single cell proteins) or SCB (single cell biomass). For example, the yield of protein synthesis in different organisms has been estimated bovine = 1, soybean = 10, yeast = 10^5, bacteria = 10^{11}. This high intensity is due to several causes, such as the ratio S/V increase. Other benefits: increased nutritional value, proteins with essential AA; they are produced in small spaces, synthesis is continuous in bioreactors; does not block agricultural land; they use cheap nutritional substrates, sometimes residues from different industries. For anabolism, MO has to find the nutrients in the medium, the source of C (energy) and a source of N, essential elements for the synthesis of specific molecules: proteins, nucleic acids.

Since the environment may not always be able to provide all the necessary compounds, MO uses a series of intermediate compounds resulting from the energy metabolism: pyruvic acid, acetyl-CoA, α-ketoglutarate, glyceraldehyde-3-P. The MO performs the nutrition process by assimilating sources of C and N.

The concept of autotrophy is thus explained, including among the autotrophs both the MO that assimilates CO_2 through the Calvin cycle but also the MO which can assimilate the compounds with 1 atom of C (CH_4, CH_3-OH, CH_3-NH_2), i.e., MO methylotroph, which they assimilate via the ribulose monophosphate ring and the serine cycle. In fact, MO is optionally autotrophs, and in turn the heterotrophs have the ability to adapt to the use of inorganic compounds when organic ones are missing (García-Gómez et al., 2003).

9.4.1 DECONTAMINATION BY THE BIODEGRADATION OF SOIL ORGANIC SUBSTANCES IN THE CARBON CYCLE

Carbon is incorporated into all organic compounds, and those occupy the most important place in the composition of all organisms on earth. The primary source of carbon for all organisms is atmospheric CO_2. The atmosphere contains a total of $2,300 \times 10^9$ tons of CO_2, representing 0.03% of the air volume. Seawater and oceans contain $130,000 \times 10^9$ tonnes of CO_2 in the form of HCO_3. CO_2 is transformed into organic carbon under the action of autotrophic organisms.

The amount of CO_2 fixed by photoautotrophic organisms (green plants, photosautotrophs MO) is incomparably higher than the amount of CO_2 fixed by chemioautotrophic organisms (bacteria). Therefore, from a quantitative point of view, photosynthesis is more important. Green plants provide the organic carbon needed for heterotrophic creatures (animals, heterotrophic MO, heterotrophic plants). The vegetal carpet consumes 60×10^9 tons of CO_2 per year, or about 2.5% of the amount of CO_2 in the air. If atmospheric CO_2 would not regenerate, it would be exhausted in about 40 years.

Plant and animal life would become impossible. Atmospheric CO_2, however, is not exhausted but is completely regenerated because organic carbon in the vegetal, animal, and microbial debris is converted to CO_2 under the action of MO. The contribution of animals and plants to the production of CO_2 through breathing is of secondary importance compared to the role played by MO in this process: MO is responsible for 80–90% of the total amount of organically oxidized organic substances.

The fundamental importance for life of CO_2 production by MO can be illustrated as follows: the $130,000 \times 10^9$ tons of CO_2 in seas and oceans could meet the current needs of photosynthesis for only about 2,000 years; the amount of CO_2 that would come from the burning of all oil reserves, methane, and coal deposits ($100,000 \times 10^9$ tonnes) would allow photosynthesis to remain at the current level for only about 1,500 years (Hou et al., 2013).

The amount of CO_2 in the atmosphere shows a steady growth trend, explained by: burning fuels in increasing quantities, especially in industry; this way, at least 6×10^9 t of CO_2 per year are produced; deforestation and land degradation on large lands; due to this situation, the total amount of CO_2 fixed by photosynthesis was reduced. CO_2 fixation by green plants and CO_2 formation under MO action are the basic steps of the carbon cycle.

The carbon cycle is carried out simultaneously and in combination with the oxygen and hydrogen cycle: by photosynthesis of green plants from CO_2 and H_2O organic compounds containing H and releasing O_2 are formed; during the total decomposition of organic substances, C, and H are released by the participation of O in the form of CO_2 and H_2O, respectively. The cycles of other elements (N, P, S, etc.), are also related to the carbon cycle.

All of these microbial species are omnivorous, capable of degradation of all organic substances of biological origin. Some synthetic, xenobiotic organic compounds (non-existent compounds created by humans) are not biodegradable: a number of pesticides, detergents, pharmaceuticals, some plastics, which are the greatest danger to environmental pollution (Tilche and Galatola, 2008).

9.4.2   DECOMPOSITION OF ORGANIC COMPOUNDS INTO SOIL

MO plays a major role in the decomposition of organic matter in the soil. It also includes soil fauna and some mechanical degradation under the influence of abiotic factors: precipitation, temperature changes, etc. For the MO, the organic substances they decompose serve as sources of carbon and energy.

Only part of the carbon used is assimilated as cellular matter, the rest of the carbon is released as CO_2 or incompletely oxidized products (organic acids, alcohol, methane, etc.), are formed. The decomposition of native organic substances from the soil: humic and non-humic substances are accomplished by bacteria in the *Pseudomonadaceae* family and actinomycetes (*Nocardia*) (Abraham and Gajendiran, 2019).

The decomposition of organic substances added to the soil is studied with vegetal and animal debris, plant extracts or pure chemical compounds (glucoses, cellulose, lignin, etc.), added to the soil. The decomposition is followed by the determination of degassed CO_2, and in the case of pure compounds, methods of their quantitative analysis or intermediate degradation products are applied.

During decomposition of plant debris, at first the water-soluble substances (mono and oligosaccharides, etc.), are decomposed, followed by starch, hemicelluloses, pectic substances and cellulose. Lignin is the most resistant substance, its decomposition being very slow.

During the decomposition of vegetal debris, the number of -OH groups decreases, and the number of -COOH groups increases as well as the cationic exchange capacity. In parallel, changes in color, volume, and mechanical strength of the mechanical remains (e.g., wheat straw after 1 to 2 months of decomposition on the field become gray or black, their strength decreases, fiber strength is becoming smaller) (Ding et al., 2016).

9.4.3 MICROFLORA OF ORGANIC MATTER DECOMPOSITION

There are three successive microbial populations:

1. **Primary Microflora:** It is made up of members of a physiological group able to decompose the added organic substance. Ex. Cellulose or plant debris rich in cellulose-cellulosic MO, proteins-proteolytic MO.

2. **Secondary Microflora:** It develops on the basis of incompletely oxidized organic compounds resulting from the action of the primary microflora.

3. **Tertiary Microflora:** It develops on the basis of dead cells of MO from the primary and secondary microflora (Khan and Malik, 2018).

9.4.4 BIOCHEMISTRY OF DECOMPOSITION OF ORGANIC COMPOUNDS ADDED TO SOIL

It varies depending on the nature of the substance. For example, the mechanism of cellulose decomposition differs from that of lignin decomposition. The factors influencing the decomposition of organic substances are as follows:

• **Type of Soil:** Intensity of decomposition is higher in brown soils than in chernozems.

• **Vegetation:** Decomposing is more intense in virgin soils, at the beginning of cultivation, than in soils cultivated for a long time.

• **Organic Substances:** They may have stimulatory, indifferent, or inhibitory effects on biodegradation. For example, a method of study is by adding vegetable debris, manure, C14 organic substances, compared to the control soil sample, incubation, CO_2 weighing with C14 and released non-radioactive C. The results show that from samples to which vegetation residues were added was released more non-radioactive CO_2, having a priming effect of stimulation of the decomposition of organic substances in the soil.

- They can stimulate humification processes by enhancing the growth of MOs that use them. The chemical composition of organic compounds has a decisive influence on the rate of decomposition. Glucose decomposes more easily than cellulose, and cellulose lighter than lignin. Lignin-deficient vegetal residues decompose faster than those lignin-rich. Vegetal residues rich in nitrogen are decomposed more easily than nitrogen-rich ones (e.g., wheat straw containing 0.5% N). The remains of young plants that are richer in nitrogen are more susceptible than the remains of old plants.

- **Inorganic Substances:** Mineral nitrogen compounds intensify the decomposition of organic compounds that are deficient or poor in nitrogen. Clay minerals adsorbing macromolecular organic substances (proteins, polysaccharides, etc.), and enzymes that act on these macromolecules reduce the intensity of decomposition. Bentonite, illite, and kaolinite inhibit decomposition.

- **pH:** Decomposition of organic substances is more intense in neutral or slightly alkaline soils than in acid soils. The lime treatment of acidic soils has the effect of intensifying the decomposition of organic substances.

- **Oxygen:** Decomposition is more intense under aerobic conditions than under anaerobic conditions.

- **Humidity:** The decomposition is maximum at 60–80% of the soil water capacity.

- **Temperature:** Decomposition increases as temperature rises to 28–40 °C.

- **Depth:** Depending on the depth, decomposition intensity decreases.

- **Seasonal Variations:** Decomposition is more intense in spring and autumn (Risberg et al., 2017).

9.4.4.1 CELLULOSE DECOMPOSITION

Cellulose decomposition is the most important process for the biological cycle of carbon, since cellulose is the most abundant organic constituent of plant debris and is the most important source of carbon and energy for soil MO. Cellulose is a linear polysaccharide consisting of glucose residues bound by β1, 4 bonds. The number of glucose residues varies between 1,400 and 10,000 (occasionally 15,000) and the molecular weight of the cellulose is 200,000–2,400,000 depending on the plant species.

The cellulose content of plants varies with their species and age, and with different organs. In grasses and pulses, young pulp represents 15% of dry weight and in wood material over 50%. For most crop plants the value is around 15–45% of dry weight (Ai et al., 2018).

> **Cellulolytic Microorganisms:** They are isolated on a nutrient medium containing cellulose as the only source of carbon and energy. Grains or soil dilutions are used. Cellulose is used as: filter paper, cotton wool, precipitated cellulose, cellophane, etc.

> Organic research uses cloths or cotton or cellophane pieces buried in the soil. Cellulo-dextrins are also used in the cellulolytic mechanism. These are oligoglucides, containing some 1,4-linked b-glucose residues. *Cellulolytic* MO may be obligatory or optionally cellulolytic. They may be cryophiles, mesophiles, and thermophiles, aerobic or anaerobic. *Cellulolytic* Mesothelial aerobic MO may be bacteria, actinomycetes, fungi, some protozoa.

> However, the most important are the fungi. Examples of bacteria (optional cellulolytic): *Cytophaga, Sporocytophaga* (form dry microchips); Mixobacteria: *Sorangium compositum, Sorangium cellulosum, Polyangium cellulosum; Pseudomonas erythra, Ps. ephemerocyanea, Ps. Lasia; Cellulomonas; Bacillus: B. vagans,* and *B. Soli.*

> **Examples of Cellulolytic Aerobic Mesothelial Actynomicetes:** *Streptomyces cellulosae, S. violaceus, S. hygroscopicus, Micromonospora chalcea, Nocardia cellulans, N. vaccinii. Chaetomium* has all cellulolytic species, and *Aspergillus* has some species, *Penicillium citrinum,* has some strains, and *Aspergillus luchuensis* (attacks by nondegreased rockwool). Very active fungal strains are the following: *Myrothecium verrucaria, Trichoderma virida, Aspergyllus terreus, Aspergyllus fumigatus* (Virunanon et al., 2008).

➤ **Examples of Protozoa:** *Hartmanella.* The composition of MO groups is influenced by soil pH. In strongly acidic soils (pH below 5.5) fungal strains will predominate, in acidic soils (pH 5.5–6.5) predominate fungi and *Cytophaga,* and in the neutral and slightly alkaline soils (pH above 6.5) the fungi and *Cellvibrio* predominate.

➤ **Examples of Meso-Anaerobic Cellulolytic Anaerobe MO:** Bacterial strains are the most important, obligatory or optional anaerobic and cellulolytic: *Clostridium sp.;* Actinomycetes: *Micromonospora;* Fungi: *Merulius.* Examples of aerobic thermophilic cellulolytic MO: *Bacillus calfactor,* and *Streptomyces thermophilus.* They are found mostly in soils that are supplemented with manure. Examples of anaerobic thermophilic cellulolytic *Clostridium thermocellum,* and *Clostridium thermocellulolyticum* (Fujii and Shintoh, 2006).

1. **Cellulose Decomposition Biochemistry:**

   i. **Cellulose Hydrolysis:** Native cellulose is not soluble in water. Cellulose is degraded by an extracellular enzyme system (cellulose) that diffuses into the medium or remains in the cell-to-cell contact area (at *Cytophaga*). Cellulose may be constitutive enzyme or induced in the presence of cellulose. In the presence of cellulase, the cellulose is biodegraded in celobiose and glucose.

   ii. **Mechanism of Enzymatic Hydrolysis:** Some cellulases attack β-glucosidic linkages within the cellulose (endocellular) chains. Cellulose dextrins are formed. Other cellulases (exocelulases) attack the β-glucosidic linkages at the non-reducing end of the chain, while sequentially cutting a cellobiose molecule.

   iii. **Decomposition of Reducing Sugars:** Reducing sugars, especially glucose and celibiose, serve as a source of carbon and energy for cellulolytic MO. Only one part assimilates in the form of cellular substances; another part is released as final decomposition products.

   Under aerobic conditions, CO_2, H_2O, and small amounts of organic acids occur. Under anaerobic conditions, the final products are: CO_2, H_2O, H_2 and organic acids in large quantities (acetic, formic, lactic, butyric) and small amounts of ethyl alcohol (Scott et al., 2016).

9.4.4.2   HEMICELLULOSES DECOMPOSITION

Hemicellulose is a polysaccharide that accompanies cellulose in green plants. They are pentozanes or hexoses. The pentosans include xylan and araban. Hexoses include mannels, glucomannans, galactans. The amount of hemicellulose in vegetal residues varies between 6–30% of the dry weight by species and age of the plant.

The decomposition of hemicelluloses is proved by the following experiment: Purified hemicelluloses are added to the soil; the soil is incubated and then CO_2 is determined. The result is increasing CO_2 release due to the decomposition of hemicelluloses. Examples of hemicellulolytic MO: fungi: *Rhizopus, Apergillus, Penicillium*; actinomycetes; bacteria: *Bacillus, Pseudomonas, Clostridium, Azotobacter, Sporocytophaga myxococcoides*.

The biochemistry of hemicellulose decomposition follows the following steps:

- **Hydrolysis of Hemicelluloses:** under the action of extracellular enzymes is called hemicellulose. Ex. Xylanase is endo-type: hydrolytically cleaves β-xyloside linkages inside xylan chains, producing oligoxylates and xylose.

- **Decomposition of Reducing Sugars:** under aerobic conditions, the final products are CO_2 and H_2O (Petre et al., 2001).

- **Decomposition of Pectic Substances:** Pectic substances cement the cellulose fibers into cell walls of plants. The pectic substances are: pectic acid, pectin, pectin, pectin, pectinate, native pectin or protopectin. Pectin acid is an α-1,4-galacturonic polymer. There are two types of pectin: pectin H, with a high degree of esterification, over 50% of the -COOH groups; pectin L with low esterification, less than 50%.

Branches of pectic acid may bind to some monosaccharides, galactose, arabinose, xylose, some polysaccharides: galactan, araban, cellulose. MO pectinolytics. In practice, they produce the biologic melting of textile plants (flax, hemp). It can be isolated from the soil, in liquid nutrient media (containing purified pectin as the sole source of C and energy) or in water media with flax straws, by cultivation of soil grains or soil dilutions.

Several examples of pectinolytic Bacteria are: Anaerobic (principal pectic decomposing agents)Clostridium aurantibutyricum, Clostridium felsineum. *Aerobic:* Bacillus polymixa, Bacillus macerans, Pseudomonas, Arthrobacter, Corynebacterium, etc. *Pectinolytic actinomycetes:* Streptomyces, Micromonospora. *Pectinolytic fungi: Aspergillus, Fusarium, Alternaria, Botrytis,* Monilia, etc., (Kim et al., 2017).

➢ **Biochemistry of Pectic Substances Decomposition:** Hydrolysis of pectic substances—it acts successively on three inducible extracellular enzymes:

- Protopectinase (it cleaves the bonds of Ca, Mg, phosphoric acid, releases pectin, monosaccharides (galactose, arabinose, xylose), polysaccharides (galactan, araban, cellulose);

- Poligalacturonase (Pectinaza-PG) is a polysaccharide acting on α-1,4 bonds at the non-reducing side of the pectin chain and pectic acid;

- Pectinesterase (PE)-acts on the methylene groups (-COOCH$_3$ in pectin) and forms pectic acid and CH$_3$OH.

From galactan, araban, cellulose, enzymes produce galactose, arabinose, xylose, glucose. Certain pectins also release acetic acid. The products of protopectin total hydrolysis are: galacturonic acid, galactose, arabinose, glucose xylose, methyl alcohol, acetic acid, o-phosphoric acid, salts of these acids (Haque et al., 2016).

9.4.4.3 DEPOLYMERIZATION OF PECTIC SUBSTANCES UNDER THE ACTION OF LYSES

Decomposition of hydrolysis products:

1. **Anaerobic:** Fermentation of hydrolysis products takes place resulting in galactose, arabinose, xylose glucose; Galacturonic acid does not ferment; the fermentation may be: acetonobutyric (producing acetone CH$_3$-CO-CH$_3$ and n-butyric acid CH$_3$-CH$_2$-CH$_2$-COOH); acetonitrile (producing CH$_3$-CO-CH$_3$acetone and CH$_3$-CH$_2$-CH$_2$-CH$_2$-COOH).

2. Aerobic: The hydrolysis products are oxidized to CO_2 and H_2O. The oxidation rate is in the order of: acetic acid, glucose > galactose > arabinose > xylose > galacturonic acid.

9.4.4.4 STARCH DECOMPOSITION

As a reserve nutritional substance, starch is very common in higher plants and some MO. Starch decomposition in soil is faster than for cellulose, hemicelluloses, and pectic substances. The starch contains two components:

- Amylose, linear glucose polymer, with α-1,4 bonds and degree of polymerization 200–1000; and

- Amylopectin, like glycogen, is a branched polymer of glucose holding a degree of polymerization of 1,000–30,000. In the linear parts, the bonds are of the α-1,4 type, and the branches are produced by α1,6-type bonds. It contains phosphoric acid residues, Ca, Mg. Starch has 10–30% amylose and 70–90% amylopectin in most plants. In wrinkled peas, contains 70% amylose and 30% amylopectin. In *Zea mays* var cheratin, the starch contains only amylopectin.

Amylolytic MO—are isolated from the soil, on liquid or solid nutrient media (containing agar, silica gel) by sowing with soil grains or soil dilutions. Starch decomposition is done with Lugol solution. Amylolytic bacteria are *Bacillus, Clostridium, Chromobacterium, Flavobacterium, Pseudomonas, Serratia, Cytophaga*, etc., *Azotobacter chroorococcum, Clostridium pasteurianum*. Amylolytic actinomycetes: *Streptomyces, Nocardia, Micromonospora*. Pectinolytic fungi: *Aspergillus, Fusarium, Rhyzopus*, etc. (Sarkar et al., 2010).

> ➤ **Starch Decomposition Biochemistry:** Starch hydrolysis is done with constitutive or induced enzymes: amylases and maltase. Amilases can be:

- α amylase (found in plants and amylolytic MO), endo amylases (cleaves α-1,4 bonds inside chains). They produce: maltodextrins (maltoheptaose, maltotriose) and dextrins with some maltose and glucose.

- β-amylase (in plants) is an exo-type enzyme (cleaves α-1,4 bonds from the non-linear end of linear chains), produces maltose and free dextrins.

- α and β amylase, together, form: maltose, glucose + limit dextrin, maltotriose.

Maltase is found in herbs and amylolytic MO. It has action against maltose and maltodextrin α-1,4 bonds, producing glucose.

Limit dextrinase (in MO) cleaves the α-1,6 bound of the limit dextrins. The enzyme (in plants) works similarly to limit dextrinase. Maltase from *Clostridium acetobutyl* and amyloglucosidase (glucoamylase) from *Aspergillus niger* hydrolyzate α-1,4 and α-1,6 bonds from starch producing glucose. The decomposition of reducing sugars, glucose, maltose is either aerobic → CO_2 and H_2O or anaerobic → by Butyric Fermentation (Steger et al., 2005).

➤ **Decomposition of Fructans:** Fructosans represent polysaccharides from fructose residues. The upper plants include: inulin (found especially in composites-Inula, Dahlia, Helianthus tuberosus); fleine (from Phleum), triticin, irisine, asparagosin. Inulin is a linear fructan β 2,1 fructan, fructose residues with β-2,1 linkages. Fleine has fructose residues with β-2,6 linkages. Decomposing MO for fructans are bacteria: *Pseudomonas, Arthrobacter, Clostridium*. Actinomycetes: Streptomyces and fungi: *Penicillium, Aspergillus, Fusarium, Sacharomyces*.

➤ **Biochemistry of Fructan Decomposition:** Hydrolysis is done with the contribution of fructanohydrolases, extracellular enzymes. *Aspergillus fumigatus* contains 2.1 fructanohydrolase, resulting inulobiosis. *Streptomyces* biosyntheses 2,6-fructanohydrolase resulting in levanbiosis. *Penicillium funiculisum* and *Fusarium moniliforme*, contain 2.1 and 2.6 fructanohydrolase, which produces fructose.

Decomposition of reducing sugars is Aerobic → CO_2 and H_2O or anaerobic → acids and alcohols (Galindo-Leva et al., 2016).

9.4.4.5 DECOMPOSITION OF VEGETAL GUMS

Vegetal gums are polysaccharides that are produced in healthy plants or when the tissue is damaged. It swells in water and gives viscous solutions. Arabic gum produced by *Acacia* is made up of arabinose, galactose, rhamnose, and glucuronic acid residues. Decomposing bacteria for vegetal gums are *Pseudomonas, Bacillus, Cytophaga* (Di Marco et al., 2017).

9.4.4.6   DECOMPOSITION OF CHITIN

Chitin is a nitrogen-containing polysaccharide, similar to cellulose. It is found in the composition of the outer shell of the invertebrates and in the cell wall of many fungi-basidiomycetes and ascomycetes. It is a linear polymer of β-1,4-linked β-glucose residues.

Chitosan, the diacetylated derivative of chitin, may be associated with chitin. It is very chemically resistant and cleanses only under the action of concentrated mineral acids. When found in soil, it degrades quite quickly due to chitinoclastic, aerobic, and anaerobic MO.

Aerobic and anaerobic chitinoclastic MO are found in soil. For example, 1 g of soil contains 10^6 chitinoclatic MO, of which: 90–99% are actinomycetes (*Streptomyces, Nocardia, Micromonospora*), 2–10% aerobic bacteria (*Pseudomonas, Bacillus, Cytophaga*, etc.), and anaerobic bacteria (*Clostridium*), 1% fungi (*Mortierella, Trichoderma*, etc.). After composting the soil with chitin, the number of chitin clastic MO increases greatly, to $700 \cdot 10^6$/g of soil, of which 90% are actinomycetes (Wang et al., 2018).

➢ **Biochemistry of Chitin Decomposition:** Chitin hydrolysis:

- **Chitinase Induced:** Constitutive or induced extracellular enzyme resulting in chitobiosis, chitotriosis, N-acetyl glucosamine.

- **Chitobiasis Induced:** Hydrolytically cleaves chitobiosis and chitotriosis resulting in N-acetyl glucosamine, from which acetic acid and glucosamine are formed.

From the metabolisation of hydrolysis products the results are in acetic acid → CO_2 and H_2O (aerobic degradation products), glucose, and ammonia, which are precursors for chitinoclastic MO as source of C, N, and energy (Lv et al., 2016.). The decomposition of polysaccharides done by bacteria, fungi, algae, and lichens involves the following steps:

1. *Dextran* produced by bacterial strains such as: *Leuconostoc dextranicum, L. mesenteroides, L. citovorum*, is a branched-chain glucose (dextrose) polymer, with linear α-1,6 linkages. Branching occurs by α-1,3 and α-1,4 bonds. In the soil is hydrolytically decomposed by bacteria, actinomycetes, fungi, under the action of the dextranase.

2. *Levan* is synthesized by many bacterial species (especially *Bacillus*), a branched fructose (levulosis) polymer, with linear linkages β-2,6

and β-2,1 at branches. Levanases are catalyzing the depolymerization of lavender by bacteria, actinomycetes, fungi.

3. *Sugar glucan* is found in the cell wall of *Saccharomyces cerevisiae*, formed by β-1,3 linked glucoside residues.

4. *Nigeran* is produced by *Aspergillus niger*, and is linear glucose polymer.

5. *Laminarine* is produced by brown algae *Laminaria*. It is a glucose polymer.

6. *Alginic acid* is produced by brown algae. It is a linear copolymer of mannuric acid and guluronic acid, b-1,4 linked monosaccharides. It is decomposed into soil or seawater by bacteria: *Alginomonas alginica, A. nonfermentana,* and *Beneckea alginica.*

7. *Agar-agar* produced by red algae (*Gelidium*), is, and a mixture of agarose (consisting of galactose residues and 3.6 anhydrogalactose with b-1,4 and β-1,3 alternative linkages) and agaropectin (consisting of galactose residues, 3,6 anhydrogalactose, galacturonic acid and sulfuric acid). Most of the MO does not decompose agar-agar. However, several bacterial species capable of hydrolyzing agar were isolated from seawater and seaweed in decomposition. These bacterial strains only exceptionally decompose agar-agar from soil.

8. *Lichenine*, present in *Cetraria islandica*, is a glucose polysaccharide, with β-1,4 and β-1,3 bonds. It is degraded in the soil by the lichenase (Haque et al., 2016).

9.4.4.7 LIGNIN DECOMPOSITION

Lignin occupies the third place, after cellulose and hemicellulose, in terms of quantity, in the composition of green plants. Grasses and young legumes contain 3–6% lignin from the dry weight, and the mature ones, 15–20%; the wood material contains 35% lignin.

Lignin-like substances also exist in some soil borne fungi, representing 20% of the dry mycelial weight (*Aspergillus, Alternaria, Damatium*). Lignin is the most resistant organic compound from vegetable debris. It has the lowest decomposition rate, depending on the species, age, and temperature.

For example, lignin from clover and maize residue decomposes faster than wheat straw. Polysaccharides favor lignin decomposition. Substances rich in N, abundant in young plants, favor the decomposition of lignin, also. Lignin degrades faster under aerobic conditions than under anaerobic conditions.

Lignin decomposition is faster in mesophilic (but also thermophilic) conditions. It is studied with soil samples, vegetal remains (combs, leaves, straw) or purified lignin. It is determined the amount of lignin before and after incubation, or on the basis of analysis of CH_3O-groups, the number of which decreases during decomposition.

Ligninoclast MO are: fungi (especially those producing white rot of wood and attacking cellulose). For example, *Polyporus versicolor* initially attacks lignin, then cellulose. *Armillaria vera* initially attacks cellulose then lignin. *Ganoderma applanatum* simultaneously attacks lignin and cellulose. The bacteria act under anaerobic and thermophilic conditions (Chen et al., 2012).

➢ **Biochemistry of Lignin Decomposition:** The depolymerization of macromolecular substances with ligninase-lignase enzyme is done under the action of MO. Due to the low decomposition rate, it is assumed that the intermediate products slowly form and oxidize quickly. Having an aromatic structure, lignin shows as intermediates in its decomposition, low molecular weight aromatic compounds (e.g., decaying fungi produce vanillin and vanilic acid). An example for modeling the decomposition of lignin by simple aromatic substances instead of lignin is given below:

• *Polyporus versicolor* culture filtrate acts oxidantly on α-conidendrine, siringillic aldehyde and vanillin. It also acts on lignin which becomes more soluble. *Flavobacterium* grows on α-conidendrine media as the sole source of C and energy, producing vanillin and oxidation-protocatechic acid. Numerous soils borne fungi oxidize vanillin, vanillinic acid, siringillic aldehyde, siringilic acid and other phenols, with the dissolution of the benzene ring.

• Lignin decomposition begins with depolymerization in phenols derived from C6-C3 phenylpropane. The oxidation of the 3C side chain follows, forming aldehydes and aromatic acids. Subsequently, the methoxy groups are attacked, demethylation, and/or demethoxylation of aromatic acids takes place.

• Phenolic acids (e.g., protocatechic acid) are formed, whereby polyphenols (e.g., pyrocatechin) are formed by decarboxylation. The dissolution of the benzene ring is the result of protechic acid formation, its decarboxylation into pyrocatechin and its oxidation into cis-cis-mononic acid. Polyphenols resulting from lignin decomposition and those synthesized by MO may be partially oxidized into p- or o-quinones under the action of polyphenoloxidase.

- Polyphenols and quinones are polymerized especially in the presence of amino compounds (AA, peptides) giving rise to humic substances. Those can fix carbohydrates (e.g., condensation of polyphenols and amine compounds-glycine in humic substances) (Xu et al., 2018).

9.4.4.8   LIPID DECOMPOSITION

In the carbon cycle can be found: triglycerides, sterols (cephalins, lecithins with N).

1. **Triglyceride Decomposition:** The tryglicerides are esters of glycerol with fatty acids. Lipolytic MO are as follows: aerobic bacteria: *Pseudomonas fluorescens, Bacterium megaterium, Mycobacterium*; anaerobic bacteria: *Clostridium sporogenes*; actinomycetes; fungi.

    i. **Biochemistry of Triglyceride Decomposition:** Triglyceride hydrolysis is done under the action of lipase with three molecules of H_2O resulting glycerol and fatty acid. Decomposition of hydrolysis products can be described as follows: Glycerol decomposes rapidly on the Embden-Meyerhof-Parnas pathway, coupled with the Krebs cycle, to CO and H_2O. Fatty acids decompose predominantly by β-oxidation. The process is carried out at a low speed in the soil (e.g., 200 g of soil + 4.5 g of neutral fat, incubated for 1-year results in 22.9% decomposition and 12 years-only 38% decomposition).

    Fatty acids tend to accumulate in the soil and become toxic to plants, (e.g., carbon sulfide added to soil sterilization, has the effect of eliminating soil fatigue because it alters the physiological state of fatty acids, producing their decomposition) (Sasaki et al., 2011).

2. **Decomposition of Sterols:** Cholesterol is degraded by actinomycetes *Nocardia, Pseudomonas, Bacillus*, molds, and yeasts.

9.4.4.9 DECOMPOSITION OF ALIPHATIC HYDROCARBONS

Aliphatic hydrocarbons are CH_3-CH_3(ethane), CH_2=CH_2(ethylene), CH_3-CH_2-CH_3(propane), CH_3-CH=CH_2(propylene), CH_3-CH_2-CH_2-CH_3 (n-butane). Etilene belongs to the group of natural plant growth inhibitors, being important in the environment.

Methane decomposition is a process of oxidation under the action of specialized MO such as: methylotroph MO: *Methylomonas, Methylococcus, Methylosinus*; fungi: *Cephalosporium, Penicillium janthienellum.*

$$CH_4 \rightarrow CH_3\text{-}OH \rightarrow HCHO \rightarrow HCOOH \rightarrow CO_2 + H_2O$$

Methane → methanol → formaldehyde → formic acid → CO_2 + H_2O

Decomposition of ethane, propane, n-butane is done by oxidation under the action of MO from the soil: mycobacteria, nocards, pseudomonas, flavobacteria, micrococi. Decomposition of higher hydrocarbons such as paraffins, higher hydrocarbons from crude oil, benzene, petroleum is also an interesting topic.

Some petroleum products are used to dissolve pesticides and this is how they reach in the soil. Some higher hydrocarbons are formed in the soil. Decomposition MO are as follows: aerobic bacteria: *Mycobacterium paraffinicum, Corynebacterium petrophillum, Pseudomonas desmolytica, Ps. fluorescens*; Anaerobic bacteria: *Desulfovibrio desulfuricans*; Actinomycetes: *Nocardia paraffinae*; Yeast: *Candida lipolytica, C. tropicali.*

> ➢ **Mechanism of Oxidation:** Oxidation of the CH_3 group in COOH at one or both ends to form monobasic or bibasic fatty acid which then degrades by β-Oxidation. For example, paraffin:

$$CH_3\text{-}(CH_2)n\text{-}CH_2\text{-}CH_3 \rightarrow \text{monobasic fatty acid:}$$
$$CH_3\text{-}(CH_2)n\text{-}CH_2\text{-}COOH \rightarrow \text{fatty acid: } HOOC\text{-}(CH_2)n\text{-}CH_2\text{-}COOH$$

Oxidation of the carbon adjacent to the terminal carbon, to form the alcohol. It is oxidized to methyl ketone (Nuñal et al., 2014).

9.4.4.10 DECOMPOSITION OF AROMATIC HYDROCARBONS

Decomposition of aromatic hydrocarbons was discovered during the study of soil sterilization with benzene or toluene. It was observed that benzene and toluene disappear from the soil after a while. Decomposition MO are as follows: bacteria (*Bacterium benzols, Bacterium toluols, Mycobacterium, Pseudomonas*), fungi, and actinomycetes. The hydrocarbons that oxidize in the soil are: Benzene, toluene, o-xylene (monocyclic), naphthalene, anthracene, phenanthrene, polycyclic.

> ➢ **The Phases of Microbial Soil Oxidation in Monocyclic Hydrocarbons:** Hydroxylation of the cycle (hidroxilaze-catalyzed) is also named monoxidation, i.e., introducing into the ring a single O atom in the presence of an H (H_2X) atom. The role of the H donor is $NADH^+H^+$ or $NADPH^+H^+$. Chlorine, the nitro and the sulfonic groups of some aromatic compounds are replaced by H or OH. However, generally, chlorine, nitro derivatives, and the sulfonic derivatives are resistant to microbial decomposition. The result of the first step is a diphenolic compound (often pyrocatechin or protocatechic acid). The diphenolic compound is oxidized with the ring cleavage, which is of three types:

> - **Type I:** Cleavage at the ortho position (between the hydroxylated C atoms). Under the action of a dioxinase introducing 2 O atoms. It results in a cyclic peroxide, which by restructuring gives rise to a cis-cis-mucic acid aliphatic compound. For the protocatechic acid will result 3-carboxy-cis-cis-muconic acid.

> - **Type II:** Breaking the bond between a hydroxylated and a nonhydroxylated atom in meta position. From this process will result 2-hydroxymauconic acid semialdehyde.

> - **Type III:** The linkage between a hydroxylated C atom and a C atom carrying a side chain or a carboxyl is cleaved.

Ex. Homogentizic acid → 4-maleilacetoacetic acid. Ex. main intermediates:

acid cis,cis-muconic and 3-carboxi-cis-cis acid, cis muconic →
β-β-cetoadipic acid → acetyl-CoA + succinyl-CoA.

2-hidroxi-4-carboxi muconic acid semialdehyde → oxaloacetic acid +
pyruvic acid

$$HOOC-CH_2-CO-COOH+CH_3-CO-COOH$$

4-maleic acetoacetic acid → acetoacetic acid + maleic acid
$$HOOC-CH_2-CO-CH_3+HOOC-CH=CH-COOH$$

Products resulting from the splitting of the diphenolic cycle are metabolized by several intermediate products. The final oxidation products are CO_2 and H_2O.

For polycyclic hydrocarbons, the mechanism of oxidation is similar in soil, with the exception that cyclisation cleavage takes place in stage I resulting in a monocyclic compound (pyrocatechin). Asphalt and synthetic graphite are degraded by microorganisms existing in the soil (Dutta et al., 2018).

9.4.4.11   DECOMPOSITION OF PHENOLS, AROMATIC ACIDS

Phenol, crezoles, benzoic acid are antiseptic substances which decompose under MO if their concentration is not too high (0.01–0.1%). Decomposing MO are: bacteria (*Micobacterium, Pseudomonas fluorescens, Vibrio cyclosites)*; actinomycetes (*Nocardia)*; fungi. The mechanism of oxidation of phenols and aromatic acids is similar to that of oxidation of aromatic hydrocarbons.

> **Rubber Decomposition:** Rubber decomposition takes place under the action of fungi, bacteria, and actinomycetes. Natural rubber degrades more easily than synthetic rubber (Liu et al., 2017).

> **Plastics Decomposition:** Polyethylene and polypropylene are plastics that cannot be degraded by MO. However, there are several fungal strains (such as *Phanerochaete chrysosporium, Lentinus tigrinus, Aspergillus niger,* and *Aspergillus sydowi*i) that are able to use PVC as a source of C and energy (Roohi, Bano et al., 2017).

9.4.4.12   PESTICIDES DECOMPOSITION

Herbicides, insecticides, fungicides, nematicides, are all pesticides with high soil remanence. Some pesticides decompose in nature through non-biological processes under the action of visible light and ultraviolet rays or under the action of non-enzymatic catalysts (Cu^{2+}, carbonates, Fe, Mn, CO, MnO_2, clay minerals).

In the soil, many pesticides decompose biologically under the action of MO. On the other hand, MO are able to biodegrade several pesticides.

Conditions favoring the growth of MO coincide with those necessary for the rapid spread of soil pesticides, and there is a positive correlation with the MO number in the soil. Sterilization of the soil leads to a reduction in the rate of pesticide decomposition or this does not take place. Inhibition of breathing processes in the dark with NaN_3.Na azide, NaF-Na fluoride prevents the decomposition of 2,4-D pesticides.

MO was isolated in pure cultures. These MO used the pesticides as sole source of C and energy. Mixed cultures demonstrated MO synergism. By cometabolism, pesticide decomposition occurs through oxidation, by MO that does not have pesticides as a source of energy and C (N), degradating the pesticides by co-oxidiation.

The first demonstration of metabolic decomposition of pesticides was carried out in 1951 on *Arthrobacter globiformis*, 2,4-D decomposing. Other bacteria, actinomycetes, molds, yeast produce pesticide decomposition. Algae and protozoa interfere with this process. The MO pesticide degradation capacity is limited. Some pesticides are decomposed slowly (for example, DDT persists aerobically in soil for 24 years) (Héritier et al., 2017).

> **Biochemistry of Microbial Decomposition of Pesticides:** Enzymatic catalyzed base reactions that may occur are: oxidation, dealkylation, dehalogenation, reduction, hydrolysis of amines and esters, cleavage of aromatic cycles, conjugation, and condensation.

> **Oxidation:** Under the action of mono-oxygenases, in the presence of O_2 and $NADH^++H^+$, leads to the hydroxylation of aromatic pesticides, after which aromatic compounds become more polar and more water-soluble and capable of cleavage. The aliphatic side chains are removed. Organic acids formed from aliphatic chains decompose by β-oxidation.

> **Dealkylation:** Due to the mono-oxygenases, in the presence of O_2 and $NADH^++H^+$ $NADPH^++H^++$ removes the alkyl groups attached to the N, O, and S atoms. The C-linked alkyl groups are generally resistant to microbial decomposition (e.g., triazine herbicides).

> **Dehalogenation:**

  • **Hydrolytic Dehalogenation:** The halogen atom is replaced by the hydroxyl (OH) group.

  • **Reductive Dehalogenation:** The halogens are replaced by H.

  • **Dehydrohalogenation:** Removes H and halogen, forming double bonds.

MO-mediated reduction consists (a) reduction of the nitro ($-NO_2$) group to the amino ($-NH_2$) and (b) reduction of quinones to phenols.

Hydrolysis of amines and esters is done under the action of amidases and esterases. Conjugation and condensation is done by coupling of the pesticide molecule or its partial decomposition with natural compounds (including AA, carbohydrates, etc.).

The result is the production of methylated compounds, formyl, conjugated AA, glycoside. Thus, the pesticide is removed, but only temporarily, because the resulting conjugate can easily break. Condensation is a synthesis of the pesticide or degradation compound. Decomposing MO are: bacteria (*Arthrobacter, Achromobacter, Corynebacterium, Pseudomonas, Flavobacterium*); actinomycetes (*Nocardia*), fungi (*Aspergillus*).

2,4-dichlorophenoxyacetic (2,4D) acid is an herbicide which, in normal doses, decomposes into soil in 4–18 weeks. Decomposing bacteria for 2,4D are *Arthrobacter*.

The decomposing mechanism is as follows: 2,4D→2,4 dichlorophenol→3,5 dichloropirocatechin→2,4-dichloro-cis-cis-muconic acid by cleaving the ring into the ortho position.

Lindane is the γ-isomer of hexachlorocyclohexane, and is degraded by dehydrohalogenation in y-2,3,4,5,6-pentachlorocyclohex-1-ene $\rightarrow CO_2$ + H_2O + Cl.

Escherichia coli and *Clostridium* slowly metabolize lindane, one of the metabolic products being benzene:

hexachlorocyclohexane → tetrachlorobenzene + chlorobenzene + benzene + Cl⁻

Chlorophenols from fungicides are decomposed by several bacteria and fungi. *Trichoderma viride*, by methylation of pentachlorophenol, produces pentachloranisole.

2,4,5-trichlorophenol → pentachlorophenol → pentachloranisol.

The 1,2-dibromoethane nematocide is dehalogenated, the MO releasing ethylene and Br.

$$Br-CH_2-CH_2-Br \rightarrow CH_2 = CH_2 + 2Br.$$

The effect of pesticides on soil borne MO can be direct or indirect. The direct influence is exerted by changing the physiological activity of decomposing MO populations. Indirectly, the pesticides are affecting other MO, plants, and animals. If pesticides are applied in normal doses, there are no long-term quantitative and qualitative changes in MO communities in the soil. For future development, in the research studies, pesticide doses of 10–1000 times higher than normal are used (Murillo-Zamora et al., 2017).

Changes in the microbial community of the soil may be as follows: endosporid bacteria (*Bacillus, Clostridium*), fungi, and protozoa that produce resistance forms are less sensitive to high pesticide concentrations; their vegetative forms may be sensitive, however: *Bacillus cereus* is susceptible to 2,4-D. 2,4-D produces temporary restriction of fungal growth.

Fungicides change the composition of the microflora, and some fungal and bacterial species become dominant. *Trichoderma viride* is dominant in soil recolonization after fungicide treatment. *Trichoderma viride* can suppress the development of other fungi such as *Armilaria, Pythium, Rhizoctonia,* and *Phytophthora.*

The pentachloronitrobenzene fungicide inhibits *Pythium* and *Fusarium* antagonists, which increases the severity of the plant diseases. Reduction of total MO of soil, especially for: *Nitrosomonas, Nitrobacter, Azotobacter, Rhizobium,* and *Thiobacillus.* Microbial activities, as an indicator of the effect of pesticides, include: CO_2 production; O_2 consumption; nitrification, free, and symbiotic fixation of N_2; cellulose degradation, ammonification.

For example, the effect on nitrification and N_2 fixation can be described: The propanil herbicide (from 50 ppm to 30 ppm) inhibits nitrification (from 50 ppm to 30 ppm) at a depth of 5 cm in soil. The paraquat herbicide (over 500 ppm) inhibits nitrification. The triazine herbicides inhibit Phase I and Phase II differently from nitrification. Fungicides reduce the rate of nitrification. Heptachlor and aldrin (50–100 ppm) inhibit *Melilotus* nodules, depending on the soil (Holmsgaard et al., 2017).

9.5   THE NITROGEN CYCLES

MO regulates nitrogen reserves in soil and their accessibility to higher plants by assimilation and immobilization of nitrogen by: symbiotic and non-simiobiotic nitrogen fixation, nitrification, denitrification, ammonification as follows:

- Green plants assimilate nitrogen in mineral form (ammonium salts, nitrates). Legumes also use N_2 by MO symbionts from root nodules (the symbiotic fixation of N_2).

- Ammonia is oxidized to nitrate under the action of nitrifying MO (nitrification).

- Nitrates can be reduced to N_2 (denitrification).

- MO from soil and water decompose the organic compounds with nitrogen, releasing it as ammonia (ammonification).

Ammonification is the process of mineralization of organic nitrogen compounds, which releases nitrogen as ammonia. Nitrogen from organic compounds, otherwise inaccessible, becomes accessible due to ammonification. Ammonification occurs with proteins, AA, nucleic acids, and nucleotides, nucleosides, purines, and pyrimidine bases, lipids, and carbohydrates containing nitrogen, creatinine, hippuric acid, urea, alkaloids, etc., (Wu et al., 2018). Decomposition of proteins from plant, animal, and MO residues.

Several examples of proteolytic MO are as follows: proteolytic bacteria: aerobic, sporulated: *Bacillus cereus var. mycoides, B. subtilis, B. megaterium, B. thermoproteoliticus*; nonsporulated: *Serratia marcescens, Arthrobacter*; optional anaerobic: *Proteus vulgaris, Pseudomonas fluorescens*; obligatory anaerobic: *Clostridium putrefaciens*. Proteolytic bacteria *Bacillus cereus var. mycoides*, is the most active in the culture media, while in the soil, under natural conditions, the most active is *Proteus vulgaris*. Proteolytic actinomycetes: *Streptomzces violaceus, Micromonospora chalcea;* Proteolytic mushrooms: *Penicillium, Aspergillus, Mucor, Rhyzopus, Alternaria*.

9.5.1 BIOCHEMISTRY OF PROTEIN DECOMPOSITION

Biochemistry of protein decomposition has two stages: hydrolysis and dezamination and decarboxylation of AA (Martínez-Santos et al., 2018).

Hydrolysis is an extra and intracellular process, comprising:

- Hydrolysis of protein towards peptides, with proteinases;

- Hydrolysis of peptides towards AA, with peptidases.

- Proteinases are:

- **Serine Proteinases:** They have serine and histidine active in the center. They are alkaline proteinases, owing an optimal pH of 9–11. An example is the proteinase produced by *Aspergillus, B. subtilis*-subtilisin.

- **Thiol-Proteinases:** They have cysteine active center (streptococcal thiol-proteinase).

- **Acid Proteinases (Carboxyl-Proteinase):** with optimal pH below 5, produced by *Aspergillus niger, Penicillium janthinelum, Mucor pusillus*.

- **Metallo-Proteinases:** It contain Zn^{2+} neutral, produced by *Bacillus subtilis*.

- The peptidases are:

- Aminopeptidases (removes only one amino acid residue from the N terminus of the chain, e.g., *Aeromonas proteolitica*);

- Carboxipeptidases (removes one amino acid from the C-terminal end of the chain, e.g., *Serin carboxipeptidase-Penicillium, Aspergillus*, metal carboxypeptidase).

- Dipeptidases (hydrolyses dipeptides, e.g., *Mycobacterium phlei*);

- Peptideases for dipeptides at the N-terminus and, respectively, the C-terminus of the peptide chain (Roohi et al., 2017).

Dezamination and decarboxilation of AA is an intracellular process.

Oxidative dezamination is done with amino acid dehydrogenases + NAD or NADP coenzyme. It can also be done with amino acid-oxidases (flavin enzymes) + FAD coenzyme, with Cu or vitamin B12 enzymes. During hydrolytic dezamination the NH_2 group of AA is not hydrolyzed by MO. The occurrence of α-hydroxyacids is explained by the reduction of α-ketoacids formed by the oxidative deamination pathway, while the primary alcohols appear by the dehydroxylation of α-hydroxy.

Amino acid decomposition produces toxic substances:

- Decarboxylation gives putrescine, cadaverine from ornithine;

- Histamine is produced from histidine.

- Tyramine, which is toxic at high doses results from tyrosine (Graves et al., 2016).

The decomposition of nucleic acids is done with ammonia release. Decomposition is by: nucleotide depolymerization, nucleotide hydrolysis with nucleoside and o-phosphate formation, hydrolysis of purine or pyrimidine base nucleosides and ribose or deoxyribose, deamination of nitrogenous bases. Decomposing MO are the following: *Bacillus, Clostridium, Arthrobacter*; actinomycetes: Streptomyces; fungi: *Aspergillus, Penicillium, Cephalosporium, Mucor, Fusarium*, etc.

9.5.2 DECOMPOSITION OF NITROGEN-CONTAINING LIPIDS

Cephalines and lecithins from organic residues are enzymatically hydrolyzed resulting in glycerol, fatty acid, o-phosphate, and a nitrogen compound: colamine or choline. Artrobacteria decompose choline to glycine by two successive oxidations and 3 demethylations.

9.5.3 DECOMPOSITION OF NITROGEN CARBOHYDRATE

Chitin, murein-mucopeptide or peptidoglucan from the cellular basal wall of eubacteria is an aminopolysaccharide. It decomposes into soil, the final products being:

$$NH_3 + 2CO_2 + H_2O$$

> **Creatinine Decomposition:**
> Creatinine hydrolysis $\rightarrow$ creatine $\rightarrow$ urea and sarcosine $\rightarrow$ glycine demethylation $\rightarrow NH_3 + 2CO_2 + H_2O$
> **Urea Decomposition:** The decomposition of urea is done in the soil under the action of urobacteria, actinomycetes, and fungi. In 1866, Pasteur named the decaying bacterium: *Torula ammoniacale*. *Bacillus pasteuri* and *Sporosarcina ureae* are alkaline pH resistant. MO decompose soil urea originating from nitrogen metabolism in mammals (urine), from some fungi, or formed during the decomposition of purine and pyrimidine bases, arginine, and creatinine, from synthetic fertilizers. Urea accumulated in the soil is decomposed to ammonia and carbonic acid.

9.5.4 DECOMPOSITION OF CALCIUM CYANAMIDE

Ca Cyanamide is a nitrogen-containing fertilizer containing calcium carbide.

$$Ca\ Cyanamide + H_2O \rightarrow Cyanoamide\ (N = C-NH_2) + CaO$$

$$Cyanamide + H_2O \rightarrow Urea$$

$$Urea + H_2O\ and\ Urea \rightarrow CO_2 + 2NH_3$$

9.5.5 DECOMPOSITION OF ALKALOIDS

Alkaloids originate from plant remains: coffee-caffeine, tea-theophylline, cocoa-theobromine. Other alkaloids are: nicotine, atropine, quinine. Soil-borne MO that are able to degrade alkaloids are bacteria: *Pseudomonas putida*, *Ps fluorescens*, *Bacillus coagulans*; fungi: *Penicilium*. Methylxanthine is degassed by hydrolysis to methanol and xanthine-which degrades oxidatively. Other nitrogen compounds are also decomposed by soil borne MO (vitamins, antibiotics, chlorophylls, drugs, pesticides, artificial dyes) (Yan et al., 2016).

9.5.6 AMMONIA TRANSFORMATIONS

Abiotic transformations that occur can be described as follows: ammonia adsorbed by clay minerals neutralizes acids in the soil, volatilizes to alkaline pH and is then leached from the soil. The biotic transformations: NH_3 and ammonium salts are assimilated by MO.

NH_3 is oxidized to nitric acid and MO nitrates. Nitrification is the oxidation of ammonia to nitrites, in the soil. Denitrification is the reduction of nitrite to organic N.

The biological bonding of molecular nitrogen is run by N_2 fixative bacteria: *Azotobacter, Azomonas, Klebsiella, Bacillus, Clostridium, Cyanobacteria*. Green plants assimilate nitrogen in mineral form by transforming it into organic nitrogen. MO from soil and water decompose organic compounds with nitrogen, releasing nitrogen as ammonia. Ammonia is oxidized to nitrate under the action of nitrifying MO. Nitrates can be reduced to N_2 (denitrification) (Li et al., 2015).

9.5.7 BIODEGRADATION OF ORGANIC WASTE AND RESIDUES: THE ISSUE OF DIOXINS

Uncontrolled combustion on the platforms, even the domestic waste-incinerator, produces large amounts of dioxins and furans, which own high toxicity and which are concentrated in the environment (water, air, soil) and in trophic, aquatic, or terrestrial chains. Specialty literature from Canada, UK, Belgium, the Netherlands, the USA, etc., reported the fact that the incidence of cancer and toxic phenomena in humans and animals is significantly higher in neighboring of incinerated waste platforms.

Under these circumstances, it is necessary to reuse all the waste, no matter how much it would cost to adjust their quality for further recirculation. Agriculture, the largest consumer of organic waste, can become the user of biodegradable waste, which can provide an energy base for increased productions and environmental cleanliness. The methods of recycling waste in agriculture and related sectors are multiple: soil fertilization, energy recycling, recycling as a direct or indirect source of animal feed, etc., (Wu et al., 2015).

9.6 THE IMPORTANCE OF TECHNOLOGIES FOR THE BIODEGRADATION (STABILIZATION) OF ORGANIC WASTE AND RESIDUES IN AGRICULTURE

Vegetal waste, manure, semi-liquid manure, effluents from the food industry, sewage sludge from sewage treatment plants in all sectors can be used as raw, unprocessed, on well-sized land surfaces correlated with the self-purification of the soil.

There are many advantages, especially economic: low investment costs, low transport costs, as well as reduced storage costs. The use of raw waste and especially untreated manure and domestic fecal water, has a number of disadvantages that limit this method of valorification:

- The huge environmental pollution potential: over 70 compounds and gases that produce odor pollution, soil, and groundwater pollution, insect biotope, rodents;

- High infective and infesting potential;

- The need for large land plots (0.5 ha/u.a.m);

- High humidity residues require expensive means of transport;

- The use of liquid manure cannot be done all year long and only in certain phases of vegetation;

Unexpected situations may occur, such as pump failure, quarantine, frost, etc. A hygienic soil can purify about 2000 kg of C.O.D. (C.O.D. = chemical oxygen demand for oxidation of organic substances) per hectare per day, degrading organic matter, fixing P and K. N is nitrified and converted to soluble nitrates (but which can leach in the groundwaters).

The agronomic limits of fertilization with manure also consist of other aspects: mineral imbalances; scarification of vegetation; restrictions on temperature and vegetation area; pH restrictions; type of soil. For these reasons, organic waste must be processed for efficient, non-polluting reuse, with the main focus on the following desideratum: economic desideratum: full recovery as organic fertilizer without nitrogen loss; volume reduction; reducing transportation costs; Sanitary and veterinary-sanitary desideratum: inactivation of pathogenic biotic agents for humans and animals; environmental pollution prevention.

Other desideratum: Inactivation of weed seeds; soil degradation prevention; waste recycling as a direct and indirect source of animal feed; energy recycling; biodegradation (in the sense of stabilization of organic waste and residues; natural (spontaneous) biodegradation of organic waste (Li et al., 2019).

Immediately after their release (urine, feces) or after formation (plant waste, animals, wastewater, decanting sludge, other organic waste), either stored in solid piles or pools and drying pads, all categories of waste and organic waste are subject to environmental factors that reduce their volume (quantity) and structural complexity. The environmental factors that influence or carry out biodegradation processes of organic matter are:

- **Physical Factors:** Temperature, humidity, light radiation, wind, and ultraviolet radiation;

- **Chemical Factors:** Oxygen, carbon dioxide, environmental pH, C: N ratio;

- **Biological Factors (Microbiocenosis and Macrobiocenosis):** Bacteria, viruses, fungi, protozoa, coprophage insects, other partially-coagulating organisms;

- The quality of the organic matter of biodegradable waste;

- Dominant relationships between microbiocenose species (competition, predation, parasitism, commensalism, synergism, neutrality, antibiotic);

- The presence and concentration of factors favoring the natural biodegradation process;

- Presence and concentration of natural biodegradation factors (anti-biotics, disinfectants, stabilizers).

Depending on the type of biodegradation, the final products can be: $H_2O + CO_2 +$ energy if biodegradation is complete and minerals, reintroduced into matter circuits in nature, or end products (intermediates) if the process is controlled (compost, methane, energy, etc.) (Xie et al., 2019).

9.7 MICROBIOLOGY OF BIODEGRADATION

The chemical composition of organic waste is a complex oxidation-inducing material, subjected to the action of MO, which uses the waste as a source of C, O, N, H, S, Fe, energy. At the time of their release, especially feces and urine, they contain a rich and varied microbial component of enteral origin whose activity is quickly taken up by microbiocenosis from the storage medium. At the surface of the landfills, aerobic heterotrophic microflora is installed and furthermore, in depth, anaerobic heterotrophic microflora, then protozoa appear.

Later on, coprophage insects lay their eggs, a large number of larvae, worms grow, and then birds and mammals consume also a varying amount of residue.

The microbiocenosis of waste can be autotrophic and heterotrophic, aerobic, anaerobic, photosynthesis, and chemosynthesis (Table 9.2). Aerobic flora dominant in manure and other organic waste: *Bacillus vulgaris, B. subtilis, B. mesentericus, B. graveolens, Bacterium fluorescens, B. enteriditis, Escherichia coli, Bacterium vulgare, Micrococcus luteus, Micr. candicans, Micr. sulfureus, Micr. piogenes, Flava saracia, Streptococcus pyogenes, Cytophaga, Cellovibrio,* etc., as well as nitrifying and denitrifying bacteria, mixobacteria, and frequent pathogenic bacteria. The dominant anaerobic bacteria are *Bacterium cellulosae hydrogenicus, Metanobacterium,* etc. The fungal strains involved in the biodegradation processes are very well represented: *Monilia, Penicillium, Aspergillus, Rhyparobius, Aschopanus, Pilaira, Pilobolus, Sordaria, Circinella, Cladosporum,* etc. Saprophytes: flagellate (*Bodo endax, Monas*), amoeba (*A. limax*), ciliate (*Lembus pusillus, Cyclidium glaucoma, Uronema*).

TABLE 9.2   Aerobic and Anaerobic Microbial Processes in Degradation of Residues

The Attacked Substrate	Final Product	MO
Aerobic Degradation		
Aliphatic hydrocarbons	CO_2 H_2O	A great variety of bacteria, protozoa,
Aromatic hydrocarbons	CO_2 H_2O	and aerobic heterotrophic fungi
Carbohydrates, sugars	CO_2 H_2O	
Fats, fatty acids,	CO_2 H_2O	
Protein, amino acids	H_2O H_2S NH_3 CO_2	
S, S_2^-	SO_4	*Begiatoa, Thiotrix, Thiobacillus*
Fe^{++}	Fe^{+++}	*Ferobacillus*
H_2	H_2O	*Hydrogenomonas*
NH_3	NO_2^- NO_3^-	*Nitrosomonas, Nitrobacter*
Anaerobic Degradation		
Aliphatic hydrocarbons	CO_2 fatty acids	A very large number of anaerobic and
Carbohydrates	CO_2 fatty acids	optionally anaerobic heterotrophic
Fats	CO_2 fatty acids	microbes, microaerophilic fungi
Proteins	NH_3, amines, fatty acids	
NO_3^-	N_2O N_2	
Fe^{+++}	Fe^{+++}	
H^+	H_2	
CO_2	fatty acids, CH_4	*Metanobacterium, Metanosarcina, Metanococcus*
SO_4	S	*Disulfovibrio, Pseudomonas, Clostridium*

The cumulative effect of abiotic and biotic factors (this complex micro-biocenosis) is the gradual reduction of the amount and complexity of organic residues, the organic matter being simplified to CO_2, H_2O, energy, and mineral matter to stable elements, both components being introduced into other biosphere material circuits (Jędrczak and Suchowska, 2018).

9.7.1 THE MAIN METABOLIC PATHWAYS FOR THE DEGRADATION OF ORGANIC WASTE

In the oxido-reduction relations, the aerobe microflora uses oxygen as a hydrogen acceptor, while the anaerobe one calls for other acceptors: NO_3^-, H^+, Fe^{+++}, SO_4^-, photosynthetic MO uses light energy in oxidation-reduction processes, but all three MO groups coexist in the processes of natural degradation of residues.

bacterial catalysis: Organic waste + $O_2 \rightarrow$ M anaerobic + CO_2 +H_2O

bacterial catalysis: Organic waste + hydrogen acceptors$\rightarrow CO_2 + CH_4 + H_2O$

photonic catalysis: Organic waste + $CO_2 + H_2O \rightarrow$ MO photosynthesis + CO_2

9.7.2   BIOCHEMISTRY OF ORGANIC MATTER DEGRADATION

The natural degradation of biodegradable waste is essentially accomplished by MO from the environment in a phased process that begins with the hydrolysis of polymers and the consumption of monomers, the growth and development of biomass due to the transfer of matter, and ultimately the autolysis of biomass along with the results of the various catabolisms included in other biosphere circuits. The phases of biodegradation succeed, overlap, and coexist, not being possible to separate the beginning or end moments, but always predominate one or the other.

Under natural conditions, the duration of biodegradation is long, 1–2 years, being longer and incomplete in an aerobic, and shorter and advanced in an aerobic environment. In their development, biodegradation processes, whether in solid waste dumps, in the liquid waste or soil (for all categories of organic waste), go through two obligatory phases: primary transformations consisting of substrate mobilization, i.e., enzymatic decomposition of substrate and secondary transformations, uptake by the MO of the hydrolyzed products of the first phase (Bátori et al., 201.).

9.7.3   PRIMARY WASTE TRANSFORMATIONS: MOBILIZATION (DEGRADATION, DECOMPOSITION) OF THE SUBSTRATE

Mobilization of the substrate consists of a set of biochemical reactions, by which the organic waste is prepared for assimilation by micro- and macroorganisms, mainly by the enzymatic equipment they possess.

> ➢ **Polysaccharide Decomposition:** It deals with: cellulose, hemicelluloses, pectins, lignin, chitin, starch:
>
>   • **Cellulose:** Decomposition through MO hydrolysis has been known since 1850. The importance of cellulose in the biosphere is extraordinary given the following data: the total amount

of cellulose in the composition of living organisms on Earth reaches 700 billion tonnes, and in organic soil and peat is 1400 billion tons.

- Most cellulolytic organisms use cellulose carbon exclusively with celluloses, enzymes that contain two factors: one that solubilizes and one depolymerizes and releases oligosides. The process of cellulose degradation appears as a hydrolytic process depending on cellulase and celobase hydrolase. The result of degradation is glucose. Some examples of cellulolytic MO are: *Cythophaga, Myxococcus, Cellvibrio, Cellfalcicula, Actinomyces* sp., *Bacillus cellulosae dissolvens, Bacterium sp., Clostridium sp., Pseudomonas,* and many fungi: *Trichoderma Fusarium* sp., *Aspergillus* sp.

- **Hemicelulloses (Polysaccharides with 5 and 6 Carbon Atoms):** Galactans, xylans, decompose gradually after 8 days.

- **Pectins:** These are a major component of manure in ruminants and coarse vegetal waste. They are attacked by specific MOS (*Pseudomonas, Prunicola*) possessing two enzymes: pectinesterase and polygalacturonidase. The degradation products are acetic acid, butyric acid, formic acid, lactic acid (LA), hydrogen. In aerobic decomposition the final products are: $H_2O + CO_2$.

- **Lignin:** It is a component of coarse vegetable waste and is decomposed by MO with polyphenoloxidase enzymes. Through genetic engineering, scientists are currently working on producing bacteria capable of metabolizing lignin from the coarse fodder.

- **Chitin:** It is attacked by bacteria producing chitinases: *B. chitinovorus, Pseudomonas, Flavobacterium, Micrococcus, Klebsiella, E. coli,* etc.

- **Starch:** It is especially found in ruminant manure, wastewater from the food industry and sewage sludge. In the stored residues, the MO that produce α-amylase and β amylase attack the starch and cleave it to monosaccharides. Among the amylase producing MO there are: *B. subtilis, Pseudomonas, Clostridium acetobutilicum, B. macerans,* etc.

➢ **Protein Decomposition:** Proteins are the most important component of waste and residues. Their biodegradation is under the action
of proteolytic enzymes or proteases.

Degraded proteins from manure are extensively hydrolyzed by
trypsin and chomotripsin. Besides proteins, other organic nitrogen
compounds are found in organic wastes (OWs): purine bases,
nucleosides, nucleotides, amines, amides, etc.; their decomposition
is accomplished by the MO that have amiases- and from such cleav-
ages, ammonia is produced (Grübel et al., 2018).

9.7.4   SECONDARY TRANSFORMATION OF WASTE

➢ **Metabolisation of Hydrolyzate by MO:** The primary decomposition of organic waste under the action of extracellular enzymes
results in molecules of substance that constitute carbon, nitrogen,
and other nutrients needed for MO growth. They are multiplying at
high speed and managed to prove their presence by increasing the
local temperature.

The MO number increases from several 100 million to a few billion
per gram of waste. It is clear that multiplication of biomass produces
an increase in biomass (bacterial and fungal), which means, in
fact, secondary transformations of organic matter. Parallel to the
decomposition of the old organic matter, new organic matter is
formed by biosynthesis, meaning a microbial protein more resistant
to biodegradation. In addition to the syntheses governed by the vital
activity of the MO, other syntheses takes place, led by enzymes
released in the mass of waste. This is how is explained the synthesis
of phosphorus organic compounds by the esterification of glycerol
with orthophosphoric acid, the synthesis of levanes from fructose
residues under the action of levansucrase. In addition, pure chemical
syntheses are produced. The pathways of MO usage of the products
resulting from the enzymatic hydrolysis of residues and wastes are
complex and varied (Mishra et al., 2018).

9.7.5   BIOMASS TRANSFORMATIONS

From what has been shown so far, it results that the primary and secondary
transformations of organic waste result in an increase in the mass of MO
(especially bacteria and fungi), in parallel with the loss of energy and volatile

matter. Hydrolysis and substrate assimilation are followed by MO death and their degradation, and these two phases occur simultaneously. To quantify the result, one can analyze the quantity and quality of residuals remaining after a determined time at the place of storage or processing.

Under natural conditions, MO mimics the situation in controlled ecosystem (controlled cultivation): have an adaptation phase, then a log growth phase, plateau phase (steady growth), negative growth phase or decline and phase of collapse when the nutrient substrate is exhausted. However, this curve is the result of the interaction of a large number of influential factors known and defined when designing biotechnology for waste degradation and recycling. The natural biodegradation of organic waste is basically an oxide redox set that results in a gradual reduction in organic matter content, parallel to the concentration of stable inorganic matter (replicated on the biosphere circuit). All organics (especially volatile) are considered environmental pollution factors (Fang et al., 2018).

9.7.6 CONTROLLED BIODEGRADATION OF ORGANIC WASTE

Composting waste residues for recycling as fertilizer—in support of the need to transform organic waste into stabilized, hygienic, very valuable and non-hazardous fertilizers, there are many technical, economic, hygienic, and sanitary arguments to prevent air, natural water and soil pollution. Controlled biodegradation of organic waste for processing and use as fertilizer is generally known as composting.

Composting involves simultaneous or sequential physicochemical, biochemical, and microbiological transformations of OWs and residues, from their original state, to different humification stages, resulting in a new product: compost.

Depending on the conditions of aeration, temperature, and humidity, the organic substance in the plant, animal, artificial waste, evolves to a new state quite stable in biodegradation, having a C: N ratio similar or close to that of humus in the soil. In literature, fermentation rather than "composting" is used more often.

Biochemistry and microbiology of waste composting are similar to processes of spontaneous biodegradation, distinguishing from them by human intervention, controlling the process, controlling the duration and the influential factors. The raw materials for composting can take various forms: straw, leaves, wood, orchard waste, human, and animal manure, urban waste, food waste, household waste, pulp, textiles, wool, etc.). Depending on the degree of composting, intermediate phases appear, such as:

- Raw compost, in which materials of origin are readily recognizable;
- Semi-finished compost;
- Finalized compost, much converted to humus;
- Earth compost (mound).

The chemical composition of a qualitative composts: about 50% organic substance; pH 7–8; total nitrogen 0.2–4%; phosphorus 0.5–2%; K, Ca, Mn, Mg about 1%; Good quality compost has a pleasant smell, homogeneous appearance, humidity below 40%.

Composting time takes around 3–4 months, 6 months or one year, depending on the process used (Zhang et al., 2018).

9.7.7   BIOTECHNOLOGY FOR COMPOSTING

Depending on the quality and quantity of the waste, the technical endowment, and the intended purpose, several composting biotechnologies can be used, in the household or industrial system.

9.7.7.1 BIOTECHNOLOGY OF COMPOSTING IN THE HOUSEHOLD SYSTEM

➢ **Extensive Composting:** Plant and animal waste are thrown into disordered piles, placed near shelters or on the land area to be fertilized. It takes about a year, the composting is incomplete, fertilizer elements are lost, and the maximum temperature of 40–45°C does not inactivate pathogens, parasites, and weed seeds. Environmental pollution is massive.

➢ **Intensive Aerobic Composting:** The origins of these aerobic composting biotechnologies were described by Steiner (biodynamic process) and Howard (indore process). Practically, all available organic waste categories can be used: manure, straw, weed, leaves, seas, algae, waste from orchards, vineyard residues, vegetables, hops, sawdust, etc. It is recommended to crush the vegetal debris. The optimal C:N ratio is 33:1.

The layers are laid loose and are aerated every 2–3 weeks. The composting is performed on platforms at the surface of the land, with a maximum height of 2 m, or on 1.2 m wide and 0.9 m deep beams. The duration of the process is approx. 3 months. In small household farms, landfill (heap) and other wastes may have the following dimensions: a large base of 2.5–4 m, and a small base of 1 m, and a height of 1.8–2 m.

Aerobic biodegradation is intensely exothermic, as the temperature rises to 60–65 °C or 70–80 °C. At these temperatures the pathogen, parasite eggs, and larvae, weed seeds are inactivated. As a fertilizer, the resulting compost has a lower value due to the high nitrogen losses (Song et al., 2018).

➢ **Anaerobic Intensive Composting:** In the platform or compost pile the air is removed; for this purpose, waste residues with high humidity, over 70% is deposited in layers over 2 m thick. In the absence of oxygen, the anaerobic microflora will decompose slowly and incompletely cellulose, hemicellulose, pectins, and partially lignin.

Anaerobic composting can also be carried out on semi-buried or buried platforms. The duration of the composting is 6 months-one year, the fertilizing value of the compost is high, and the nitrogen losses are small. The temperature is lower: 20–25°C in the winter and 30–35°C in the summer, thus not being capable of inactivating pathogenic MO, which must be destroyed by other mechanisms (Lin et al., 2019).

➢ **Mixed Aerobic-Anaerobic Composting:** Mixed aerobic-anaerobic composting (KRANTZ method) combines aerobic and anaerobic composting. On the platform, the layers of 0.8–1.0 m of organic waste are allowed to aerate for 3–4 days (loosened) for aerobic fermentation, after which new layers are added, up to a final height of 3–4 m. Afterwards they are covered with a layer of soil. It takes 4–5 months, and the method is cumbersome and does not ensure secure destruction of the pathogenic MO (maximum temperature achieved is 55°C).

➤ **Other Composting Biotechniques in the Household System:** Besides the described aerobic or anaerobic classic methods, which mainly refer to composting for the recycling of solid manure obtained in livestock farms, there are other biotechnological variants used for decanting sludge, household waste, human feces, water marshes, aquatic plants and unconventional vegetal waste (other than straw, chaff, degraded hay).

• **Biotechnology of Sewage Sludge Composting from Sewage Treatment Plants:** Compost batch was produced both by spontaneous biodegradation and controlled biodegradation, by addition of microbial bioproducts from selected bacteria and fungi cultures.

The sludge mixed with straw and other vegetal waste in a proportion of 2–4 parts of straw to 1 part of the sludge to correct the C: N ratio. Spontaneous and especially controlled biodegradation was carried out in good conditions with physical changes (temperature, humidity), chemical changes (organic matter, nitrogen, nitrate) and bacteriological changes characteristics of composting. To achieve good quality compost, different plant mixtures + biopreparations of selected bacteria, actinomycetes, and fungi were used in a perfectly controlled biotechnology (Mu et al., 2017).

• **Biotechnology for Composting Household, Urban, and Industrial Waste:** The use of household, urban, and industrial residues in agriculture is related to the spread of metropolitan areas and their strong development. The composition of the residues is very variable, including readily biodegradable elements (food debris), others with slow fermentation (paper, cardboard) and other virtually non-degradable biological agents (slag, ash, metals, etc.). Currently, large-scale composting of household waste with sewage sludge from treatment plants is widely applied. The process is used with good results in Switzerland, France, Sweden, USA. "ALCYON BYOTERMIQUE" is a French procedure of ecological recycling of household waste, producing compost, biogas, and BIOSOL-HUMUS, globally recognized.

- **Biotechnology of Aerobic Composting of Complex Mixtures of Waste and Various Wastes:** In many countries of the world (China, India, USA, Mexico), mixtures of vegetal waste, aquatic plants, animal, and human waste, mud, domestic waste and sludge from domestic wastewater treatment plants are composted without any demanding equipment or technique. This method of composting has been particularly important in countries where human waste and household fecaloid residues have been used as raw material for fertilization, resulting in the alarming spread of parasitic infestations in humans.

- **Biotechnology of Composting of Straw and Other Vegetal Waste for the Production of "Artificial Garbage:"** Straw and other vegetal waste are rich in cellulose, hemicellulose, and lignin, but poor in nitrogen. In order for cellulosic MO to hydrolyze the substrate, nitrogen is needed. This is why straw and other wastes are treated with ammonium salts, nitrates, cyanamide, urea, calcium, phosphorus, and potassium salts, soak, and build a pile in which composting takes about 3 months. Good quality compost is called "artificial garbage."

- **Processes for Speeding up the Biodegradation in Compost Piles:** In order to reduce the biotransformation duration of organic substrate used for composting, good results were achieved through the use of additives: introducing into the composting pile of some biopreparations-macerated products of medicinal plants (chamomile, nettle, dandelion, valerian, horse tail), which by a hormonal-like mechanism can direct the intensity and direction of biodegradation. Such preparations are marketed in countries such as France, Germany, Switzerland, and the Netherlands, etc. Other methods are: the use of a mixture of selected cultures of bacteria, fungi, and actinomycetes; introducing enzyme concentrates into the compost pile, Correcting the C:N ratio to 33:1 or higher; control of aeration and moisture in the compost pile; pH correction, etc. (Viaene et al., 2016).

- **Biotechnologies in Industrial System:** The household waste composting system and unprocessed waste management in agriculture was perpetuated in all countries of the world until around 1920. Later on, at the margins of the large urban areas a lot of waste residues accumulated. There was huge sludge of decanting domestic and industrial fecaloid waters. Their degradation by burning has become too costly and highly polluting.

Large industrial zootechnical complexes have drowned neighboring areas in manure. At that moment, the regulatory framework began to set new environmental protection legislation.

As a result, the biodegradation, deodorization, and sanitation of all categories of waste and sludge through aerobic composting have become the most effective method. Until the emergence of biotechnology for composting, the aerobic composting was exclusively used by both municipalities and zootechnical farms. After discovering the possibilities of obtaining new, high-value products through composting, private entrepreneurs became interested in these methods.

The beginning of enterprises for the recovery of sludge and household waste was around 1945. Their profitability was proven, and large capital entered the circuit, many industrial enterprises appeared, technological processes and lines, equipments, and installations were patented, competitive relations for the concession of the right on the processing of waste and sludge.

Biotechnology focusing on forced composting, performed as follows: after separating hard bodies and plastics, the urban waste is mixed (or not) with organic sludge, adjust C:N to 25:1 with manure and other organic residues and it is introduced into containers, where they are continuously oxygenated, and thermophilic microbial cultures are sometimes added. In a few hours the temperature rises to 50–60°C, the mixture deodorizes quickly and the compost is finished in 2–14 days. The mixture is dried under hot air until the humidity reaches 30%, and briquettes and packs are then produced.

Composting is not performed towards the formation of humic compounds. Biotechnology that, after the completion of the composting, uses the second phase of production, meaning the industrial technique. The process is carried out as follows: the waste is deposited in the beams or concrete platforms, the heaps being 1.5–3 m high. The heaps are aerated and the material reaches a moisture content of 50–60%, the C:N ratio is 20 to 25:1. After 3–4 months, compost enters the industrial techno-logical line.

The final conditioning of the compost is performed mechani-cally: the metals, the glass, the plastics, the boulders are sorted, the rest is ground and sieved through the 10 cm and afterwards 2 cm sieves, after which the final product is packed in bags, or delivered in bulk according to requirements (Eklind et al., 2004). Among the bio-products obtained on an industrial scale, which are found on the European market, we mention: "BIOCAMPO"– a compost produced by the Italian company "ITALCAMPO"; "BIO-VEGETAL"-organic-biological fertilizer produced by the Italian company "TERSANPUGLIA" and "SUDITALIA." Recycling of organic waste by composting and its use as a fertil-izer is a long-term economic and ecological strategy. Associ-ating large organic waste producers with an agroecosystem can bring significant amounts of knowledge into the energy balance of the farmland at affordable prices. Consideration should also be given to the size of agroecosystems according to the avail-ability of organic waste in the surrounding area and, if they lack sufficient resources, the establishment of the zootechnical sector is the most favorable remedy (Korniłłowicz-et al., 2010).

9.8 CONCLUSIONS AND RECOMMENDATIONS

The issue of maintaining the biological and energy balance in ecosystems, ensuring the need for organic matter and, at the same time, removing the waste and waste environment, presents two important, strongly motivated issues:

The continuous requirement of soil in relation to organic matter "exported" in massive amounts, through the primary production of ecosystems.

In the case of intensive agroecosystems, consumption, and the need for organic matter are even higher than in spontaneous ecosystems. Soil is concerned with specialists in many countries, threatening the agronomic potential of many regions. Decreasing the percentage of organic matter in the soil is of concern to specialists in many countries, threatening the agronomic potential of many regions.

The large amount of waste and residues collected from various practical activities (agriculture, food industry, households activities, etc.), and which threaten the hygiene of the environment. In order to ensure the biological and energy balance in nature, to discard the environment from the burden of waste, the ideal solution is to recycle the waste products, and recirculate substances and elements into natural cycles.

The issue is of vital importance for the survival of society, given:

- Large expenditures for the disposal of household and industrial waste and residues, representing a major problem of large metropolitan areas;

- Reducing land available for waste disposal, with increasingly pesimistic forecasts for the coming decades;

- Urban and domestic waste cannot be burned altogether, despite the large furnaces built next to each metropolis;

- Waste storage platforms have become a biotope favorable for pathogenic MO, insects, rodents, odorous pollution outbreaks, bringing up the most unpleasant areas, with a psychologically depressive effect.

Chemical analyzes estimate the amount of pollution from the agricultural waste, but they do not reflect the environmental consequences resulting from their mobilization, accumulation along the food chain and especially their impact on the key metabolic processes in the soil. Biological methods, in turn, reflect the impact on soil organisms, highlighting intensification/inhibition of activities in stress conditions. Given the fact that the soil is subject to strong anthropogenic influences, it is of major importance to determine the effect of pollutants from agricultural waste on edaphic communities of MO.

KEYWORDS

- **actinomycetes**
- **bacteria**
- **biochemistry**
- **biodegradation**
- **environmental factors**
- **microorganisms**

REFERENCES

Abraham, J., & Gajendiran, A., (2019). Biodegradation of fipronil and its metabolite fipronil sulfone by *Streptomyces rochei* strain AJAG7 and its use in bioremediation of contaminated soil. *Pestic. Biochem. Physiol.*, *155*, 90–100.

Ai, S., Zhao, Y., Sun, Z., Gao, Y., Yan, L., Tang, H., & Wang, W., (2018). Change of bacterial community structure during cellulose degradation by the microbial consortium. *Sheng Wu Gong Cheng Xue Bao.*, *34*(11), 1794–1808.

Amodeo, C., Sofo, A., Tito, M. T., Scopa, A., Masi, S., Pascale, R., Mancini, I. M., & Caniani, D., (2018). Environmental factors influencing landfill gas biofiltration: Lab scale study on methanotrophic bacteria growth. *J. Environ. Sci. Health A Tox Hazard. Subst. Environ. Eng.*, *53*(9), 825–831.

Armstrong, D. L., Rice, C. P., Ramirez, M., & Torrents, A., (2018). Fate of four phthalate plasticizers under various wastewater treatment processes. *J. Environ. Sci. Health A Tox Hazard. Subst. Environ. Eng.*, *53*(12), 1075–1082.

Bátori, V., Åkesson, D., Zamani, A., Taherzadeh, M. J., & Sárvári, H. I., (2018). Anaerobic degradation of bioplastics: A review. *Waste Manag.*, *80*, 406–413.

Chen, Q., Marshall, M. N., Geib, S. M., Tien, M., & Richard, T. L., (2012). Effects of laccase on lignin depolymerization and enzymatic hydrolysis of ensiled corn stover. *Bioresour. Technol.*, *117*, 186–192.

Di Marco, E., Soraire, P. M., Romero, C. M., Villegas, L. B., & Martínez, M. A., (2017). Raw sugarcane bagasse as carbon source for xylanase production by *Paenibacillus* species: A potential degrader of agricultural wastes. *Environ. Sci. Pollut. Res. Int.*, *24*(23), 19057–19067.

Ding, Y., Wang, W., Liu, X., Song, X., Wang, Y., & Ullman, J. L., (2016). Intensified nitrogen removal of constructed wetland by novel integration of high rate algal pond biotechnology. *Bioresour. Technol.*, *219*, 757–761.

Dodd, A., Swanevelder, D., Zhou, N., Brady, D., Hallsworth, J. E., & Rumbold, K., (2018). *Streptomyces albulus* yields ε-poly-L-lysine and other products from salt-contaminated glycerol waste. *J. Ind. Microbiol. Biotechnol.*, *45*(12), 1083–1090.

Dourou, M., Aggeli, D., Papanikolaou, S., & Aggelis, G., (2018). Critical steps in carbon metabolism affecting lipid accumulation and their regulation in oleaginous microorganisms. *Appl. Microbiol. Biotechnol.*, *102*(6), 2509–2523.

Dutta, K., Shityakov, S., Khalifa, I., Mal, A., Moulik, S. P., Panda, A. K., & Ghosh, C., (2018). Effects of secondary carbon supplement on biofilm-mediated biodegradation of naphthalene by mutated naphthalene 1,2-dioxygenase encoded by Pseudomonas putida strain KD9. *J. Hazard. Mater.*, *357*, 187–197.

Eklind, Y., Hjelm, O., Kothéus, M., & Kirchmann, H., (2004). Formation of chloromethoxybenzaldehyde during composting of organic household waste. *Chemosphere*, *56*(5), 475–480.

El-Naggar, A., Shaheen, S. M., Hseu, Z. Y., Wang, S. L., Ok, Y. S., & Rinklebe, J., (2019). Release dynamics of As, Co, and Mo in a biochar treated soil under pre-definite redox conditions. *Sci. Total Environ.*, *657*, 686–695.

Fang, H., Zhang, H., Han, L., Mei, J., Ge, Q., Long, Z., & Yu, Y., (2018). Exploring bacterial communities and biodegradation genes in activated sludge from pesticide wastewater treatment plants via metagenomic analysis. *Environ. Pollut.*, *243*(Pt B), 1206–1216.

Fujii, K., & Shintoh, Y., (2006). Degradation of mikan (Japanese mandarin orange) peel by a novel Penicillium species with cellulolytic and pectinolytic activity. *J. Appl. Microbiol.*, *101*(5), 1169–1176.

Galindo-Leva, L. Á., Hughes, S. R., López-Núñez, J. C., Jarodsky, J. M., Erickson, A., Lindquist, M. R., Cox, E. J., et al., (2016). Growth, ethanol production, and inulinase activity on various inulin substrates by mutant *Kluyveromyces marxianus* strains NRRL Y-50798 and NRRL Y-50799. *J. Ind. Microbiol. Biotechnol.*, *43*(7), 927–939.

Ganser, B., Bundschuh, M., Werner, I., Homazava, N., Vermeirssen, E. L. M., Moschet, C., & Kienle, C., (2019). Wastewater alters feeding rate but not vitellogenin level of *Gammarus fossarum*(Amphipoda). *Sci. Total Environ.*, *657*, 1246–1252.

García-Gómez, A., Bernal, M. P., & Roig, A., (2003). Carbon mineralization and plant growth in soil amended with compost samples at different degrees of maturity. *Waste Manag. Res.*, *21*(2), 161–171.

Geed, S. R., Prasad, S., Kureel, M. K., Singh, R. S., & Rai, B. N., (2018). Biodegradation of wastewater in alternating aerobic-anoxic lab scale pilot plant by *Alcaligenes* sp. S3 isolated from agricultural field. *J. Environ. Manage.*, *214*, 408–415.

Ghaffar, I., Imtiaz, A., Hussain, A., Javid, A., Jabeen, F., Akmal, M., & Qazi, J. I., (2018). Microbial production and industrial applications of keratinases: An overview. *Int. Microbiol.*, *21*(4), 163–174.

Graves, C. J., Makrides, E. J., Schmidt, V. T., Giblin, A. E., Cardon, Z. G., & Rand, D. M., (2016). Functional responses of salt marsh microbial communities to long-term nutrient enrichment. *Appl. Environ. Microbiol.*, *82*(9), 2862–2871.

Grübel, K., Wacławek, S., Machnicka, A., & Nowicka, E., (2018). Synergetic disintegration of waste activated sludge: Improvement of the anaerobic digestion and hygienization of sludge. *J. Environ. Sci. Health A Tox Hazard. Subst. Environ. Eng.*, *53*(12), 1067–1074.

Haque, M. A., Kachrimanidou, V., Koutinas, A., & Lin, C. S. K., (2016). Valorization of bakery waste for biocolorant and enzyme production by *Monascus purpureus*. *J. Biotechnol.*, *231*, 55–64.

Héritier, L., Duval, D., Galinier, R., Meistertzheim, A. L., & Verneau, O., (2017). Oxidative stress induced by glyphosate-based herbicide on freshwater turtles. *Environ. Toxicol. Chem.*, *36*(12), 3343–3350.

Holmsgaard, P. N., Dealtry, S., Dunon, V., Heuer, H., Hansen, L. H., Springael, D., Smalla, K., et al., (2017). Response of the bacterial community in an on-farm biopurification system, to which diverse pesticides are introduced over an agricultural season. *Environ. Pollut.*, *229*, 854–862.

Hou, W., Gu, B., Lin, Q., Gu, J., & Han, B. P., (2013). Stable isotope composition of suspended particulate organic matter in twenty reservoirs from Guangdong, southern China: Implications for pelagic carbon and nitrogen cycling. *Water Res.*, *47*(11), 3610–3623.

Huang, Y., Pan, H., Wang, Q., Ge, Y., Liu, W., & Christie, P., (2019). Enrichment of the soil microbial community in the bioremediation of a petroleum-contaminated soil amended with rice straw or sawdust. *Chemosphere*, *224*, 265–271.

Hultberg, M., & Bodin, H., (2018). Effects of fungal-assisted algal harvesting through bio pellet formation on pesticides in water. *Biodegradation*, *29*(6), 557–565.

Jędrczak, A., & Suchowska-Kisielewicz, M., (2018). A comparison of waste stability indices for mechanical biological waste treatment and composting plants. *Int. J. Environ. Res. Public Health*, *15*(11). pii: E2585. doi: 10.3390/ijerph15112585.

Jin, D., Kong, X., Jia, M., Yu, X., Wang, X., Zhuang, X., Deng, Y., & Bai, Z., (2017). *Gordonia phthalatica* sp. nov., a di-n-butyl phthalate-degrading bacterium isolated from activated sludge. *Int. J. Syst. Evol. Microbiol., 67*(12), 5128–5133.

Kandasamy, S., Dananjeyan, B., Krishnamurthy, K., & Benckiser, G., (2015). Aerobic cyanide degradation by bacterial isolates from cassava factory wastewater. *Braz. J. Microbiol., 46*(3), 659–666.

Khan, S., & Malik, A., (2018). Toxicity evaluation of textile effluents and role of native soil bacterium in biodegradation of a textile dye. *Environ. Sci. Pollut. Res. Int., 25*(5), 4446–4458.

Kim, H. M., Song, Y., Wi, S. G., & Bae, H. J., (2017). Production of D-tagatose and bioethanol from onion waste by an integrating bioprocess. *J. Biotechnol., 260*, 84–90.

Kim, J. S., Lee, Y. H., Kim, Y. I., Ahmadi, F., Oh, Y. K., Park, J. M., & Kwak, W. S., (2016). Effect of microbial inoculant or molasses on fermentative quality and aerobic stability of sawdust-based spent mushroom substrate. *Bioresour. Technol., 216*, 188–195.

Korniłłowicz-Kowalska, T., & Bohacz, J., (2010). Dynamics of growth and succession of bacterial and fungal communities during composting of feather waste. *Bioresour. Technol., 101*(4), 1268–1276.

Li, C., Yang, J., Wang, X., Wang, E., Li, B., He, R., & Yuan, H., (2015). Removal of nitrogen by heterotrophic nitrification-aerobic denitrification of a phosphate accumulating bacterium *Pseudomonas stutzeri* YG-24. *Bioresour. Technol., 182*, 18–25.

Li, D., Sun, J., Cao, Q., Chen, Y., Liu, X., & Ran, Y., (2019). Recovery of unstable digestion of vegetable waste by adding trace elements using the bicarbonate alkalinity to total alkalinity ratio as an early warning indicator. *Biodegradation, 30*(1), 87–100.

Li, L., Guo, C., Fan, S., Lv, J., Zhang, Y., Xu, Y., & Xu, J., (2018). Dynamic transport of antibiotics and antibiotic resistance genes under different treatment processes in a typical pharmaceutical wastewater treatment plant. *Environ. Sci. Pollut. Res. Int., 25*(30), 30191–30198.

Li, Y., Liu, H., Li, G., Luo, W., & Sun, Y., (2018). Manure digestate storage under different conditions: Chemical characteristics and contaminant residuals. *Sci. Total Environ., 639*, 19–25.

Lin, L., Shah, A., Keener, H., & Li, Y., (2019). Techno-economic analyses of solid-state anaerobic digestion and composting of yard trimmings. *Waste Manag., 85*, 405–416.

Liu, Q., Singh, V. P., Fu, Z., Wang, J., & Hu, (2017). An anoxic-aerobic system for simultaneous biodegradation of phenol and ammonia in a sequencing batch reactor. *Environ. Sci. Pollut. Res. Int., 24*(12), 11789–11799.

Lu, H., Han, T., Zhang, G., Ma, S., Zhang, Y., Li, B., & Cao, W., (2018). Natural light-micro aerobic condition for PSB wastewater treatment: A flexible, simple, and effective resource recovery wastewater treatment process. *Environ. Technol., 39*(1), 74–82.

Lu, H., Zhang, G., Lu, Y., Zhang, Y., Li, B., & Cao, W., (2016). Using co-metabolism to accelerate synthetic starch wastewater degradation and nutrient recovery in photosynthetic bacterial wastewater treatment technology. *Environ. Technol., 37*(7), 775–784.

Lv, M., Hu, Y., Gänzle, M. G., Lin, J., Wang, C., & Cai, J., (2016). Preparation of chitooligosaccharides from fungal waste mycelium by recombinant chitinase. *Carbohydr. Res., 430*, 1–7.

Lyu, T., He, K., Dong, R., & Wu, S., (2018). The intensified constructed wetlands are promising for treatment of ammonia stripped effluent: Nitrogen transformations and removal pathways. *Environ. Pollut., 236*, 273–282.

Martínez-Santos, M., Lanzén, A., Unda-Calvo, J., Martín, I., Garbisu, C., & Ruiz-Romera, E., (2018). Treated and untreated wastewater effluents alter river sediment bacterial communities involved in nitrogen and sulfur cycling. *Sci. Total Environ.*, *633*, 1051–1061.

Mishra, S., Singh, P. K., Dash, S., & Pattnaik, R., (2018). Microbial pretreatment of lignocellulosic biomass for enhanced biomethanation and waste management. *3 Biotech.*, *8*(11), 458. doi: 10.1007/s13205-018-1480-z.

Mu, D., Horowitz, N., Casey, M., & Jones, K., (2017). Environmental and economic analysis of an in-vessel food waste composting system at Kean University in the U.S. *Waste Manag.*, *59*, 476–486.

Murillo-Zamora, S., Castro-Gutiérrez, V., Masís-Mora, M., Lizano-Fallas, V., & Rodríguez-Rodríguez, C. E., (2017). Elimination of fungicides in bio-purification systems: Effect of fungal bioaugmentation on removal performance and microbial community structure. *Chemosphere*, *186*, 625–634.

Ning, Z., Zhang, H., Li, W., Zhang, R., Liu, G., & Chen, C., (2018). Anaerobic digestion of lipid-rich swine slaughterhouse waste: Methane production performance, long-chain fatty acids profile and predominant microorganisms. *Bioresour. Technol.*, *269*, 426–433.

Nuñal, S. N., Santander, D. E., Leon, S. M., Bacolod, E., Koyama, J., Uno, S., Hidaka, M., et al., (2014). Bioremediation of heavily oil-polluted seawater by a bacterial consortium immobilized in cocopeat and rice hull powder. *Biocontrol. Sci.*, *19*(1), 11–22.

Petousi, I., Daskalakis, G., Fountoulakis, M. S., Lydakis, D., Fletcher, L., Stentiford, E. I., & Manios, T., (2019). Effects of treated wastewater irrigation on the establishment of young grapevines. *Sci. Total Environ.*, *658*, 485–492.

Petre, M., Teodorescu, M. E., Zarnea, G., Adrian, P., Gheorghiu, E., & Gheordunescu, V., (2001). Microbial degradation of cellulose wastes in continuous bioreactors. *Meded Rijksuniv Gent Fak Landbouwkd Toegep Biol. Wet.*, *66*(3a), 195–198.

Risberg, K., Cederlund, H., Pell, M., Arthurson, V., & Schnürer, A., (2017). Comparative characterization of digestate versus pig slurry and cow manure - Chemical composition and effects on soil microbial activity. *Waste Manag.*, *61*, 529–538.

Roohi, B. K., Kuddus, M., Zaheer, M. R., Zia, Q., Khan, M. F., Ashraf, G. M., Gupta, A., & Aliev, G., (2017). Microbial enzymatic degradation of biodegradable plastics. *Curr. Pharm. Biotechnol.*, *18*(5), 429–440.

Roohi, M., Riaz, M., Arif, M. S., Shahzad, S. M., Yasmeen, T., Ashraf, M. A., Riaz, M. A., & Mian, I. A., (2017). Low C/N ratio raw textile wastewater reduced labile C and enhanced organic-inorganic N and enzymatic activities in a semiarid alkaline soil. *Environ. Sci. Pollut. Res. Int.*, *24*(4), 3456–3469.

Sarkar, S., Banerjee, R., Chanda, S., Das, P., Ganguly, S., & Pal, S., (2010). Effectiveness of inoculation with isolated *Geobacillus* strains in the thermophilic stage of vegetable waste composting. *Bioresour. Technol.*, *101*(8), 2892–2895.

Sasaki, D., Hori, T., Haruta, S., Ueno, Y., Ishii, M., & Igarashi, Y., (2011). Methanogenic pathway and community structure in a thermophilic anaerobic digestion process of organic solid waste. *J. Biosci. Bioeng.*, *111*(1), 41–46.

Sayen, S., Rocha, C., Silva, C., Vulliet, E., Guillon, E., & Almeida, C. M. R., (2019). Enrofloxacin and copper plant uptake by *Phragmites australis* from a liquid digestate: Single versus combined application. *Sci. Total Environ.*, *664*, 188–202.

Scott, B. R., Huang, H. Z., Frickman, J., Halvorsen, R., & Johansen, K. S., (2016). Catalase improves saccharification of lignocellulose by reducing lytic polysaccharide monooxygenase-associated enzyme inactivation. *Biotechnol. Lett.*, *38*(3), 425–434.

Song, C., Li, M., Qi, H., Zhang, Y., Liu, D., Xia, X., Pan, H., & Xi, B., (2018). Impact of anti-acidification microbial consortium on carbohydrate metabolism of key microbes during food waste composting. *Bioresour. Technol., 259*, 1–9.

Song, C., Zhang, Y., Xia, X., Qi, H., Li, M., Pan, H., & Xi, B., (2018). Effect of inoculation with a microbial consortium that degrades organic acids on the composting efficiency of food waste. *Microb. Biotechnol., 11*(6), 1124–1136.

Steger, K., Eklind, Y., Olsson, J., & Sundh, I., (2005). Microbial community growth and utilization of carbon constituents during thermophilic composting at different oxygen levels. *Microb. Ecol., 50*(2), 163–171.

Tilche, A., & Galatola, M., (2008). The potential of bio-methane as bio-fuel/bio-energy for reducing greenhouse gas emissions: A qualitative assessment for Europe in a life cycle perspective. *Water Sci. Technol., 57*(11), 1683–1692.

Ugwuanyi, J. O., Harvey, L. M., & McNeil, B., (2005). Effect of digestion temperature and pH on treatment efficiency and evolution of volatile fatty acids during thermophilic aerobic digestion of model high strength agricultural waste. *Bioresour. Technol., 96*(6), 707–719.

Valenti, F., Porto, S. M. C., Selvaggi, R., & Pecorino, B., (2018). Evaluation of biomethane potential from by-products and agricultural residues co-digestion in southern Italy. *J. Environ. Manage., 223*, 834–840.

Viaene, J., Van, L. J., Vandecasteele, B., Willekens, K., Bijttebier, J., Ruysschaert, G., De Neve, S., & Reubens, B., (2016). Opportunities and barriers to on-farm composting and compost application: A case study from northwestern Europe. *Waste Manag., 48*, 181–192.

Virunanon, C., Chantaroopamai, S., Denduangbaripant, J., & Chulalaksananukul, W., (2008). Solventogenic-cellulolytic clostridia from 4-step-screening process in agricultural waste and cow intestinal tract. *Anaerobe., 14*(2), 109–117.

Wang, D., Li, A., Han, H., Liu, T., & Yang, Q., (2018). A potent chitinase from Bacillus subtilis for the efficient bioconversion of chitin-containing wastes. *Int. J. Biol. Macromol., 116*, 863–868.

Wu, N., Zeng, M., Zhu, B., Zhang, W., Liu, H., Yang, L., & Wang, L., (2018). Impacts of different morphologies of anammox bacteria on nitrogen removal performance of a hybrid bioreactor: Suspended sludge, biofilm and gel beads. *Chemosphere, 208*, 460–468.

Wu, P., Xu, L., Wang, J., Huang, Z., Zhang, J., & Shen, Y., (2015). Partial nitrification and denitrifying phosphorus removal in a pilot-scale ABR/MBR combined process. *Appl. Biochem. Biotechnol., 177*(5), 1003–1012.

Xie, J., Duan, X., Feng, L., Yan, Y., Wang, F., Dong, H., Jia, R., & Zhou, Q., (2019). Influence of sulfadiazine on anaerobic fermentation of waste activated sludge for volatile fatty acids production: Focusing on microbial responses. *Chemosphere, 219*, 305–312.

Xu, Q., Dai, R., Ruan, Y., Rensing, C., Liu, M., Guo, S., Ling, N., & Shen, Q., (2018). Probing active microbes involved in Bt-containing rice straw decomposition. *Appl. Microbiol. Biotechnol., 102*(23), 10273–10284.

Yan, L., Zhang, S., Hao, G., Zhang, X., Ren, Y., Wen, Y., Guo, Y., & Zhang, Y., (2016). Simultaneous nitrification and denitrification by EPSs in aerobic granular sludge enhanced nitrogen removal of ammonium-nitrogen-rich wastewater. *Bioresour. Technol., 202*, 101–106.

Zhang, D., Luo, W., Li, Y., Wang, G., & Li, G., (2018). Performance of co-composting sewage sludge and organic fraction of municipal solid waste at different proportions. *Bioresour. Technol., 250*, 853–859.

CHAPTER 10

Values of Biofertilizers

HILAL AHMAD GANAIE,[1] NASEER UE DIN SHAH,[2] FALAK MUSHTAQ,[2] and JASBIR KOUR[2]

[1] Department of Zoology, Government Degree College (Boys), Pulwama – 192301, Jammu and Kashmir, India, E-mail: hilalganie@hotmail.com

[1] Cytogenetics and Molecular Biology Research Laboratory, Center of Research for Development (CORD), University of Kashmir, Srinagar – 190006, Jammu and Kashmir, India

ABSTRACT

The use of bio-fertilizers is one of the administration rehearses which keep up or increment the substance of the natural issue and improve soil ripeness in arable soils. While, a few outcomes have been gotten in connection to the impact of bio-composts on natural issue content, less in thought about the fragmentary arrangement of humus. Biofertilizers are the common manures that possess living material known as inoculum, which may be microscopic organisms, algae; parasites alone or in mix. These microbial inoculants raise the accessibility of supplements directly to creatures that grows on the soil viz. plants. The job that bio composts play in horticulture has an incredible noteworthiness, especially in the present setting of expanded expense of concoction manures and their unsafe consequences for soil health.

10.1 INTRODUCTION

Biofertilizers contain living microbes which, when associated to plant surfaces, seeds, or soil; colonize the soil region having the influences of plant roots (rhizosphere), within the plant and the development of the plant by increasing the supply or accessibility of necessary supplement to the plant to which they are in association or simply the host plant. These biofertilizers perform nitrogen obsession and make supplements accessible to the plants. They additionally solubilize the nutrients like phosphorus thereby stimulating the development of the plant all the way through the blend of these advancing substances. The use of biofertilizers has lessened the use of the manufactured pesticides and fertilizers. The microscopic organisms present in these biofertilizers assemble soil natural issue by re-establishing the soil's characteristic supplement cycle (Valarini et al., 2003). The healthy flora of soil plants can be developed with the utilization of bio manures as they improve the manageability and the soundness of the soil too. Henceforth a favored logical term utilized for such helpful microscopic organisms is "Plant-Growth-Promoting Rhizobacteria" (PGPR). In this manner, they demonstrate to be incredibly significant by improving the soil readiness thereby fulfilling plant nutrients which are prerequisites and provide inorganic nutrients via these microscopic organism and their side effects (Melero et al., 2006; Liu et al., 2010). Thus, biofertilizers can't be viewed as hurtful to the living soil as they are without any synthetic compounds.

Current agribusiness procedures incorporate utilizing hybrid seeds and high yielding assortments of plants that are very receptive to huge portions of compound manures and water system. Utilizing engineered manures unpredictably has prompted contamination and defilement of soil and water bowls. This has denied the dirt of basic supplements and natural issue. It has prompted exhaustion of advantageous small-scale life forms and creepy crawlies in a roundabout way decreasing soil richness and making crops increasingly inclined to illnesses (Chander et al., 1997). It has been evaluated that constantly 2020, to achieve the concentrated on formation of 321 million tons of sustenance grain, the prerequisite of enhancement will be 28.8 million tons, while their availability will be simply 21.6 million tons. This will bring about a shortage of about 7.2 million tons which can prompt exhaustion of feedstock/non-renewable energy sources (vitality emergency) and expanding cost of composts influencing the little and minimal ranchers. This would heighten the draining degrees of soil fruitfulness because of augmenting hole between supplement evacuation and supplies. Compound manures which have been utilized widely since the green upset have drained soil wellbeing by making the dirt environment non-inhabitable for soil smaller scale greenery and miniaturized scale fauna (Lal, 2011; Krasowicz et al., 2011). It is this vegetation which keeps up soil richness and gives fundamental and vital supplements to plants. Biofertilizers are really the items containing at least one type of microorganisms. These microorganisms can activate healthfully significant parts from non-useful into the useful forms through usual procedures, for example, nitrogen obsession, phosphate solubilization, discharge of plant development advancing substances or cellulose and biodegradation in soil, fertilizer, and different situations.

Biofertilizers give "eco-accommodating" natural agro-input. Different types of biofertilizers are being used since earlier times, e.g., *Azotobacter, Rhizobium, Azospirilium,* andblue-green algae(BGA). *Rhizobium* inoculants are used for the plants bearing nodules at their roots or simply leguminous crops. *Azotobacter,* another biofertilizer, is mainly used for different types of grains and oilseeds (wheat, maize, mustard, cotton, potato) and other vegetable yields. *Azospirillum* immunizations are suggested mostly for sorghum, millets, maize, sugarcane, and wheat. BGA, *Anabaena* or *Nostoc* or *Aulosira* or *Tolypothrix,* fixes climatic nitrogen and are utilized as vaccinations for paddy harvest which have been developed in swamp conditions as well as in uplands (Gonet and Dębska, 2006; Li et al., 2014). *Anabaena* in relationship with *Azolla* (water greenery) contributes nitrogen up to 60 kg per hectare per season and furthermore enhances soils with natural issue. Different kinds of microscopic organisms, known as phosphate-solubilizing microbes, for example, *"Pantoea agglomerans"* strain P5 or *"Pseudomonas putida"* strain P13, can solubilize the unsolvable phosphate from natural and inorganic phosphate sources. Truth is told, since immobilization of phosphate by mineral particles, for example, iron, Aluminum, and Calcium or natural acids, the amount of accessible phosphate (Pi) in the soil is quite lower than plants need for their proper growth. What's more, compound Pi composts are likewise immobilized in the dirt, and under 20% are consumed by plants. Consequently, decrease in Pi assets, on the one hand, and ecological contaminations coming about because of both creation and utilization of substance Pi manure; then again, have just requested the utilization of phosphate-solubilizing microbes or phosphate biofertilizers. Even though bio-manures have been known for a long time, generally slight study has been carried out to trace their belongings (or non-impacts) on harvest generation and to give proof of their possible consequences for properties of the soil and procedures, particularly in exceptional companion explored logical diaries (Dinesh et al., 2010; Khaliq et al., 2006; Mayer et al., 2010; Piotrowska et al., 2012). The hypothetical mechanism of action of biofertilizers in the root cells of the plants is shown in Figure 10.1.

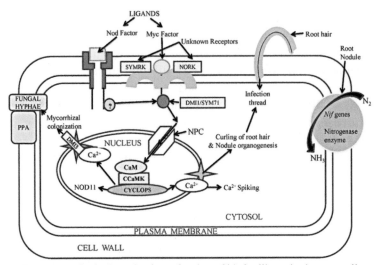

FIGURE 10.1   Hypothetical mechanism of action of biofertilizers in the root cell.

Source: Reprinted with permission fron Bhardwaj et al., 2014. © Springer Nature.

10.2   BIOFERTILIZER: THE NEED OF THE HOUR IN AGRICULTURE

At present, there is a growing concern around the world about ecological hazards and threats to sustainable farming. The long haul utilization of bio-manures has been demonstrated to be discreet, eco-accommodating, progressively effective, profitable, and open to minor and little farmers over the utilization of chemical composts. The requirement for the utilization of bio composts emerges fundamentally for two reasons. The first reason is being the utilization of manures prompts expanded yield efficiency without affecting the soil productivity or texture, and second, in light of the fact that expanded use of synthetic fertilizer leads to environmental problems.

10.3   CLASSIFICATION OF BIOFERTILIZERS

Biofertilizer production exploits the association of microorganisms with crop plants. The biofertilizers are grouped into the following different categories based on their function and nature:

1. ***Rhizobium*:** It is a soil bacterium, which colonizes vegetable roots and fixes air nitrogen beneficially for the plant. The morphology and physiology of this biofertilizer change from its free-living structure to the bacteroid. *Rhizobium* is the best biofertilizers as indicated as

long as the amount of nitrogen fixed by this is concerned. These biofertilizers join various genera and are significantly unequivocal to outline handle in vegetables, insinuated as cross vaccination gathering (Khosro and Yousef, 2012).

2. *Azotobacter*: *A. chroococcum*, among the few types of *Azotobacter*, are present in the arable lands where they are fit for fixing the atmospheric nitrogen (2–15 mg of N_2 fixed per gram of carbon source). These species produces bounteous ooze which aides in soil accumulation (Halim, 2009). The quantities of *A. chroococcum* in soils of India infrequently surpass 10^5 per gram soil because of lack of normal issue and the proximity of opposing microscopic organisms in the soil.

3. *Azospirillum*: *Azospirillum lipoferum* and *A. brasilense* (*Spirillum lipoferumin* prior writing) are vital microorganisms that live in the soils, the rhizosphere of the plant and between intercellular spaces of root cortex of graminaceous plants. They create co-operative advantageous association with graminaceous plants. Aside from nitrogen obsession, IAA generation (development advancing substance), malady opposition, and dry season resistance are a portion of the extra advantages of immunization with *Azospirillum*.

4. **Blue-Green Algae (Cyanobacteria):** Both the forms of *Cyanobacteria*, i.e., free-living as well as symbiotic, have been used in India for the production of rice. Once a great amount was spent in its advertisement as a biofertilizer for rice crop, it has not by and by pulled in the concern of rice cultivators throughout the country. However, the current approach of having BGA as a biofertilizer poses a restriction itself.

5. *Azolla*: It is a free-gliding water plant that floats in surface of water. *Azolla* fixes environmental nitrogen by means of the bacteria which fixes the nitrogen in association with the *Cyanobacteria* (*Anabaena azollae*). It is being used as fertilizer for paddy cultivation. Azolla contributes about 40–60 kg of nitrogen per hectare per rice crop.

6. **Phosphate Solubilizing Microorganisms (PSM):** A few soil microbes and growths, remarkably types of *Penicillium, Aspergillus, Pseudomonas, Bacillus*, and so forth emit natural acids thereby

minimizing the pH to achieve disintegration of undissolved phosphates in soil (Gupta, 2004). Expanded production of potato and wheat were exhibited because of vaccination of peat-based societies of *Bacillus polymyxa* and *Pseudomonas striata.*

7. **AM Fungi:** The exchange of supplements, basically phosphorus and furthermore zinc and sulfur from the dirt milleu to the phones of the root cortex is intervened by intracellular commit parasitic endosymbionts of the genera *Acaulospora, Endogone, Sclerocysts, Gigaspora,* and *Glomus* which have vesicles for capacity of supplements and arbuscles for piping these supplements into the root framework. The most common class among all seems, by all accounts, to be Glomus, which has a few animal varieties appropriated in soil.

8. **Silicate Solubilizing Bacteria (SSB):** These microorganisms are equipped for corrupting silicates and aluminum silicates. Throughout the digestion of microorganisms a few natural acids are delivered which assume a double job in silicate enduring. They provide H^+ particles to their surrounding and advance hydrolysis and the natural acids like citrus, oxalic corrosive, Keto acids and hydroxy carbolic acids which structure edifices with cations, advance their evacuation and maintenance in the medium in a disintegrated state.

9. **Plant Growth Promoting Rhizobacteria (PGPR):** The gathering of microorganisms that live in the regions of the roots or in soil rhizosphere are alluded to as plant development advancing rhizobacteria (PGPR). The PGPR inoculants advance development through concealment of plant ailment (named Bioprotectants), improved supplement procurement (named Biofertilizers), or phytohormone generation (named Bio-stimulants). Types of *Bacillus* and *Pseudomonas* can create phytohormones or development controllers that reason bountiful fine root development which thusly increment the absorptive surface of plant pulls for take-up of water and supplements. These PGPR are alluded to as bio-stimulants and the phytohormones created by them incorporate indole-acidic corrosive, gibberellins, cytokinins, and inhibitors of ethylene generation. The biofertilizers recommended for the crops are listed in Table 10.1.

TABLE 10.1 Biofertilizers Recommended for Various Crops

Biofertilizer	Amount	Crops
Rhizobium + Phosphotika	200 g each per 10 kg of seed as seed treatment.	Pigeon pea, black gram, green gram, cowpea, groundnut, and soybean
Azotobacter + Phosphotika	200 gm each per 10 kg of seed as seed treatment	Wheat, sorghum, maize, cotton, mustard
Azospirillum + Phosphotika	5 kg each/ha.	Transplanted rice

Source: https://www.bio-fit.eu/.

10.4 APPLICATION OF BIOFERTILIZERS TO THE CROPS

10.4.1 TREATMENT OF THE SEED

Every bundle (200 g) of inoculant is blend with 200 ml of rice slop or jaggery arrangement. The seeds required per hectare are blended in the slurry in order to have a uniform covering of the inoculants over the seeds. They are then shade dried for 30 minutes. The treated seeds ought to be utilized within 24 hours. One bundle of inoculant is sufficient to treat to 10 kgs of the seeds. *Rhizobium, Azospirillum, Azotobacter,* and *Phosphobacteria* is connected as a seed treatment.

10.4.2 DIPPING OF THE ROOT SEEDLING

This strategy is utilized for the crops which are transplanted. Five bundles (1.0 kg) of the inoculants are required for one ha which are blended with 40 liters of water. The root part of the seedlings is plunged in the answers for 5 to 10 minutes and afterward transplanted. Azospirillum is utilized for seedling root plunge especially for rice.

10.4.3 TREATMENT OF THE SOIL

For treating the soil, 4 kg of the prescribed biofertilizers are mixed in 200 kg of manure and kept medium-term. This blend is fused in the soil at the season of planting the crops.

10.5 VALUES OF BIOFERTILIZERS

10.5.1 COST-EFFECTIVE AND EASY APPLICATION TECHNIQUES

Biofertilizers are savvy when contrasted with substance composts. They vary from substance and natural manures since they do not legitimately provide any nutrient to crops and incorporate uncommon microorganisms and growths which have moderately low establishment price. The utilization of biofertilizers can enhance the profitability per unit region in a generally brief time period. Their assembling and use expenses are low. Their application devours littler measures of vitality. That is their lower expenses can be straightforwardly converted into gainful advantages for ranchers. In this sense, utilization of natural composts can bring profits by a monetary perspective, since biofertilizers are a financially savvy and inexhaustible wellspring of plant supplements to substitute the concoction manures for supportable horticulture.

Most generally, biofertilizers are utilized in powder, transporter based structure. The transporter more often than not is lignite. Lignite has high natural issue substance and hold over 200% moisture. The property of high water substance upgrades the development of the microbes in the biofertilizers. It comprises of granules which are 1–2 mm in size and are produced by using tank bed mud (TBC) and heated at 200°C in a stifle heater, which cleans the matter and offers porosity to the granules. The prepared granules are absorbed as suspension by the microscopic organisms in an appropriate medium. The soil granules are dried in the air 25°C under the conditions which are free from microbes. These granules contain around 10^9 microscopic organisms for every gram of granules. These granules are reasonable for field application alongside seeds. Be that as it may, the amount of biofertilizer is higher than used in seed treatment applications.

10.5.2   BIOFERTILIZERS ENHANCE 15–35% YIELD

Production of "biofertilizer" is a mechanical advancement which has the possibility to build harvest yield as well as improve soil condition and decrease generation cost. Biofertilizers can be viewed as a superior enhancement to substance composts as when these biofertilizers are connected with soil inoculants or with seeds, they duplicate and take an interest in the supplement cycle, in this manner profiting the yield profitability. Biofertilizers can possibly improve harvest produces through naturally improved supplement supplies. The biofertilizers provide nutrients to the plants. It is accounted for that biofertilizers increase crop production by 20–30% and invigorates the development of the plant. The proficiency of biofertilizer utilization to the crops is the main reason that adds to the expansion of the harvest production.

Different models demonstrate that biofertilizers decidedly influence the harvest yield. For example, Vital N®, a natural biofertilizer containing

Azospirillium, prompts broad development in foundations of yields like orchids, banana, rice, corn, garlic, and onion. *Azospirillium* produces indole-3-acetic acid (IAA), bringing about higher development yield.

Reports have demonstrated that the general execution of potato harvests is emphatically impacted by use of green fertilizers (*Cowpea* and *Crotolaria* sp.) giving 30% production enhancements. The expanded profitability esteems check the proficiency of biofertilizers in horticultural creation. Then again, physicochemical properties of the dirt are improved and ecological effects because of the drawn out utilization of concoction manures are slowly alleviated.

Moreover, 10% increments in the crop production/ha have been watched for harvests that have been treated with *"Arbuscular mycorrhizal"* (AM) organisms. It has likewise brought about expanded opposition of the plants to the activity of pathogenic microorganisms. Moreover, the joined utilization of AM with nitrogen fixing microbes or manure gives better production execution, usually double, and enriches the qualities of individual plants.

A preliminary examining the practicality of biofertilizers models dependent on local microscopic organisms from paddy harvests detailed 10% increments in production creation by utilizing the blends, from 7,625 kg per hectare to 8,500 kg per hectare. Results are higher incomes and increment efficiency by utilizing biofertilizers.

The utilization of *"Azolla-anabaena"* as a biofertilizer in paddy fields of north regions of Italy permitted acquiring production by fixing nearly about 40 kg of nitrogen per hectare within a time period of three months and confirming increments in the development rate of paddy. Moreover, higher opposition of a portion of the paddy varieties to the nearness of herbicide Propanil was confirmed.

10.5.3 BIOFERTILIZERS PROVIDE NITROGEN AND GROWTH HORMONES TO THE PLANTS

The bio-fertilizer adds nitrogen to the soil in order to make it available to the plants. These biofertilizers supply synthetic nitrogen about 25%. In this manner, nitrifying microorganisms assume a significant job in nitrogen supply by changing over climatic nitrogen into natural structures which can be utilized by the plants (Barak, 1999). Utilization of natural N_2-obsession innovation can add to abatement in the N compost application and to the decrease of ecological dangers. *Azotobacter* (free-living nitrifying bacteria) assumes a significant job in the "nitrogen cycle" in environment because of its assorted metabolic potential. This microorganism can likewise combine and discharge impressive measures of organically dynamic substance, among which the nutrients riboflavin, thiamine, pantothenic acid, biotin, and nicotinic acid; the plant-development hormones heteroxins, gibberellins. These naturally dynamic nutrients assist in alteration of the supplement take-up by the plants. *Azospirillum,* another free-living nitrifying bacterium, produces indole acetic acid (IAA) which is a plant-development advancing substances and indole butyric acid (IBA) which enhances the pace of nutrient take-up by the roots of the plants, bringing about the improvement of plant produce.

It is outstanding that most plants structure cooperative in relationship with the *"Arbuscular mycorrhizal fungi"* (AMF) and thereby acting as bio-ameliorators. These AMF can possibly impressively upgrade the rhizospheric soil attributes. This, thus, prompts improved soil structure and advances plant development under ordinary just as focused on conditions. Studies have uncovered that AMF-incited upgrade in supplement take-up advances different naturally significant nutrients. Some of the significant among them are the hormones of the plants which include auxins and gibberlic acid, that assume a novel job in development of plant under both normal and abnormal conditions (stress conditions). Plants vaccinated with AMF additionally have high action of cytokinins and IAA which result in enhanced development and advancement of the plant.

10.5.4   BIOFERTILIZERS ENHANCE SOIL FERTILITY AND ARE POLLUTION FREE

The utilization of the "biofertilizers" isn't just practical; it additionally enlarges the issue of ecological contamination. They are ecologically cordial on the grounds that their utilization counteracts harming the regular assets as well as liberates the plants of accelerated synthetic composts. Biofertilizers advance the decrease of natural effects related with the over the top utilization of concoction preparation. Consequently, their utilization in natural cultivating, feasible agribusiness, green cultivating, and non-contamination cultivating add to usage of sound condition strategies at national, provincial, and worldwide level.

A wide range of harvests developed in various agro-ecologies can profit by the utilization of these biofertilizers. Persistent utilization of bio-fertilizers empowers the microorganisms to develop in the soil and aides in keeping up soil fruitfulness adding to economical agribusiness.

Biofertilizers keep the dirt condition wealthy in a wide range of miniaturized scale and full-scale supplements by means of different procedures as nitrogen obsession, potassium, and phosphate solubilization or mineralization, arrival of plant-development directing substances, generation of antimicrobials and bio-degradation of natural issue in the soil. Utilization of biofertilizers is profitable in shielding the dirt from corruption. Biofertilizers can prepare supplements that support the improvement of organic exercises in soils. Along these lines, they anticipate small-scale supplement lacks in plants and assurance better supplement take-up and expanded resilience to dampness stress and dry season, everything that emphatically add to richness of the soil.

10.5.5 ANTIBIOTIC SECRETION AND PESTICIDE

The biofertilizer use can have ill effects and can act as a biological control for various disease-causing organisms of plants. Hence, they can prove beneficial to the microbiology of the soil by suppressing these disease-causing organisms or by competing out the harmful microorganisms which are present in the soil.

The strategies that are adopted to control the various species of the fungus through biological controls consist of using the biologically digested products from these in order to control the disease-causing organisms and the pests that are the targets of humans. As the biofertilizers produce antibiotics and siderophores, they may prove harmful to the fungi, foliar, insects, and even to the disease-causing bacteria that reside in the rhizosphere of the roots.

Arbuscular mycorrhizal parasites (AMF) can possibly decrease harm brought about by soil-borne disease-causing organisms, nematodes, and microorganisms. Meta-investigation has demonstrated that AMF by and large declines the impacts of parasitic pathogens. Various mechanisms have been put forward, which explain that mycorrhizal fungi play a significant role. Since the plants with good phosphorous content are often less sensitive to disease-causing pathogens, hence nutritional aspect plays an important role. The mechanisms other than nutritional also play an important role as it has been found that non-mycorrhizal and mycorrhizal plants that have the same amount of indigenous concentration of phosphorous are affected by the disease-causing pathogens differentially. These various non-nutritional mechanisms of the plants include activation of its defense system, patterns of exudation and associated variations in populations of mycor-rhizosphere, increase in the cell wall lignifications and the struggle for colony formation and sites of infection.

Besides the mycorrhizal species, various endophytes of the fungi such as *Sebacinales* (*Basidiomycota*, with *Piriformospora indica* as a model organism) and *Trichoderma* spp. (*Ascomycota*) have attracted the attention of the scientific community during recent times. These fungal endophytes have the ability to complete a part of their life cycle apart from the host plant, the colonization of their roots, and providing nutrients to their host plant by the mechanisms that are not yet clear. Nowadays, these fungi are getting attention from the scientists as they can multiply easily *in vitro* conditions through plant inoculants and act as experimental models to find out the mechanisms how the nutrients are transferred from these endo-symbionts of the fungi to their host plants.

The endophytes of the fungi like *Trichoderma* spp. are being widely studied for their bio-pesticidal (mycoparasitic) and bio-control (inducer of disease resistance) potential. These have also been studied extensively as enzyme sources by the industrial biotechnologists. Presently, based on evidences, it is hypothesized that *Trichoderma* spp. also brings about various responses in the plants. Some of the important responses among them include the abiotic stress tolerance, efficiency of the use of nutrients, morphogenesis, and organic growth. Based on the above-mentioned effects, endophytes of the fungi might be considered not only as bio-stimulants but also as bio-pesticides.

10.5.6 BIOFERTILIZERS IMPROVE THE CHEMICAL AND PHYSICAL PROPERTIES OF THE SOIL

The maintenance of good structure and improved stability of soil stimulates and promotes root growth. Biofertilizers aid in achieving the good physical condition of the soil by improving the constitution and aggregation of particles of the soil, minimizing compaction and increasing the spaces between the pores, soil aeration, water penetration and help in reducing the soil erosion thus allowing better tilth. Moreover, they serve as a major source of the food for microorganisms and help to keep the soil alive. Biofertilizers also help in the physicochemical stability of soils through replenishing the nutrients present within soil, keeping the elements free in order to facilitate their uptake by the root system and enhanced capacity of exchange of these nutrients in the soil.

The formation of soil structure and its maintenance gets influenced by various properties of the soil, architecture of the roots and practices of the management. The excessive use of fertilizers and machines are mainly liable for the deprivation of the soil structure. In all ecosystems, mycorrhizal fungi are largely responsible for the maintenance of good structure of the soil. The *mycorrhizal fungi* help in maintenance of high-quality structure of the soil by means of the processes described below:

- The hyphae that grows externally within the soil forms a skeletal structure that helps in holding the particles of the soil together;

- External hyphae produce a favorable environment that is fit for formation and enlargement of micro-aggregates;

- Direct tapping of resources of carbon of plant to the soil. Thereby stimulates soil aggregates formation; the carbon present in the soil is vital to form organic ingredients which are important to keep the soil particles cemented.

The hyphae of AM fungi have longer residence time in soil and hence are very essential for the above phenomenon than the "hyphae of sapro-trophic fungi." Additionally, AM parasites produce glomalin about 12–45 mg/cm^3. Glomalin is a particular protein present in the soil which is having an obscure bio-chemical nature. It has a more extended living arrangement time period in the soil than has hyphae, taking into consideration an extended steady commitment to soil total adjustment. The time of habitation for the hyphae is changed considerably from a few days even to months. The habituation time for glomalin changes from 6 to 42 years. The glomalin steadily stick the hyphae into the soil. The component is the arrangement of a 'sticky' string-sack of hyphae that prompts the soundness of totals.

10.5.7   BIOFERTILIZERS ENHANCE PRODUCTION DURING POOR IRRIGATION

The bio-fertilizers improve the soil's moisture and holding the nutrient ability and also boost soil drainage and moisture absorption, particularly in those with nutrient deficiencies structural deficiencies. They improve tolerance to drought and stress caused by excessive moisture. In this manner, even in plantations lacking adequate natural water supply or irrigation, the crop yield improves. AM association, for example, increases the conductivity of water from the roots even at reduced soil water potential. This enhancement is one among the other variables that contribute to better absorption of water by the plant. The leaf wilting disease that occurs due to drying of the soil has not been reported in plants that are in association with mycorrhiza unless the potential of the soil water is significantly low. The drought tolerance induced by Mycorrhiza may be related with colonization of AM factors such as enhanced leaf water and turgor potential and stomatal function and maintenance of transpiration, enhanced hydraulic conductivity, and enhanced length and growth of the roots.

10.5.8   BIOFERTILIZERS ARE ECO-FRIENDLY

A significant decrease in environmental pollution and enhancement of agro-ecological reliability is the most significant and contributing function of biofertilizers. Compared to chemical fertilizers, biofertilizers are eco-friendly organic agro-inputs. They do not damage biodiversity and are environmentally useful as they allow decreased use of chemical fertilizers in crop production globally. Because of their eco-friendly characteristics, the demand of the bio-fertilizers has risen over the past decade. Their operations affect the soil ecosystem and provide the plants with additional substances. Biofertilizers improve health of the plants and contribute to the ecology of soil by providing constant supply of balanced micronutrients to crops and eliminating plant disease. The supply of food and the stimulated growth of useful microscopic organisms contribute to maintenance of the biological equilibrium. Eventually, biofertilizers are scheduled to supplement and replace standard chemical fertilizers, leading to financial and environmental advantages.

10.6   CONSTRAINTS IN BIOFERTILIZER PRODUCTION TECHNOLOGY

The unpredictability of the results after application is a significant feature common to most biofertilizers. The output depends on the manufacturing technology of the biofertilizer and thus consistent improvement in the output of biofertilizers is of essential significance.

While bio-fertilizer machinery is a cost effective and environmentally friendly technology, its application, or execution is restricted by several limitations. These limitations are unawareness regarding the quality of technology, infrastructure, finance, environment, human resources. The various limitations influence biofertilizer manufacturing technology, marketing, and use.

10.6.1 TECHNOLOGICAL CONSTRAINTS

Despite substantial development in biofertilizer innovation throughout the years, the advancement in the field of biofertilizer generation innovation isn't satisfactory. Mechanical requirements looked by both natural and customary ranchers in selection of natural cultivating practices are centered around the accompanying perspectives.

10.6.2 STRAINS FOR PRODUCTION

Using improper, less effective strains for biofertilizer manufacturing can result in inadequate microorganism population and is an important constraint. One of the main limitations is the lack of region-specific strains, because the bio-fertilizers are not merely particular to crops, but are also specific to the soil.

In addition, the strains that are chosen must be compatible and have the capacity to survive in broth and inoculant carriers over other strains in a variety of ecological circumstances. The elevated amount of contaminants may be another issue. Therefore, an excellent bio-fertilizer should have a better efficient strain in a suitable population and must be free from other disease-causing microbes. Moreover, biofertilizers cannot be used in challenging soil (acidic, saline, and alkaline) owing to decreased effectiveness. During times of high temperature, the implementation of bio-fertilizers is also unsuccessful. The biofertilizers are not applied to the soils with unfavorable phosphorous. Lastly, the biofertilizers get mutated during the process of fermentation, hence raising the produce and cost for quality control. In order to eliminate these undesirable changes, wide research is the need of the hour.

10.6.3 TECHNICAL PERSONNEL

Technically unqualified, inexperienced, and inadequate staff contributes to technological problems with the bio-fertilizer industry. The deficiency of data related to this technique and expertise on the use of bio-fertilizers is a major high-intensity limitation, as farmers are not offered adequate guidance on the implementation elements. Poor organization of the execution procedure and absence of free time when these biofertilizers are applied during sowing season; absence of understanding of technology of inoculation by additional staff and farmers is another major issue. Most marketing sales staff is not familiar with adequate inoculation methods. Biofertilizers, basically living organisms, require adequate equipment for handling, transportation, and storage.

10.6.4   QUALITY PRODUCTION UNITS

The deficiency of skilled scientific staff in manufacturing unit may result in wrong handling and manipulation during the process of manufacturing.

10.6.5   QUALITY OF CARRIER MATERIAL

The effective biofertilizer implementation can be hampered due to non-availability of carrier material that is of excellent quality or the use of distinct carrier components by different manufacturers without understanding the worth of the products. The unavailability of an appropriate carrier in which bacteria can be multiplied is a significant cause for restriction of biofertilizers shelf life. Selection of a carrier material must be done on account of the accessibility and price at the manufacturing site. A good quality carrier must have an excellent ability to hold moisture, be free of poisonous substances, and be sterilizable and easily adjustable to pH 6.5–7.0. Under the unfavorable climatic circumstances when both the weather extremes soil prevails, no appropriate carrier material has yet been recognized to support biofertilizer development. Better bacterial growth is achieved in the sterile carrier and gamma irradiation is the best technique of sterilization.

The microbes are having a shelf life of just 6 months in the carrier-based biofertilizers. UV rays and temperatures above 30°C are not tolerated by the microbes. These microbes 'population density at the moment of manufacturing is only 10^8 cfu/ml. This count diminishes every day. Therefore, biofertilizers based on carriers prove not to be efficient and common amongst the farmers. The probable steps to alleviate the above-mentioned drawbacks include the use of carriers that are sterile and the installation of a centralized unit of sterilizing equipment; selection of common carrier products based on accessibility and recommendation to manufacturers in distinct nations.

The so-called "liquid biofertilizers" are the option. Liquid bio-fertilizers are distinct liquid formulations that contain not only the preferred microbes with their nutrients, but also the particular protective cells or chemical substances that support the formation of cysts or resting spores for extended the shelf-life and tolerance towards extreme circumstances. The microbes present in this type of biofertilizer have a shelf life of only two years with a number of 10^9 cfu/ml and this count is kept constant. These can tolerate not only the high temperature (55°C) but also can tolerate ultraviolet radiations. Being fluid formulations, they are very simple and easy to apply in the field. They are used with power sprayers, hand sprayers and fertigation tanks, etc. To develop an appropriate alternative formulations, i.e., liquid inoculants or granular formulations for all bio-inoculants, the new formulations require standardization of the media, inoculation method, etc.

10.6.6 INOCULANT QUALITY

Manufacture of inoculants lacking the knowledge of fundamental techniques of microbiology threatens the good quality of the inoculants, hence the effectiveness of the inoculants. If the coat of the seeds is removed due to the scratching with the biofertilizer solution can lead to reduced germination. Insufficient product formulation may be a severe obstacle to the marketing of these biofertilizers. Therefore, for the good quality input of the biofertilizer, it is need to have improvement in their innovation. The following factors must be taken into account when formulating high-quality inoculants:

- The inoculants should be first identified according to the location of the crop for the particular soil that must fix nitrogen, phosphorous, and zinc solubilizing and absorbing according to distinct climatic circumstances.

- Application of suitable techniques biotechnology for enhancement of the strains;

- Exchange of culture between nations with comparable climate; and

- Assessing their output for a specific crop for the better strains;

- In order to prevent the culture from mutants, monitoring of the culture should be done.

10.6.7 INOCULANT SHELF-LIFE

Efficient storage is required for a brief shelf life (generally 6 months). This brief shelf life usually discourages ventures from generating a large quantity of the inoculants that they sell in the markets. The other factor being that farmers also cannot purchase large quantities of the same as they cannot store them for long periods. The countries which import most of their biofertilizers face the problems of shelf life and storage settings due to poor adaptation to local environmental conditions. For example, for nations where temperatures are generally quite high, the bio-fertilizers which need cold storage for an expanded shelf life usually are not appropriate. It is, therefore, not amazing that these products shall not fulfill the standards of quality, likely due to the loss of viability in the inappropriate circumstances of storage. Hence, taking into account the product shelf life, the product formulation is critical with regard to the handling and storage circumstances.

The issues in growth of biofertilizer industry are generally related with their low demand because of the absence of knowledge and comprehension of these bio-fertilizers. The production continues to be a challenge in many instances, not only due to the price, but in addition to the limited demand and mediocre distribution systems that might be connected with the specific prerequisite storage and handling conditions. The shelf life of the item, the quality of the carrier equipment, the storage circumstances, handling (e.g., transport), and the existence of contaminants influence the efficiency of the sector and, consequently, the rate of adoption. In order to guarantee product viability over a substantial period of time, it is therefore essential to enhance the shelf life of biofertilizers that is formulated locally under different storage circumstances.

10.7   INFRASTRUCTURAL CONSTRAINTS

10.7.1   PRODUCTION FACILITIES

A significant constraint of infrastructural is the non-availability of appropriate manufacturing equipment. Furthermore, insufficient accessibility of inputs and input unavailability at a proper point in time pose another issue. An excellent strategy in manufacturing units is to employ the skill of microbiologists monitoring the production and to develop facilities for cold storage in manufacturing centers. Biofertilization suffers from insufficient marketing equipment and lack of periodic data on biofertilizer use, which imposes danger and uncertainty among the farmers.

10.7.2   EQUIPMENT FACILITIES

This deficiency of vital machinery, supply of power, etc., leads to increased labor, as the method of manufacturing is slow and time-consuming in this situation.

10.7.3   STORAGE SPACE, PRODUCTION, AND LABORATORY USED FOR BIOFERTILIZERS

The production laboratory, manufacturing, storing of biofertilizer, etc., space accessibility is incredibly essential. In order to enlarge the manufacturing of biofertilizers, there is a need for additional land to grow, for instance, green manure plants. Other significant problems are the absence of subsidy provision and trade in biofertilizers at a fair cost. However, growing demand for biofertilizers and farmers knowledge of using them has facilitate the production of biofertilizers and entrepreneurs should be encouraged to enter the biofertilizers production.

10.7.4   LACK OF STORAGE FACILITIES FOR INOCULANT

The shortage of cold storage facilities for packets of inoculant is the threat for the production of quality biofertilizers as these must be kept in cool places that must be far from warm winds and direct heat of sun. The insufficient facilities for storage can expose elevated temperatures to biofertilizers, which are unfavorable circumstances.

10.8   FINANCIAL CONSTRAINTS

10.8.1   INADEQUATE FUNDS

Inadequate funds and issues in obtaining bank loans are among major problems. The use and cost of inorganic fertilizers is constantly growing but the effectiveness of their use remains small and the pressure on their implementation comes from regulatory/environmental issues. Alternatively, renewable biofertilizers give high effectiveness of use, comparatively low cost and minimal effect on the environment. Their funding is getting better at the moment.

10.8.2   SALE RETURNS

By selling products in smaller manufacturing units, the biofertilizer sector is susceptible to lower yields. This is a significant issue that needs to be addressed as the organization and operation of big manufacturing installations are multifaceted owing to the science, financial, environmental, and social issues that need to be addressed.

10.9   PHYSICAL AND ENVIRONMENTAL ISSUES

10.9.1   SEASONAL DEMAND

The demand for bio-fertilizers is of a seasonal nature, and hence they are required in a particular season and as a result the manufacturing and supply of biofertilizers takes place for 2–3 months in a year. The manufacturing of bio-fertilizer have a challenge in designing enhanced formulations that are adapted circumstances of a specific area and delivering the biofertilizers in a manner that meets the regional and seasonal changes in responses of crops. Therefore, a comprehensive study of the technology is needed to create formulations that can meet these demands. The manufacturers will be unable to take advantage of the complete biofertilizer potential without such studies.

10.9.2   CROPPING OPERATIONS

Application of biofertilizers is usually dependent on other crop operations that require concurrent activity. It is also necessary to consider the brief sowing span or transferring the seedlings to a specific location. Biofertilizers ought to be implemented according to a suggested technique at suitable doses. Any use of bad quality adhesives and high doses of chemicals that protect the plant will reduce the efficiency of the implementation of biofertilizer.

10.9.3 SOIL CHARACTERISTICS

Soil characteristics *viz.*, drought, salinity, waterlogging, acidity, etc., are of essential significance. The soils having temperature or low moisture content, high acidity or alkalinity, scarce Phosphorus and Molybdenum availability, and elevated indigenous population or bacteriophageal presence should all be considered as they impact development of microbes and, in turn, crop production. For example, biofertilizer field performance, e.g., *Rhizobium* inoculants, is impacted not only by genotype of crop plant and strains of the microbes, but as well by ecological circumstances (i.e., climate and soil) and agronomic management.

10.10 QUALITY CONSTRAINTS AND HUMAN RESOURCES

10.10.1 COMPETENCE OF THE STAFF

The production and implementation of biofertilizers can be compromised by inadequate human, economic, and material resources. A severe issue is the lack of technically skilled personnel in the manufacturing units. This constraint is directly related to the absence of adequate training and technical skills for biofertilizer manufacturing. By improving the ability of the humans and technical handling of biofertilizer, the good quality has also been recognized as crucial for proper marketing of biofertilizer. Consequently, it appears vital to support public policies to guarantee that only good featured biofertilizers are sold with authorization.

10.10.2 EDUCATIONAL AND TRAINING IN BIOFERTILIZERS

Overall, the primary issue is the absence of adequate organic farming training and insufficient understanding of organic farming functionaries in the sector. In addition, lack of adequate training in manufacturing techniques and abilities on enhanced biofertilizer manufacturing methods; lack of knowledge of the concentration, time period, and process of implementation of biofertilizers; absence of information on various pesticides are other significant problems that need to be taken into account in terms of humans and limitations in their quality. Technical training for producers on manufacturing and quality control; providing manufacturers with technical guidance and initiatives; organizational training for extension employees and farmers in order to popularize the technology; arranging for improved and broad dissemination of data are the steps that must be taken into account.

10.10.3 PRODUCTION TECHNIQUES

The most significant problems occur owing to the manufacturer's ignorance of product quality due to the absence of quality specifications and requirements from both manufacturing and consumer leadership.

Government assistance for biofertilizer manufacturing and use can lead to promising outcomes. Thus, through public assistance, different Asian nations have accomplished enhanced use of biofertilizers. In Thailand, for instance, biofertilizer manufacturing and use increased dramatically as a consequence of the Ministry of Agriculture's assistance for the industry. In India, a comparable public initiative has been noted.

Many nations have required domestic biotechnology organizations to tackle the problems of biosafety to guarantee that products are secure for humans, environment crops, and the livestock, even as establishing an enable innovation environment. The investment trends in the manufacturing of biofertilizers show favorable outcomes. However, the manufacturing resource generation is very restricted considering the danger that is posed by the brief shelf life and the absence of warranty of biofertilizer off-take.

10.10.4 QUALITY SPECIFICATIONS AND QUICK QUALITY CONTROL METHODS

To ensure compliance with prescribed norms, product safety and efficacy, quality control and regulation of biofertilizers is essential. Selling biofertilizers of bad quality through the black marketing methods results in faith loss among the farmers. The low quality of biofertilizers must be anticipated on the market if the framework is not well defined for quality control, leading in bad results on the ground. It is essential that companies adhere to established quality standards to guarantee that only appropriate quality products are permitted on the market. Recurrent product surveillance on the market is essential to guarantee the quality of the product in the entire marketing chain.

An evaluation of biofertilizer products disclosed that owing to the lack of active constituents or the existence of pollutants, a large amount of product formulations did not match product labels. Implementing quality standards could lead to considerably mitigating this limitation. The process of approval should be facilitated in order to have good quality of the biofertilizers. The lack of quality regulations and rapid quality control techniques is the reason why the manufacturing and requirements of biofertilizers are susceptible to compromise. The first commercially manufactured inoculant was produced in 1952 in South Africa. However, owing to bad quality goods on the market, an autonomous quality control scheme was implemented in the 1970s to guarantee that the inoculants generated in other nations could match the highest quality goods.

Inter-country quality standards could promote provincial buys and sells of the biofertilizers. One strategy is to bring into line norms with those that use biofertilizers in nations with important history, including Australia, Canada, France, India, New Zealand, and South Africa. This will improve customer security while facilitating cross-border trade. In these countries, for example, inoculants based on *rhizobium* should have at least 5×10^7–10^9 colony forming units (cfu) of the active microorganism strain per gram of the product of biofertilizer. In the meantime, at 10^5 dilutions, there should be no contaminants. Biofertilizer sector self-regulation has been developed in New Zealand, the USA, Thailand, Canada, China, Australia, as well as most EU nations. The sector is paying for quality control here. The state plays a part in controlling the quality of biofertilizers in nations such as Canada, France, and Uruguay. In France, for example, in spite of the long history of bio-fertilizer utilization in agricultural manufacturing, producers are still needed to produce adequate information to promote novel products 'quality, effectiveness, and security.

10.10.4.1   REGULATION

One of the biggest contributors to low availability and product acceptance is the lack of efficient regulation on biofertilizers. Research aimed at improving the agricultural use of these bio-fertilizers is frequently interrupted by lack of consciousness, humans, and infrastructure, in addition to the lack of a legislative and framework of policy that supports them. Because of insufficient policy and regulatory framework, the prospective advantages of biofertilizers can stay mainly unexploited. The poor regulation may lead to low demand for biofertilizers.

The potential of using biofertilizers can be considerably revealed by effective regulatory settings. In order to grow the market for the biofertilizers and thereby enhancing their trade, it is the need of the hour to have efficient regulations for the improving their quality and to discard the ones with poor quality. Lack of adequate legislative structure on product quality contributes to bad facilitation of biofertilizer manufacturing, delivery, and use.

The challenging processes for registering fresh products are another barrier in the use of biofertilizers. Poor management of the registration process of chemical fertilizers and other nutrients like biofertilizers can increase barriers to development and restrict access to new products that would otherwise enhance the competitiveness of farmers. Most of the European Union, North America, and some nations in Asia have put in place adequate laws to regulate such challenges and generate a constructive company scenario for biofertilizers.

However, in many countries, there are not the guidelines from the government that should be made available for use, leading to difficulties in introducing new products for biofertilizers on the market. A common framework is needed that covers laws, policies, standards, institutional arrangements, and regulations to ensure the biofertilizer industry's prospects. The key issues that a structure will include:

- Insufficient or incomplete biofertilizer and biopesticide regulatory policies and guidelines;

- Multiple legislative mandates, often overlapping, by accountable officials;

- Limited capability including personnel, abilities, and product surveillance laboratory;

- Inadequate implementation of biofertilizer and biopesticide quality control;

- Lack of legislation, norms, and instructions specific to biofertilizers and biopesticides;

- Weak institutional arrangements with restricted cooperation between the authorities concerned.

10.10.5   TIPS TO GET GOOD RESPONSE TO BIOFERTILIZER APPLICATION

- Biofertilizer item should contain great compelling strain in fitting populace and ought to be free from pathogenic microbes.

- Selection of suitable mixture of bio-fertilizers and they should be used before their expiry date.

- Use recommended technique for application and apply at a suitable time according to the data gave on the mark.

- For the treatment of the seeds, sufficient glue ought to be utilized for better outcomes.

- For tricky soils utilize restorative techniques like lime or gypsum pelleting of seeds or remedy of soil pH by utilization of lime.

- One should make sure to provide of phosphorus and different supplements.

10.10.6 PRECAUTIONS TO TAKE WHILE USING BIOFERTILIZERS

- Biofertilizer bundles should be put away in cool and dry spot far from direct daylight and warmth.

- Right mixes of biofertilizers must be utilized.

- As Rhizobium is crop explicit, one should use for the predefined crop as it were. Do not mix other chemicals along with the biofertilizer.

- Other synthetics ought not to be blended with the biofertilizers.

- While acquiring one ought to guarantee that every bundle is given essential data like name of the item, name of the yield for which proposed, name, and address of the maker, date of assembling, date of expiry, group number and directions for use.

- The parcel must be utilized before its expiry, just for the predetermined yield and by the suggested strategy for application.

- Biofertilizers are live item and require care in the capacity.

- Both nitrogenous and phosphatic biofertilizers are to be utilized to get the best outcomes.

- It is critical to utilize biofertilizers alongside substance composts and natural excrements. Biofertilizers are not substitution of manures but rather can enhance plant supplement necessities.

10.11   ADVANTAGES OF USING BIOFERTILIZERS

Some of the advantages associated with biofertilizers include:

- They are eco-friendly just as financially effective.

- Their use prompts soil advancement and the nature of the dirt improves with time.

- Though they don't demonstrate prompt outcomes, however the outcomes appeared after some time are fabulous.

- These manures tackle barometrical nitrogen and make it straightforwardly accessible to the plants.

- They increment the phosphorous substance of the dirt by solubilizing and discharging inaccessible phosphorous.

- Biofertilizers improve root multiplication because of the arrival of development advancing hormones.

- Microorganism changes over complex supplements into straightforward supplements for the accessibility of the plants.

- Biofertilizer contains microorganisms which advance the sufficient supply of supplements to the host plants and guarantee their legitimate improvement of development and guideline in their physiology.

- The biofertilizers increase the harvest yield by 10–25%.

- Biofertilizers can likewise shield plants from soil conceived illnesses to a specific degree.

10.12   CONCLUSION

Biofertilizers being fundamental parts of natural cultivating assume an essential job in keeping up long haul soil richness and manageability by fixing barometrical di-nitrogen, assembling fixed full scale and smaller scale supplements in the dirt into structures accessible to plants. Right now, there is a hole of 10 million tons of plant supplements between the evacuation of yields and supply through synthetic manures. In the setting of both the expense and natural effect of substance composts, exorbitant dependence on compound manures isn't practicable over the long haul on account of the cost, both in household assets and remote trade engaged with setting up of manure plants and continuing the creation. In this unique circumstance, biofertilizers would be the feasible choice for farmers to build efficiency per unit area.

KEYWORDS

- **biofertilizers**
- **blue-green algae**
- **control methods**
- **nitrogen fixation**
- **plant-growth-promoting rhizobacteria**
- **rhizobium**

REFERENCES

Barak, (1999). *Essential Elements for Plant's Growth* (pp. 1–5). Published by Nature Publishers.

Bhardwaj, D., Ansari, M. W., Sahoo, R. K., & Tuteja, N., (2014). Biofertilizers function as key player in sustainable agriculture by improving soil fertility, plant tolerance and crop productivity. *Microbial Cell Factories, 13*, 66.

Chander, K., Goyal, S., Mundra, M. C., & Kapoor, K. K., (1997). Organic matter, microbial biomass and enzyme activity of soils under different crop rotation in the tropics. *Biol. Fertil. Soils, 24*, 306–310.

Dinesh, R., Srinivasan, V., Hamza, S., & Manjusha, A., (2010). Short-term incorporation of organic manures and biofertilizers influences biochemical and microbial characteristics of soils under an annual crop [Turmeric (*Curcuma longa* L.)]. *Bioresour. Technol., 101*, 4697– 4702.

Gonet, S. S., & Dębska, B., (2006). Dissolved organic carbon and dissolved nitrogen in soil under different fertilization treatments. *Plant Soil Environ., 52*(2), 55–63.

Gupta, A. K., (2004). *The Complete Technology Book on Biofertilizer and Organic Farming* (pp. 242–253). National Institute of industrial research press India.

Halim, A. N. B., (2009). *Effects of Using Enhanced Bio-Fertilizer Containing N-Fixer Bacteria on Patchouli Growth* (p. 145). Thesis Faculty of Chemical and Natural Resources, Engineering University Malaysia Pahang.

Khaliq, A., Kaleem, A. M., & Hussain, T., (2006). Effects of integrated use of organic and inorganic nutrient sources with effective micro-organ isms (EM) on seed cotton yield in Pakistan. *Bioresour. Technol., 97*, 967–972.

Khosro, M., & Yousef, S., (2012). Bacterial bio-fertilizers for sustainable crop production: A review *APRN Journal of Agricultural and Biological Science, 7*(5), 237–308.

Krasowicz, S., Oleszek, W., Horabik, J., Debicki, R., Jankowiak, J., Styczyński, T., & Jadczyszyn, J., (2011). Rational management of the soil environment in Poland. *Pol. J. Agron., 7*, 43–58.

Lal, R., (2011). Sequestering carbon in soils of agro-ecosystems. *Food Policy, 36*, 533–539.

Li, W., Pan, K. W., Wu, N., Wang, J. C., Wang, Y. J., & Zhang, L., (2014). Effect of litter type on soil microbial parameters and dissolved organic carbon in a laboratory microcosm. *Plant Soil Environ., 60*(4), 170–176.

Liu, E., Yan, C., Mei, X., He, W., Bing, S. H., Ding, L., Liu, Q., Liu, S., & Fan, T., (2010). Long-term effects of chemical fertilizers, straw, and manure on soil chemical and biological properties in northwest China. *Geoderma, 158*, 173–180.

Mayer, J., Scheid, S., Widmer, F., Fliebach, A., & Oberholzer, H. R., (2010). How effective are 'Effective microorganisms (EM)'? Results from a field study in temperate climate. *Appl. Soil Ecol., 46*, 230–239.

Melero, S., Ruiz, P. J. C., Herencia, J. F., & Madejón, E., (2006). Chemical and biochemical properties in a silty loam soil under conventional and organic management. *Soil Till. Res., 90*, 162–170.

Piotrowska, A., Długosz, J., Zamorski, R., & Bogdanowicz, P., (2012). Changes in some biological and chemical properties of an arable soil treated with the microbial biofertilizer UG max. *Pol. J. Environ. Stud., 21*(2), 455–463.

Valarini, P. J., Díaz, M. C., Gasco, J. M., Guerrero, F., & Tokeshi, H., (2003). Assessment of soil properties by organic matter and EM-Microorganisms incorporation. *R. Bras. Ci. Solo., 27*, 519–525.

CHAPTER 11

Alternatives to Synthetic Fertilizers

MUHAMMAD IJAZ,[1] IJAZ HUSSAIN,[1] MUHAMMAD TAHIR,[2]
MUHAMMAD SHAHID,[3] SAMI UL-ALLAH,[1] MOHSIN ZAFAR,[4]
IQRA RASHEED,[1] and AHMAD NAWAZ[1]

[1] Bahauddin Zakariya University, Bahadur Sub-Campus,
College of Agriculture, Layyah, Pakistan

[2] Department of Environmental Sciences, COMSATS University Islamabad,
Vehari Campus, Pakistan, E-mail: muhammad_tahir@ciitvehari.edu.pk

[3] Department of Bioinformatics and Biotechnology,
Government College University, Faisalabad – 38000, Pakistan

[4] Department of Soil and Environmental Sciences, University of the
Poonch Rawalakot, Azad – 12350, Jammu and Kashmir, Pakistan

ABSTRACT

A gradual decline in soil health due to the permanent use of artificial fertilizers affects humans, animals, and plants. Nitrogenous fertilizers cause diseases in human beings and animals by polluting groundwater and leach down to the root zone. Therefore, due to the pollution problems of synthetic fertilizers in the atmosphere and soil fertility reduction, farmers are getting interested in the use of natural fertilizers like manures, biochar, and composts.

11.1 INTRODUCTION

Organic fertilizers perform vital functions that cannot be performed by man-made artificial fertilizers. For example, the physical structure of soil can be improved by organic fertilizers through which more air gets into plant roots, which increase microbial activities in soil. Huge amounts of nutrients are also provided by organic fertilizers with substantial improvement in chemical properties of the soil (Chrispaul et al., 2010). Organic fertilizers do not leach down from soil and therefore add less to water pollution as compared to synthetic fertilizers. Several farmers have adopted extensive use of organic sources of nutrient for crop production such as green manures, animal dung, composted materials and domestic waste due to high prices of synthetic fertilizers (Fening et al., 2010).

The main problem in waste products recycling is the slow release of organic nutrients from decomposition of organic products (Fening et al., 2010). Therefore, it was suggested that the efficiency of organic systems can be increased by microorganisms (Dobereiner, 1994). This is because the microorganisms can create suitable conditions, provide mutual support to soil and plant and make it possible to compete with injurious pathogens, and promotes beneficial microbial fauna (Doran, 1995).

In an intensive agriculture system, the application of organic fertilizers is not beneficial because it minimizes crop production and creates inequality of nutrients (Ayoola and Adeniyan, 2008). Therefore, the repeated application of synthetic fertilizers results in degradation of organic matter as in continuous cropping (Agboola and Omueti, 1982). The reason behind this is that the response of the crop to fertilizers application depends upon organic matter present in soil (Bhattacharyya et al., 2009). The gradual increase in soil organic matter and nutrients supply are the major functions of organic resources in soil (Bhattacharyya et al., 2009). The role of organic fertilizers for quality plant yield is depicted in Figure 11.1.

11.2 BIOCHAR

Biochar is porous and insubstantial charcoal produced by pyrolysis (Intani et al., 2016). In the process of pyrolysis, the biochar is prepared by heating up of the biomass in an oxygen limited environment at 350°C to 600°C (Sohi et al., 2010). For example, the physical and chemical properties of biochar are analogous to charcoal typified by high surface area, less concentrations of nutrients, high amounts of carbon and cation exchange capacity in comparison with unheated biomass (Singh et al., 2010a). Woody materials such as straw and manures are included in feedstock (Intani et al., 2016). The process of pyrolysis produces wide range of products which are similar to biochar but may differ in both high temperatures and extent. The soil fertility, nutrients retention, carbon sequestration, and increase in microbial biomass can be improved by applications of biochar despite of these differences (Lehmann et al., 2011). The amendments of biochar can immobilize

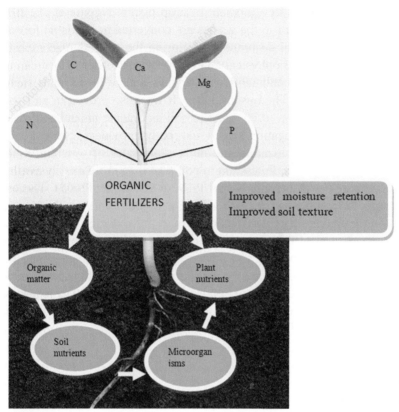

FIGURE 11.1   Role of organic fertilizers in soil fertility.

copper and also reduces its availability to micro and macro-organisms (Buss et al., 2012).

Biochar application improves the soil properties; soil quality, raises pH, water holding and, cation exchange and retention of soil nutrients (Rodriguez et al., 2009; Zheng et al., 2010; Duku et al., 2011; Paz-Ferreiro et al., 2014). Hence, application of biochar makes a potential sink for carbon and as well as due to its superior nutrient retention and sorption capacity it can increase availability of nutrients in soil (Duku et al., 2010).

11.3   COMPOST

Compost is a stabilization product produced through the aerobic decomposition of agricultural wastes (AWs) (Cai et al., 2007) and an alternative source of several important plant nutrients which are required for appropriate growth and development of crop plants (Ngwira et al., 2013). Organic nutrients present in the wastes are converted to available forms of plants through microbial decomposition during the process of composting (Ngwira et al., 2013). Biophysical and chemical properties of soil can also be improved by soil amendments with organic source of nutrients such as compost (Ndegwa and Thompson, 2001). It was also proved in an experiment that the application of compost to soil as organic amendment results in enhanced soil aggregation which indirectly provides several benefits to soil such as pH stabilization and more rapid rates of water infiltration (Erhart and Hartl, 2010). Potassium in compost is in more readily available forms compared with organic N and P (Vern Grubinger, 2005). The other advantages of composting include killing pathogens and weed seeds, and improving handling characteristics of manure by reducing its volume and weight (Rynk et al., 1992). Loss of nutrient and carbon during composting, the cost of land, equipment, and labor required for composting, and odor associated with composting are the major disadvantages of making compost (Castaldi et al., 2004; Rivero et al., 2004). It was found that 20 to 40% loss of total N and 46 to 62% loss of total C during composting of beef cattle feedlot manure occurs (Eghball et al., 1997).

11.4 MANURES

11.4.1 FARM YARD MANURE

Application of beef cattle feedlot composted manure to soil results in an increase of concentrations of soil organic matter (Eghball and Power, 1994), which ultimately increases the soil fertility and crop production. Available water content, aggregate mean weight diameter, infiltration rate and saturation hydraulic conductivity can be enhanced as well as limited bulk density are the result of the application of manure at the rate of 50–100 t/ ha (Bahremand et al., 2003). The degradation of soil can be prevented by addition of organic manures into the soil because it improves the structure of soil (Thomas et al., 1996). Therefore, application of farmyard manures has gaining vital status to provide the essential nutrients for growth and development of crop plants and making the soil healthy (Roy et al., 2006).

11.4.2 POULTRY MANURE (PM)

Due to the presence of higher levels of proteins and amino acids (AA), the poultry manure (PM) is rich in nitrogen (N) contents (Enticknap et al., 2006), it was proposed that, the PMs are consist of high concentrations of N-P-K due to which they have reduced the need and utilization of other organic sources of nutrients such as animal manures (Wilkinson et al., 2011). Therefore, PM, are considered as an important source of nutrients through organic fertilizer because of its higher levels of nutrients. Poultry farming has been increased in the last two decades due to the increased requirements of chicken meat (Wilkinson et al., 2011). This has increased the production of organic wastes (OWs) and its utilization as organic or natural fertilizers (Wilkinson et al., 2011). Therefore, due to the high concentrations of nutrients in PMs, it is considered as important organic manure as compared to other manures (Dikinya and Mufwanzala, 2010).

11.5 EGG SHELLS

The waste materials from homes, hatcheries, and other fast food industries are known as eggshells (Phil and Zhihong, 2009). Environmental pollution is also be caused by the waste disposal of eggshells. For example, the cost, odor, abrasiveness, and availability of disposal sites are the major challenging factors which are associated with the proper dumping off of eggshells. Therefore, these wastes are processed into profitable products such as organic fertilizers (Phil and Zhihong, 2009). The concentrations of magnesium, phosphorous, and calcium carbonate are 0.9, 0.9 and 98.2% respectively in eggshell composition as well as shell membranes are comprised of different percentage of proteins, fats, moisture, and ash in the ratio of 69.2, 2.7, 1.5, and 27.2%, respectively (Burley and Vadehra, 1989). The effective sources of liming are grounded eggshells, hence increases the pH of acidic soil to a neutral level and this helps in the successful cultivation of various plants in the soil because slightly acidic soils with a pH of 5.8–7.0 are preferred by most of the plant species for their growth and development (John and Paul, 2006).

11.6 WOOD ASH

Wood ash has conventionally been used as organic fertilizer as an alternative to synthetic fertilizers (Fritze et al., 2000). Use of wood is being increased day by day as energy supply source that leads to increase in wood chips production, these wood chips are used for heating of bricks in brickworks or disposed off on landfills (Vance, 1996). In agriculture and forestry, these wood ashes can be used as an organic source of nutrients. The soil pH can be enhanced by hydroxides, oxides, and carbonates of potassium, magnesium, and calcium present in wood ash and it is considered as great advantage for treating the acidic soils (Fritze et al., 2000).

Soil microbial activity and biomass can be enhanced by amendments of soil with wood ashes and pathogen fungal growth can also be limited by amendments of soils with ashes (Fritze et al., 2000). Soil microbial activity and other properties of soil mainly biochemical properties are also affected by addition of wood ash at the rate of 5–20 tonnes/ha (Perucci et al., 2006). The application of wood ash to soil results in reduced utilization of synthetic fertilizers in acidic soils for the production of maize crop as well as soil pH, cations, and yield can also be enhanced by wood ashes (Kuba et al., 2008).

11.7 GREEN MANURING

The crop which is used for soil amendment and acts as a source of nutrient for succeeding or successive crop is known as green manure crop and incorporation of the crop either in green form as well as in crop residues (CR) form is known as green manuring. In crop production economic viability can be improved through green manure approaches. Influences of harsh environment on agriculture are also limited by green manuring. Because these approaches are dependent on the interactions among the green manure, environment, and its management, therefore these approaches are slightly difficult to adopt in certain conditions. Utilization of legumes as green manures is a potentially renewable resource on the farm because it fixes nitrogen carbon biologically to several cropping systems as compared with synthetic nitrogen fertilizers. Handling and transportation costs of several organic inputs are high as compared with green manures because they do not incur such costs (Tilman et al., 2002).

The green manure residues are slow releasers of nitrogen during decomposition and can be better corresponding with the uptake of plants than sources of inorganic nitrogen (Goyal et al., 1992). It leads to reduced loss of nitrogen through leaching while, increasing the crop yield and its efficiency for uptake of nitrogen (Agustin et al., 1999) and also improves the nitrogen uptake efficiency and nutrients retention in soil (Goyal et al., 1992). The application of soil with green manures in fallow patch of land leads to reduction in soil erosion (Gaston et al., 2003) and it suppresses the weeds and specific pests of crops (Caswell et al., 1991). The habitats or resources for useful and valuable organisms may also be provided by green manures (Nicholls and Altieri, 2001). The use of green manures has been increased after the World War II for the control of pests and weeds (Smil, 2001). However, the main objective of soil amendment with green manure is to maintain the concentrations of organic matter in the soils or to increase the quantity of organic matter. Crop production can be maintained by rotation of cereals with legume green manure crops even after green manuring no synthetic fertilizer was applied in the soil (Smil, 2001).

11.8 BAGASSE

Bagasse is a waste product which is remained after the extraction of sugarcane juice in the machines. When this remaining sugarcane is burnt, then the remaining ash from this is known as Bagasse ash. High percentage of silica (SiO_2) is present in bagasse through which the existing properties of soil can also be improved. After proving the efficiency of Bagasse, it is found to be used as a replacing material because it contains huge concentrations of silica (Desai, 2014). Therefore, due to high concentrations of silica in Bagasse ash, it is considered as an important mineral additive from sugar cane industry waste (Ganesan et al., 2007). Sugar industry wastes such as Bagasse may be composed of pith and fiber (Febrer, 2002). The fiber is relatively long as well as thick. The inner pith consists of most of the sucrose with bundles of small fibers. It was also found that longer and finer fibers are present in the outer skin throughout the stem in unsystematic arrangement and they are also bound together by hemicelluloses and lignin. Bagasse can be used in ecological organic agriculture to great extent, because it has commonly low risks for the environment (Negro et al., 1999).

11.9 SLURRY

The improved form of organic manure which is applied in the form of liquid, dry, or compost is known as slurry. Plant nutrients are present in considerable concentrations in slurry than in PM, farmyard manure, cow dung, and compost (Manna and Hazra, 1996). Soil fertility can also be improved by the application of slurry. Thus, the addition of slurry to the soil reduces the utilization of synthetic fertilizers. The higher yields of vegetables can be obtained by the application of slurry in soil. Larger head size of sunflower is produced when the biogas slurry is applied to the soil at the rate of 300 g per pot (Joshi et al., 1994). The increase in the cob yield of maize was also brought about by the application of biogas slurry. Slurry production significantly exceeds in the areas of higher livestock densities per hectare of grassland (Sommer et al., 2004). Plants can easily uptake the ammonical form of nitrogen, but this form are also susceptible to volatilization of ammonia or leads to denitrification (Bussink and Oenema, 1998). Cattle slurry is comprised of nitrogen in organic and inorganic forms in equal concentrations but mainly in the form of ammonium because the slurry application has some general implications.

Several application techniques as an alternative to a broadcast application like sliding shoe and injection have been introduced to reduce the gaseous losses after the application of cattle slurry. Potential for improving the efficiency of nitrogen use can be obtained by these techniques (Misselbrook et al., 2002). The loss of phosphorous by runoff from surface followed by land application can be reduced by soil amendment with cattle slurry and poultry litter. Slurry is also applied on pasture lands to increase the production and quality of herbage on farms with limited amounts of arable lands (Sanderson, 2004). Hence, in relation to nutrient leaching, herbage production, soil fertility as well as the quality of plant species composition need an urgent research for the application of slurry as soil amendment (Liu et al., 2010).

Several nutrients and organic matter are present in huge amounts in the slurry, so it has to apply as an effective organic fertilizer to fertilize the agricultural land (Islam et al., 2010). However, different nitrogen-fixing and phosphorous solubilizing organisms also obtain energy from biogas slurry application to soil (Nkoa, 2014). Hence, crop yield and quality of soil is improved by the application of slurry, and it also increases the organic matter contents of the soil (Islam et al., 2010).

11.10   PRESS MUD

Press mud is a natural byproduct from sugar mills which can be used as an organic soil amendment to provide nutrient-rich and high-quality organic matters to the soil. High yields of crops are obtained when press mud is applied on the soil. Press mud consists of considerable concentrations of macro plant nutrients such as nitrogen, phosphorous, potassium, magnesium, and calcium and it also has some amounts of micronutrients required for plants such as zinc, iron, manganese, and copper as well as a high quantity of sugars (Laird et al., 2001). Therefore, it can be used as bio compost for enhancing the crop production and maintaining the soil fertility (Banulekha et al., 2007). Hence, the positive effects of using press mud is well established for enhancing the fertility of the soil, and it indirectly improves the production of crops (Laird et al., 2001). Press mud has assured as a most economic source for plant nutrients in the time when prices of chemical fertilizers are not affordable by most of the farmers and are increasing day by day (Diaz et al., 2001). Hence, press mud use as soil amendment results in the improvement of physical properties such as water-holding capacity, soil texture, soil aeration, soil structure and also soil porosity and chemical properties such as pH, EC, and CEC and also the biological properties of the soil (Barry et al., 2002).

11.11 BIOFERTILIZERS

The substances consisting of living microorganisms are known as biofertilizers, and they colonize the rhizosphere and anterior of the plants when applied to soil, seed, and plant surface. The availability of major nutrients to the host plants can also be enhanced by biofertilizer application. Microorganisms are present in large quantities in bio-fertilizers (Gaur, 2010). There is an important need to improve the nitrogen concentrations of organic matter in order to prepare the perfect fertilizers. It can be achieved by the addition of free-living bacteria which enhances the organic matters nitrogen content (Lima et al., 2007). Therefore, for replacing the synthetic fertilizers with bio-fertilizers, different efforts have been made in the last few couples of decades. The most important and technically advanced process for supplying the essential nutrients to crop plants is the application of biological fertilizers (Mikhak et al., 2017). Bio-fertilizers application leads to improve the crop yield through environmentally safe nutrient supplies (Malusa et al., 2016). Therefore, soil quality and production of crops can be improved by the utilization of biofertilizers (Mukhlis, 2012).

11.12 MULCHING

Process of spreading different types of mulch material (organic or inorganic) on the surface of soil called mulching. Evaporation of water from soil can be limited by the application of different mulch material. Mulching also improves the fallow land efficiency and water availability of stored water in the soil to plants is also increased. The concentration of salts can also be reduced by mulching (Li et al., 2013). Soil water holding capacity, aggregate stability can be increased as well as reduction in water evaporation can be achieved by placing the CR on the soil (Pang et al., 2010). Building up of soil organic carbon leading to enhance the productivity and quality of soil can be achieved by the returning of CR to the soil as natural mulch (Carter, 1998).

11.12.1 STRAW MULCH

The process of spreading straw on the soil as mulch material is called straw mulch. The crop productivity is limited in arid and semi-arid areas due to the shortage of water, which is the major challenge for these areas. Therefore, to promote the moisture content of the soil, it is necessary to cover the soil surface with straw as straw mulch. It leads to improve the water use efficiency and yields of crops (Li et al., 2012). Application of straw mulch deeply in soil inhibits the movement of salts from the subsoil and shallow groundwater to surface soil such as topsoil as well as acts as a barrier for water and salt transport during the process of water evaporation (Yonggan et al., 2017). Increase in salt leaching and accumulation of salts in the root zone is the result of straw layer (Li et al., 2012). Physical and chemical properties of soil can also be improved by spreading of the straw layer. In the cultivation of ginger and turmeric, rice straw mulch is used as appropriate mulch material (Wang et al., 2012). Although, the application of straw mulch leads to increase in the yield of the crop while in some other studies different results were also found such as reduced yield of winter wheat to some extent by the application of straw mulch (Li et al., 2013). Effect of straw mulch can be studied on soil properties and yield of the crop to find several methods for straw mulch application to understand the response of crop productivity and soil water dynamics to the application of straw mulch (Zou et al., 2005).

11.12.2 WOOD CHIPS

Weeds around acid-loving perennials like blueberries and in the pathways can be suppressed by wood chips and bark mulches (Huo et al., 2017). When the soils dried up during the warmer seasons, wood chips have the ability to soak up and retain water feeding it back to adjacent soil in warmer weather (Huo et al., 2017). Therefore, incorporation of previous crops stubbles and wood chips will enhance the soil health (Huo et al., 2017).

Wood chips are capable of providing many benefits such as weed suppression and delayed seedling emergence of weeds in soil (Huo et al., 2017). Therefore, these kind of natural mulches are economically and environmentally sustainable alternatives for weed controlling (Bond and Grundy, 2001). Wood chips are also beneficial for soil moisture conservation and for improving water infiltration (Faber et al., 2001). For example, different properties of soil such as soil water holding capacity, activities of the microbial population in soil, cation exchange capacity of soil, soil porosity, and soil stabilization are also the important benefits of wood chips. Wood chips can also decrease soil-borne diseases of plants (Gleason et al., 2001; Doug et al., 2002).

11.13   BIOSOLIDS

Biosolid byproducts are considered as wastes in Ireland and USA, so their application as soil amendment will be comparatively economical in these countries (Meyer et al., 2001). Surface runoff and leaching from soil surface leads to loss of several organic and inorganic nutrients where land applications are followed by periodic rainfall (Gottschall et al., 2012). Application of biosolids on land is an important way leading to entering of contaminants into the environment as shown in many previous studies (Smith, 2009).

Biosolids provide a lot of benefits to soil application and they are only disposed off in the soil. The entry of biosolids into the human food chains and risks related to them was insignificant due to minimum uptake by plants when they are applied on land. As phosphorous resources are depleting, the land application of biosolids are also becoming progressively more essential feature of sustainable nutrient management (Steen, 1998).

11.14   BONE MEAL

Meal bone meal (MBM) is an organic and essential source of important nutrients which are required for optimum growth and development of crop plants. It contains high amounts of nutrients such as nitrogen and phosphorous in the ratio of 8% and 5% respectively which makes it a significant source of nutrients for different crop plants. Bone meal reduces the requirement of crops for mineral fertilizers and it is also a fact that MBM has positive indirect influence on the environment. It also provides a source for the safe disposal of a large quantity of wastes from meat processing industries (Chen et al., 2011). They also provide an exceptional source of important elements and metals required for better growth of crops and plants when they are treated to required standards on grasslands and arable soils (Meyer et al.,

2001).Physical and chemical properties of soil can also be improved by bone meal amendment (Salomonsson et al., 1995). The potato quality can also be enhanced by soil amendment with MBM due to reduction in the risks of attack of potato scab and by also a limited population of parasitic nematodes (Fredriksson et al., 1998). Therefore, meat and bone meal has been allowed to be used as organic fertilizer for all crops except grasslands. These grasslands are used for grazing or mowing. Because the bone meal is organically bound, and it must be transformed to the inorganic form of nitrogen to make it available for the use of plants. Bone meal has low carbon to nitrogen ratio which favors faster mineralization of nitrogen than most of the other natural fertilizers (Tammeorg et al., 2012). Carbon to nitrogen ratio is used for approximation of organic matter quality in the soil to estimate nitrification and decomposition processes. As bone meal consists of adequate amounts of nitrogen, phosphorous, and calcium which makes it an attractive organic fertilizer to be used for different crops(Van et al., 2009).

11.15 COTTONSEED MEAL

The product, after removing oil from the cottonseed, is known as cottonseed meal. Cottonseed meal is absolute fertilizer which consists of 7–2.5–1.5% N-P-K respectively. It is byproduct of many cotton species like *Gossypium hirsutum* and *Gossypium herbaceum*. About 4–5 pounds of cottonseed meal should be applied for optimum productivity of flowers and vegetables per hundred square feet of land area and it should be applied into the soil before planting of crops (Roy et al., 2012). Soil texture can be improved by cottonseed meal and it also improves the growth of plants and soil health by providing humic acid to soil (Tenuta and Lazarovits, 2004). Hence, cottonseed meal aids in improving water holding capacity of soil and also protects the soil from fast erosion. It also provides different substances to light and sandy soils and also loosens the crowded soil (Martin and Gershun, 1992). The concentration of potash, phosphorous, and other important nutrients can be increased in soil environs by application of cottonseed meal (Cooperband, 2002).

11.16 ALFALFA PELLETS

Alfalfa is an excellent source of nutrients and has high digestibility with a huge amount of biological yield. It is a perennial forage legume and provides a huge quantity of nitrogen in pasture system with a lot of economic benefits and is a good alternative source of nutrients to synthetic fertilizers application (Campillo et al., 2003). Alfalfa pellets are used as an animal feed because they are an excellent source of nitrogen which is up to 5% (Agehra and Warncke, 2005).

They also contain high amounts of naturally occurring plant growth promoters such as trace minerals and triacontanol. These plant growth promoters help to improve growth and yield of several crops. An alfalfa green pellet consists of host of many micronutrients which are obligatory for growth of plants and are absent in many of predictable fertilizer products (Clark, 2008; Den et al., 2010). It enhances favorable environmental conditions for microbial activities, which results in healthy soil environs (Fageria and Baligar, 2008).

11.17   FISH EMULSIONS

Fish emulsion is a natural, organic, and liquid fertilizer which is made by the fish industry byproducts. It is also called fish fertilizer or liquid fish emulsions. Because of its high levels of nitrogen content, fish emulsions are useful for leafy green vegetables and lawns. For better crop production, fish industry byproducts are normally used as a natural fertilizer for soil amendments. In a vegetable production system, the fish meal has been utilized as a good soil amendment with a lot of success, and that fish meal is dried up protein which can be obtained from processed fish (Blatt and McRae, 1998). Fertilizers which are made from fish remains contain a considerable amount of proteins and nitrogen. These natural fertilizers also contain a strong balance of nutrients such as micro and macronutrients, essential for the growth of crops. The ratio of N-P-K of fertilizers made from fish is 10-6-2%, respectively. Continuous application of fish fertilizers as soil amendment results in rapid, healthy, and dynamic growth of crop plants (Gaskell, 1990). Fish silage and fish bone meal (FBM) either in liquid or dry form are considered as appropriate soil amendments and are also a good alternative source to inorganic fertilizers for the production of high-quality vegetables (Blatt, 1991). Therefore, the yield obtained from the application of organic nitrogen fertilizers from fish offal's was the same as yield obtained from the application of two inorganic nitrogen fertilizers (Gagnon and Berrouard, 1994).

11.18   BAT GUANO

Feces of bats consisting of high concentrations of carbon, important minerals, as well as advantageous microbes are known as bat guano. Soil fertility and soil texture can be improved by the chemical properties and microorganisms in the guano. Toxins and nematodes in the soil may also be cleared by these microbes. Different important properties of guano can be affected by species of bat, its location and as well as by the age of guano. Use of this bat guano rich manure is not getting valuable popularity in farmer's community despite having too many significant properties. Therefore, the nutrients demand of different crops are significantly provided by natural manures which are synthesized from solid or liquid wastes of livestock such as horses, camel, cattle, sheep, poultry, and bats (El-Sherif and Sarwat, 2007). As compared to other organic fertilizers, bat guano is considered as a better alternative natural fertilizer for crop production. Hence, granule yield and biological yield of maize can be considerably increased by the addition of PM and bat guano as soil amendment (Mentler et al., 2002). Bat species, age of guano, locations where bat lives and the amount of guano are major factors that affect the availability of manures of bat guano (Omak et al., 2014). Hence, for the production of crops, there is minimum knowledge and research on the application of bat guano in soil (Korine et al., 1999; Sridhar et al., 2006). The application of bat guano as organic manure is getting popularity (Sridhar et al., 2006).

11.19   CONCLUSION

The continuous degradation in soil health due to the everlasting furnishing of artificial fertilizers has momentous drawbacks on humans, animals, and plant health. Nitrogenous fertilizers cause diseases in human beings and animals by polluting groundwater and leach down to root zone. Therefore due to the pollution problems of synthetic fertilizers in the atmosphere and reduction in soil fertility, farmers are getting interest for the use of natural fertilizers like manures, biochar, and composts. A huge quantity of nutrients is also provided by organic fertilizers with substantial improvement in chemical properties and physical structure of soil. However, organic fertilizers do not leach down from soil and therefore add less to water pollution as compared to synthetic fertilizers. Organic fertilizers perform vital functions that cannot be performed by man-made artificial fertilizers. The gradual increase in soil organic matter and nutrients supply are the major functions of organic resources in the soil. Farmyard manure, compost, PM, CR, and other natural resources can be used as an alternative and supplement to the application of synthetic fertilizers. Therefore, several farmers have adopted extensive use of organic sources of nutrient for crop production such as green manures, animal dung, biochar, biofertilizers, composted materials, and domestic waste due to high prices of synthetic fertilizers and high benefits of organic fertilizers.

KEYWORDS

- biochar
- composting
- fishbone meal
- meal bone meal
- plant nutrients
- synthetics fertilizers

REFERENCES

Agehara, S., & Warncke, D. D., (2005). Soil moisture and temperature effects on nitrogen release from organic nitrogen sources. *Soil Sci. Soc. Amer. J., 69*(6), 1844–1855.

Agustin, E. O., Ortal, C. I., Pascua, S. R., Santa, C. P. C., Padre, A. T., Ventura, W. B., Obien, S. R., & Ladha, J. K., (1999). Role of indigo in improving the productivity of rainfed lowland rice-based cropping systems. *Exp. Agric., 35*, 201–210.

Ayoola, O. T., & Adeniyan, O. N., (2008). Influence of poultry manure and NPK fertilizer on yield and yield components of crops under different cropping systems in south west Nigeria. *Afr. J. Bio., 5*(15), 1386–1392.

Bahremand, M. R., Afyuni, M., Hajabbasi, M. A., & Rezainejad, Y., (2003). Short and mid-term effects of organic fertilizers on some soil physical properties. *J. Agric. Natur. Resour. Sci. Technol., 6*, 4.

Banulekha, C., (2007). *Eco-Friendly Utilization of Organic Rich Biomethanated Distillery Spent Wash and Biocompost for Maximizing the Biomass and Quality of Cumbunapier Hybrid Fodder (CO₂).* Env. Sci. Thesis, Tamil Nadu Agricultural University, Coimbatore.

Bhattacharyya, R., Prakash, V., Kundu, S., Srivastva, A. K., Gupta, H. S., & Mitra, S., (2009). Long term effects of fertilization on carbon and nitrogen sequestration and aggregate associated carbon and nitrogen in the Indian sub-Himalayas. *Nutr. Cycl. Agroecosyst., 86*, 1–16.

Blatt, C. R., & McRae, K. B., (1998). Comparison of four organic amendments with a chemical fertilizer applied to three vegetables in rotation. *Can. J. Plant Sci., 78*, 641–646.

Blatt, C. R., (1991). Comparison of several organic amendments with a chemical fertilizer for vegetable production. *Sci. Hort.*, *47*, 177–191.

Bond, W., & Grundy, A. C., (2001). Non-chemical weed management in organic farming systems. *Weed Research, 41*, 383–405.

Burley, R. W., & Vadehra, V., (1989). The eggshell and shell membranes: Properties and synthesis. In: *The Avian Egg Chemistry and Biology* (pp. 25–64). John Wiley, New York.

Buss, W., Kammann, C., & Koyro, H., (2012). Biochar reduces copper toxicity in *Chenopodium quinoa willd.* in a sandy soil. *J. Environ. Qual., 41*, 1157–1165.

Bussink, D. W., & Oenema, O., (1998). Ammonia volatilization from dairy farming systems in temperate areas: A review. *Nutr. Cycl. Agroecosyst., 51*, 19–33.

Cai, Q., Mo, C., Wu, Q., Zengand, Q., & Katsoyiannis, A., (2007). Concentration and speciation of heavy metals in six different sewage sludge-compost. *J. Hazard Mater., 147*, 1063–1072.

Campillo, R., Urquiaga, S. I., & Montenegro, A., (2003). Estimation of the biological fixation of nitrogen using the methodology. *N. Agr. Tec., 63*, 169–179.

Carter, L. M., (1998). Tillage. In: *Cotton Production* (pp. 1–14). University of California, Division of Agriculture and Natural Resources Publication.

Castaldi, P., Garau, G., & Melis, P., (2004). Influence of compost from sea weeds on heavy metal dynamics in the soil-plant system. *Fresen. Environ. Bull., 13*, 1322–1328.

Caswell, E. P., DeFrank, W. J., & Tang, C. S., (1991). Influence of non-host plants on population decline of rotylen chusreni formis. *J. Nematol., 23*, 91–98.

Chen, L., Kivela, J., Helenius, J., & Kangas, A., (2011). Meat bone meal as fertilizer for barley and oat. *Agric. Food Sci., 20*, 235–244.

Chrispaul, M., David, M. M., Joseph, A. O., & Samuel, V. O., (2010). Effective microorganism and their influence on growth and yield of pigweed (*Amaranthus dubians*). *J. Agric. Biol. Sci., 5*, 17–22.

Clark, A., (2008). *Managing Cover Crops Profitably*. Diane Publishing.

Cooperband, L., (2002). *Building Soil Organic Matter with Organic Amendments*. Centre for Integrated Agricultural Systems, College of Agricultural and Life Sciences, University of Wisconsin, Madison.

Den, H. G., Van, I. G., Beeckman, T., & De Smet, I., (2010). The roots of a new green revolution. *Trends in Plant Science, 15*(11), 600–607.

Desai, C. S., (2014). Waste production 'bagasse ash' from sugar industry can be used as stabilizing material for expansive soils. *Int. J. Res. Eng. Tech., 3*(2), 506–512.

Diaz, P. M., (1992). Consequences of compost press mud as fertilizers. DJ International Journal of Advances in Yadav DV. Utilization of press mud cakes in Indian agriculture. *Indian J. Sugarcane Technol., 7*, 1–16.

Dikinya, O., & Mufwanzala, N., (2010). Chicken manure-enhanced soil fertility and productivity: Effects of application rates. *Journal of Soil Science and Environmental Management, 1*, 46–54.

Dobereiner, J., (1994). Role of microorganisms in sustainable tropical agriculture. *Proceedings of the 2nd International Conference on Kyusei Nature Farming, (ICKNF'94)* (pp. 64–72). US. Department of Agriculture, Washington, D.C., USA.

Doran, J., (1995). Building soil quality. In: *Proceedings of the 1995 Conservation Workshop on Opportunities and Challenges in Sustainable Agriculture* (pp. 151–158). Red Deer, Alta., Canada, Alberta Conservation Tillage Society and Alberta Agriculture Conservation, Development Branch.

Doug, A. D., Randy, L. A., Robert, E. B., & Bruce, M., (2002). Weed dynamics and management strategies for cropping systems in the northern great plains. *Agron. J.*, *94*, 174–185.

Duku, H. M. u, S., & Hagan, E. B., (2010). Biochar production potentials in Ghana: A review. *Renewable and Sustainable Energy Review*, *15*, 3539–3551.

Eghball, B., & Power, J. F., (1994). Beef cattle feedlot manure management. *J. Soil Water Consrv.*, *49*, 113–122.

Eghball, B., & Power, J. F., (1999a). Composted and non-composted manure application to conventional and no-tillage systems: Corn yield and nitrogen uptake. *Agron. J.*, *91*.

Eghball, B., Power, J. F., Gilley, J. E., & Doran, J. W., (1997). Nutrient, Carbon and mass loss of beef cattle feedlot manure during composting. *J. Env. Qual.*, *26*, 189–193.

El-Sherif, M. H., & Sarwat, M. I., (2007). Physiological and chemical variations in producing Roselle plant *Hibiscus sabdariffa*L. using some organic farmyard manure. *J. Agric. Sci.*, *3*, 5, 609–616.

Erhart, E.,& Hartl, W., (2010). Compost use in organic farming. In: Lichtfouse, E., (ed.), *Genetic Engineering, Biofertilization, Soil Quality and Organic Farming* (pp. 311–345). Springer Netherlands.

Faber, B. A., Downer, A. J., & Menge, J. A., (2001). *Differential Effects of Mulch on Citrus and Avocado* (Vol. 1, p. 63). ISHS Acta Hort. 557: VII Intl. Symp. Orchard Plantation Syst. Nelson, New Zealand.

Fageria, N. K., & Baligar, V. C., (2008). Ameliorating soil acidity of tropical Oxisols by liming for sustainable crop production. *Adv. Agron.*, *99*, 345–399.

Febrer, M. C. A., (2002). Dynamics of omes aomesophilic decomposition of organic residues mixed with guinea pigs. *Engineering Agri.*, *10*, 18–30.

Fening, J. O., Ewusi-Mensah, N., & Safo, E. Y., (2010). Improving the fertilizer value of cattle manure for sustaining small holder crop production in Ghana. *J. Agron.*, *9*, 92–101.

Fredriksson, H., Salomonsson, L., & Salomonsson, A. C., (1997). Wheat cultivated with organic fertilizers and urea: Baking performance and dough properties. *Acta Agriculturae Scandinavica, Section B. Soil Plant Sci.*, *47*, 35–42.

Fredriksson, H., Salomonsson, L., Andersson, R., & Salomonsson, A. C., (1998). Effects of protein and starch characteristics on the baking properties of wheat cultivated by different strategies with organic fertilizers and urea. *Acta. Agriculturae Scandinavica, Section B. Soil Plant Sci.*, *48*, 49–57.

Fritze, H., Perkoiomaki, J., Saarela, U., Katainen, R., Tikka, P., Yrjala, K., Karp, M., et al., (2000). Effect of Cd-containing wood ash on the microflora of coniferous forest humus. *FEMS Microbiol. Ecol.*, *32*, 43–51.

Gagnon, B., & Berrouard, S., (1994). Effects of several organic fertilizers on growth of greenhouse tomato transplants. *J. Plant Sci.*, *74*, 167–168.

Ganesan, K., Rajagopal, K., & Thangavel, K., (2007). Evaluation of bagasse ash as supplementary cementitious material. *Cement and Concrete Composites*, *29*, 515–524.

Gaskell, M., (1999). *Efficient use of Organic Fertilizer Sources*. Organic farming Research Foundations, University of California cooperative Extension.

Gaston, L. A., Boquet, D. J., & Bosch, M. A., (2003). Fluometuron sorption and degradation in cores of silt loam soil from different tillage and cover crop systems. *Soil Sci. Soc. Am. J.*, *67*, 747–755.

Gaur, V., (2010). Biofertilizer-necessity for sustainability. *J. Adv. Dev.*, *1*, 7, 8.

Gleason, M., Wegulo, S., & Nonnecke, G., (2001). *Efficacy of Straw Mulch for Suppression of Anthracnose on Day-Neutral Strawberries.* ISU Ext. FG-601:48.

Gottschall, N., Topp, E., Metcalfe, C., Edwards, M., Payne, M., Kleywegt, S., Russell, P., & Lapen, D. R., (2012). Pharmaceutical and personal care products in groundwater, subsurface drainage, soil, and wheat grain, following a high single application of municipal biosolids to a field. *Chemosphere, 87*, 194–203.

Goyal, S., Mishra, M. M., Hooda, I. S., & Singh, R., (1992). Organic matter-microbial biomass relationships in field experiments under tropical conditions: Effects of inorganic fertilization and organic amendments. *Soil Biol. Biochem., 24*, 1081–1084.

Huo, L., Pang, H., Zhao, Y., Wang, J., Lu, C., & Li, Y., (2017). Buried straw layer plus plastic mulching improves soil organic carbon fractions in an arid saline soil from northwest China. *Soil Tillage Res., 165*, 286–293.

Intani, K., Latif, S., Kabir, A. R., & Müller, J., (2016). Effect of self-purging pyrolysis on yield of biochar from maize cobs, husks and leaves. *Bioresource Technology, 218*, 541–551.

Islam, M. R., Rahman, S. M. E., Rahman, M. M., Oh, D. H., & Ra, C. S., (2010). The effects of biogas slurry on the production and quality of maize fodder. *Turk. J. Agric. For., 34*, 91–99.

John, H., & Paul, K., (2006). *Can Ground Eggshells be Used as a Liming Source?* Integrated crop Management Conference, Iowa State University.

Joshi, J. R., Moncrief, F., Swan, J. B., & Malzer, G. L., (1994). *Soil Till. Res., 31*, 225.

Jurwarkar, A. S., Tnewale, P. L., Baitute, U. H., & Moghe, M., (1993). *Sustainable Crop Production Through Integrated Plant Nutrition System-Indian Experience* (Vol. 12, pp. 87–97). RAPA Publication.

Korine, I., & Arad, Z., (1999). Is the Egyptian fruit bat *Rousettus aegyptiacusa* pest in Israel? An analysis of the bat's diet and implications for its conservation. *Bio. Conservation, 88*, 301–306.

Kuba, T., Tscholl, A., Partl, C., Meyer, K., & Insam, H., (2008). Wood ash admixture to organic wastes improves compost and its performance. *Agric. Ecosyst. Environ., 127*, 43–49.

Laird, D. A., Martens, D. A., & Kingery, W. L., (2001). Nature of clay humic complexes in an agricultural soil: I. chemical, biological and spectroscopic analysis. *Soil Sci. Soc. America J., 65*, 1413–1418.

Lehmann, J., Rillig, M. C., Thies, J., Masiello, C. A., Hockaday, W. C., & Crowley, D., (2011). Biochar effects on soil biota: A review. *Soil Biol. Biochem., 43*, 1812–1836.

Li, R., Hou, X. Q., Jia, Z. K., Han, Q. F., & Yang, B. P., (2012). Effects of rainfall harvesting and mulching technologies on soil water, temperature, and maize yield in Loess Plateau region of China. *Soil Res., 50*, 105–113.

Li, S. X., Wang, Z. H., Li, S. Q., Gao, Y. J., & Tian, X. H., (2013). Effect of plastic sheet mulch, wheat straw mulch, and maize growth on water loss by evaporation in dryland areas of China. *Agric. Water Manag., 116*, 39–49.

Li, S., Wang, X., Li, Z. H., Gao, S. Q., & Tian, Y. J., (2013). Effect of plastic sheet mulch, wheat straw mulch, and maize growth on water loss by evaporation in dryland areas of China. *Agric. Water Manag., 116*, 39–49.

Lima, R. C. M., Stamford, N. P., Santos, C. E. R. S., & Dias, S. H. L., (2007). Lettuce yield chemical attribute of an oxisol function of the application of rock biofertilizers with phosphorous potassium. *Braz. J. Hortic., 25*, 224–229.

Liu, W., Zhu, Y. G., Christie, P., & Laidlaw, A. S., (2010). Botanical composition, production and nutrient status of an originally *Lolium perenne*-dominant cut grass sward receiving long-term manure applications. *Plant Soil, 326,* 355–367.

Malusa, E., Pinzari, F., & Canfora, L., (2016). Efficacy of bio-fertilizers: Challenges to improve crop production. *Microbial Inoculants in Sustainable Agricultural Productivity.* doi: 10.1007/978-81-322-2644-4_2, Springer India.

Manna, M. C., & Hazra, J. N., (1996). Comparative performance of cow dung slurry, microbial inoculums and inorganic fertilizers on maize. *J. Indian Soc. Soil Sci., 44,* 526–528.

Martin, D. L., & Gershun, G., (1992). *The Rodale Book of Composting: Easy Methods for Every Gardener.* Rodale.

Mentler, A., Partaj, T., Strauss, P., Soumah, M., & Blum, W. E., (2002). *Effect of Locally Available Organic Manure on Maize Yield in Guinea* (Vol. 13, pp. 1–8). West Africa, 17th WCSS, 14–21 August, Thailand. Sympos.

Meyer, V. F., Redente, E. F., Barbarick, K. A., & Brobst, R., (2001). Biosolids applications affect runoff water quality following forest fire. *J. Envi. Qu., 30,* 1528–1532.

Mikhak, A., Sohrabi, A., Kassaee, M. Z., & Feizian, (2017). Synthetic nanozeolite/nanohydroxy apatite as a phosphorus fertilizer for German chamomile (*Matricaria chamomilla* L.). *Industrial Crops and Products, 95,* 444–452.

Misselbrook, T. H., Smith, K. A., Johnson, R. A., & Pain, B. F., (2002). Slurry application techniques to reduce ammonia emissions: Results of some UK field-scale experiments. *Biosyst. Eng., 81,* 313–321.

Mukhlis, (2012). Effectivities of biotara and biosure in reducing chemical fertilizer use and increasing rice production in tidal swamp lands. *Agroscientiae, 19,* 170–177.

Ndegwa, P. M., & Thompson, S. A., (2001). Integrating composting and vermin-composting in treatment and bioconversion of biosolids. *Bioresour. Technol., 76,* 107–112.

Negro, M. J., Solano, M. L., Ciria, P., & Carrasco, J., (1999). Composting of sweet sorghum bagasse with other wastes. *Bioresour. Technol., 67,* 89–92.

Ngwira, A. R., Nyirenda, M., & Taylor, D., (2013). Toward sustainable agriculture: An evaluation of compost and inorganic fertilizer on soil nutrient status and productivity of three maize varieties across multiple sites in Malawi. *Agroecol. Sust. Food Systems, 37,* 859–881.

Nicholls, C. I., & Altieri, M. A., (2001). Manipulating plant biodiversity to enhance biological control of insect pests: A case study of a northern California vineyard. In: Gliessman, S. R., (ed.), *Agroecosystem Sust. Developing Practical Strategies* (pp. 29–50). CRC Press, Boca Raton, FL.

Nkoa, R., (2014). Agricultural benefits and environmental risks of soil fertilization with anaerobic digestates: A review. *Agronomy for Sustainable Development, 34,* 473–492.

Omak, B. P., Forray, F. L., Wynn, J. G., & Giurgiu, A. M., (2014). Guano-derived 13C-based paleohydro climate record from gaura cu muscă cave, SW Romania. *Environ. Earth Sci., 71,* 4061–4069.

Pang, H., Li, Y., Yang, J., & Liang, Y., (2010). Effect of brackish water irrigation and straw mulching on soil salinity and crop yields under monsoonal climatic conditions. *Agric. Water Manage, 97,* 1971–1977.

Paz-Ferreiro, H., Lu, S., Fu, A., & Gasco, G., (2014). Use of phytoremediation and biochar to remediate heavy metal polluted soils: A review. *Solid Earth, 5,* 65–75.

Perucci, P., Monaci, E., Casucci, C., & Vischetti, C., (2006). Effect of recycling wood ash on microbiological and biochemical properties of soils. *Agron. Sust. Dev., 26,* 157–165.

Phil, G., & Zhihong, M., (2009). *High Value Products from Hatchery Waste*. RIRDC publication no. 09/061. glatz.phil@saugov.sa.gov.au.

Rivero, C., Chirenje, T., Ma, L. Q., & Martinez, G., (2004). Influence of compost on soil organic matter quality under tropical conditions. *Geoderma, 123,* 355–361.

Roy, R. N., Finck, A., Blair, G. J., & Tandon, H. L. S., (2006). Plant nutrition for food security. A guide for integrated nutrient management. *FAO Fertilizer and Plant Nutrition Bulletin, 16,* 368.

Roy, S., Karim, K., Rahman, S. M. M., Aziz, Z., & Hassan, S., (2012). The fatty acid composition and properties of oil extracted from cotton (*Gossypium herbaceum*) seed of Bangladesh. *J. Sci. Ind. Res., 47,* 303–308.

Ryank, R. M., Van, D. K. G. B., Wilson, M. E., Singley, M. E., Richard, T. L., & Brinton, W. F., (1992). *On Farm Composting Northeast Regional Agricultural Engineering Service*. Ithaca, NY.

Salomonsson, L., Salomonsson, A. C., Olofsson, S., & Jonsson, A., (1995). Effects of organic fertilizers and urea when applied to winter wheat. *Acta Agriculturae Scandinavica, Section B. Soil Plant Sci., 45,* 171–180.

Sanderson, M. A., Skinner, R. H., Barker, D. J., Edwards, G. R., Tracy, B. F., & Wedin, D. A., (2004). Plant species diversity and management of temperate forage and grazing land ecosystems. *Crop Science, 44,* 1132–1144.

Singh, B., Singh, B. P., & Cowie, A. L., (2010). Characterizations and evaluation of biochars for their application as a soil amendment. *Soil Res., 48,* 516–525.

Smil, V., (2001). *Enriching the Earth: Fritz Haber, Carl Bosch, and the Transformation of World Food Production*. MIT Press, Cambridge, MA.

Smith, S. R., (2009). Organic contaminants in sewage sludge (biosolids) and their significance for agricultural recycling. *Philosophical Transactions of the Royal Society A Mathematical Physical and Engineering Science, 367.*

Sohi, S. P., Krull, E., Lopez-Capel, E., & Bol, R., (2010). A review of biochar and its use and function in soil. In: *Adv. Agron.,* (pp. 47–82). Academic Press, Burlington.

Somme, S. G., Schjorring, J. K., & Denmead, O. T., (2004). Ammonia emission from mineral fertilizers and fertilized crops. *Adv. Agron., 82,* 557–622.

Sridhar, K. R., Ashwini, K. M., Seena, S., & Sreepada, K. S., (2006). Manure qualities of guano of insectivorous cave bat *Hipposideros speoris*. *Trop. Subtrop. Agroecosyst., 6,* 103–110.

Steen, I., (1998). Phosphorus recovery-phosphorus availability in the 21st century, management of a non-renewable resource. *Phosphorus Potassium, 217.*

Tammeorg, P., Brandstaka, T., Simojoki, A., & Helenius, J., (2012). Nitrogen mineralization dynamics of meat bone meal and cattle manure as affected by the application of softwood chips biochar in soil. *Earth Envir. Sci. Transactions Royal Soc. Edinburgh., 103,* 19–30.

Tenuta, M., & Lazarovits, G., (2004). Soil properties associated with the variable effectiveness of meat and bone meal to kill microsclerotia of *Verticillium dahliae*. *Applied Soil Ecol., 25*(3) 219–236.

Thomas, G. W., Haszler, G. R., & Blevines, R. L., (1996). The effects of organic matter and tillage on maximum compatibility of soils using the proctor test. *Soil Sci., 161,* 502–508.

Tilman, D., Cassman, K. G., Matson, P. A., Naylor, R., & Polasky, S., (2002). Agricultural sustainability and intensive production practices. *Nature, 418,* 6898, 671.

Van, D. B. A., De Bolle, S., De Neve, S., & Hofman, G., (2009). Effect of tillage intensity on N mineralization of different crop residues in a temperate climate. *Soil Tillage Res.*, *103*, 316–324.

Vance, E. D., (1996). Land application of wood-fired and combination boiler ashes: An overview. *J. Environ. Qual.*, *25*, 937–944.

Wilkinson, K. G., Tee, E., Tomkins, R. B., Hepworth, G., & Premier, R., (2011). Effect of heating and aging of poultry litter on the persistence of enteric bacteria. *Poult. Sci.*, *90*, 10–18.

Wilkinson, S. R., (1979). Plant nutrient and economic value of animal manures. *J. Anim. Sci.*, *48*, 121–133.

Yadav, A., & Garg, V. K., (2016). Vermi conversion of biogas plant slurry and parthenium weed mixture to manure. *Int. J. Recycl. Org. Waste Agric.*, *5*, 301–309.

Yonggan, Z. H. A. O., Yan, L. I., Shujuan, W. A. N. G., Jing, W. A. N. G., & Lizhen, X. U., (2017). Combined application of a straw layer and flue gas desulfurization gypsum to reduce soil salinity and alkalinity. *Pedosphere*.

Zhang, A., Cui, L., Pan, G., Li, L., Hussain, Q., Zhang, X., Zheng, J., & Crowley, D., (2010). Effect of biochar amendment on yield and methane and nitrous oxide emissions from a rice paddy from Tai Lake plain, China. *Agric. Ecosystems Envir.*, *139*, 469–475.

Zheng, W., Sharma, B. K., & Rajagopalan, N., (2010). *Using Biochar as a Soil Amendment for Sustainable Agriculture*. Illinois Department of Agriculture, Illinois, USA, Field Report.

Zou, J., Huang, Y., Jiang, J., Zheng, X., & Sass, R. L., (2005). A 3-year field measurement of methane and nitrous oxide emissions from rice paddies in China: Effects of water regime, crop residue, and fertilizer application. *Global Biogeochemical Cycles*, *19*.

CHAPTER 12

Mushroom Cultivation Technology for Conversion of Agro-Industrial Wastes into Useful Products

SHAUKET AHMED PALA,[1] DIG VIJAY SINGH,[2] ABDUL HAMID WANI,[1] ROUF AHMAD BHAT,[3] and BASHIR AHMAD GANAI[4]

[1] Section of Mycology and Plant Pathology, Department of Botany, University of Kashmir, Hazratbal Srinagar – 190006, Jammu and Kashmir, India

[2] Babasaheb Bhimrao Ambedkar Central University, Lucknow – 226025, Uttar Pradesh, India

[3] Division of Environmental Sciences, Sher-e-Kashmir University of Agricultural Sciences and Technology, Srinagar – 190025, Jammu and Kashmir, India, Phone: +91-7006655833, E-mail: rufi.bhat@gmail.com

[4] Centre of Research for Development, University of Kashmir, Hazratbal Srinagar – 190006, India

ABSTRACT

Quality food, health, and environmental security are the major challenges for the modern man. The increase in the population and industrial activities has led to the generation of a variety of wastes that have negative effects on the quality of environs. The lack of resources and innovative techniques for the sustainable management of the waste generated is lacking in most countries. The use of proper bioconversion techniques will result in the production of valuable resources having multiple uses in the different fields of the environment. One of the recently developed bioconversion technique is the mushroom production using organic waste mostly produced from food industries. Cultivating mushrooms on the organic wastes (OWs) reduces the burden on disposal of the wastes. This technique helps achieve zero-emission and convert the nutrient-rich wastes into edible form, thereby preventing the wastage of nutrients well as eutrophication of water bodies. The multiple benefits of bioconversion are the conversion of nutrient rich organic waste and the production of quality food (medicinal mushroom) and additional income source for the farmers. Thus, this technique is ecofriendly and sustainable for the production of valuable products and efficient one for the conversion of organic waste.

12.1   INTRODUCTION

Growing population and increasing urbanization results in the generation of enormous quantity of waste around the world. The increasing generation of waste is one of the biggest problems is facing by modern generation, but recycling of waste is helpful in reducing the quantity of waste disposed into the water bodies or disposal sites. Recycling agro-industrial waste helps in conservation of natural resources and increases productivity by producing valuable products. Most of the waste produced is disposed off without any treatment or disposed by conventional methods (burning, dumping, or by unplanned landfilling) that have negative effects on the environment. The impacts are associated with the emission of greenhouse gases. The utilization of the conventional techniques results in the wastage of bio-resource and the emission of greenhouse gases. The increasing cost of the conventional methods for the waste disposal and their negative effect on the environment has forced to develop ecofriendly and sustainable techniques for the treatment of the waste especially organic in nature. Organic wastes (OWs) produced mostly from agro-industries poses great threat to the environment but at the same time can acts as the resource for various products like mushroom and biofertilizers. The utilization of organic waste generated from different agro-industries provides alternative substrate and helps in resource recovery, which is otherwise wasted by disposing the waste on the surface or in water bodies. Different products like mushroom, ethanol, single cell protein, secondary metabolites can be produced by different processes using organic waste as the alternative substrate. Thus Mushroom cultivation on the organic waste is prominent method for recycling of the waste (agricultural, industrial, forestry, and household wastes) and simultaneous production of the healthy food in the form of mushroom.

Mushrooms being devoid of chlorophyll, depend upon the green plant for the organic matter. The organic matter/biomass produced by green plants is used by mushrooms as the source of nutrients by secreting enzymes that can degrade complex food into a simple one. The simple compounds produced are then used easily by the mushrooms for their growth and development. The organic materials produced from industry, households, and agriculture on which mushrooms depends for nutrients are known as substrates. Disposal of the organic waste in unscientific manner can have adverse impact on the receiving environment like eutrophication of water bodies, development of the anaerobic condition, formation of algal blooms and decrease in dissolved oxygen level of water. These adverse impacts can be avoided by using proper techniques as the waste produced is rich in nutrients and can acts as the substrate for different organisms. Keeping in mind the impacts related to unscientific disposal of waste, the focus on the green techniques in increasing all over the world. The green techniques like composting for the conversion of organic waste are considered crucial for reducing the burden of enormous waste. Composting of organic waste also results in the formation of biofertilizers that can improve nutrient status and moisture retention capacity of the soil. This technique has got wide acceptance and can also acts as the crucial component of integrated waste management.

A general and simplified diagrammatic representation of mushroom cultivation on agro industrial wastes is shown in Figure 12.1.

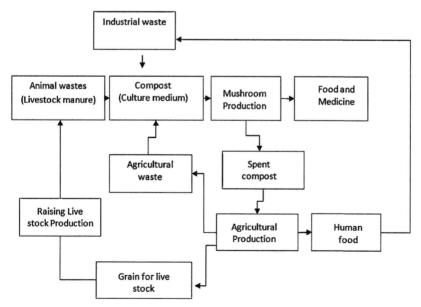

FIGURE 12.1   Simplified diagrammatic representation of mushroom cultivation on different agro industrial wastes.

In India, more than 60% of the population is dependent upon agriculture for their livelihood. Due to various agriculture activities (crop production, livestock) huge quantity of waste generated is rich in cellulose, hemicellulose, and lignin. An enormous quantity (>200 × 10^9 tons) of lignocellulosic rich waste is produced annually around the world (Zhang, 2008).

Organic waste produced from different food crops (27) is estimated to be about 4 × 10^9 tons out of which cereals annually accounts almost 3 × 10^9 tons of lignocellulosic waste (Lal, 2008). 3/4th of the total world production of organic waste is occupied by cereals (FAO, 2004) like rice, barley, wheat, maize, millet, and sorghum. Sugar cane, cotton, coconut, groundnut, sunflower can also produce considerable amount of waste that can be converted into useful products after bioconversion.

The organic waste from agro-industries contains nutrient-rich material in small quantity, but the collection and the transportation cost of the waste is high. The utilization of organic waste as a source of nutrients for mushroom cultivation is considered as resource but not the waste. Mushroom cultivation present an economically important industry that recover proteins from the organic waste by using solid-state fermentation. Solid-state fermentation uses fungi that can feed on various substrates from different agro-industries for the production of mushrooms and other beneficial compounds. Cultivation of mushrooms on organic waste can utilize waste as a resource that has no impact on human health as well as the human environment.

12.2   BENEFICIAL EXPRESSIONS OF MUSHROOM CULTIVATION ON THE AGRO-INDUSTRIAL WASTES

Human has been using mushroom as a source of food from a long time, but their production in a controlled condition and nutrient assessment was not done due to the lack of techniques. With time, the cultivation under controlled conditions has started, and their nutrient value has also been accessed. The mushrooms can be used as a source of food and also medicine. Mushrooms are digestible and can act as a rich source of proteins (20–30%). Mushrooms on comparing with animal meat on a nutrient basis, are kept below the meat but well above vegetables and fruits. Due to good nutrient content, mushrooms are also known as mycomeat or vegetable meat. Mushrooms can not only act as a good source of nutrients but also a good source of energy (fats and carbohydrates), vitamins, and other compounds that are beneficial for human health (Chang and Miles, 2004; Suman and Sharma, 2007).

A large quantity of biomass is produced that can act as a sustainable resource like solar energy. The enormous biomass (lignocellulosic) produced has no value in its original form until it is converted into other useful products. Currently, a good amount of the fund is provided for research in increasing grain, fiber, and oil production, but the limited financial resource is provided for the reuse of waste as a resource for the production of different beneficial products. The cheap method of agriculture waste disposal is burning that is followed in most of the developing countries but increase the release of different pollutant into the environment, thereby ultimately effects human health. The proper management of the lignocellulosic waste produced can act as a new resource for different microorganisms. Microorganisms feed on the waste as a nutrient source and convert the waste into valuable products like ethanol and manures (Figure 12.2). The proper management of agro-waste can also help to improve the economic condition of the farmers. Waste from cereals can be easily converted into food as mushroom cultivation directly converts waste into mushroom. The material left after cultivation of mushroom can have positive effects on the soil and also good source of nutrients. The residue can also be used as soil conditioner as it have high cation exchange capacity and also help in maintaining increases moisture retention capacity of the soil. The nitrogen rich compost contains partially decomposed lignocellulosic waste on mixing with animal or human excreta results in the production of biogas and nitrogenous sludge. The integrated approach for the conversion of the organic waste into mushroom, biogas, nitrogenous sludge, bio-fertilizers can definitely double the income of the farmers with sustainable management of the waste. Thus integrated approach not only produces different by-products but also zero waste emission.

Lignocellulosic wastes are generated in huge quantity in rural as well as urban area by agricultural and agro-industrial activities. It is a waste and contains complex compounds with no food value in original form. It can be converted into simple compounds by using chemicals like dilute hydrochloric acid and calcium chloride but using the chemical for increasing digestibility of the waste is a very costly process. In contrast, mushroom cultivation techniques have become significantly important in recent years in improving nutritional quality and upgrading the economic value of the solid OWs. Mushrooms have different enzymes that convert complex compound into simple one that can be used by mushroom for its growth and development. This demonstrates the impressive capacities of mushrooms for broad-spectrum 'biosynthesis,' which is different from 'photosynthesis' by green plants. The use of different substrates by mushroom species is totally dependent on mushroom and substrate. On examining the enzyme profile of mushroom species on a different substrate, mushroom shows varying ability in utilizing different substrate as the nutrient source.

Mushroom production is a short and efficient method of protein recovery by utilizing fungi that have a great capacity to degrade lignocellulosic materials. Two mushroom species, namely *L. edodes* and *Pleurotus* can feed on different substrates (wheat straw, cotton wastes, coffee pulp, corncobs, sunflower seed hulls wood chips, and sawdust) and convert mushroom protein. Enzymes in large quantity are produced by these two species that degrade the organic waste in order to release nutrient necessary for the growth and development of mushrooms. Petre and Petre (2013) proposed the scheme of laboratory-scale biotechnology of *Pleurotus* sp. and *Lentinula edodes* on winery and apple wastes as these produce enzymes in high quantity and are efficient in converting agro-industrial material into nutrient-rich food. The conversion efficiency and productivity is expressed in the form of biological efficiency and is calculated by the ratio of the weight of fresh mushroom to the weight of the dry substrate.

Mushrooms are also known for their absorptive nature and can easily absorb and accumulate different nutrients into their body. This absorptive power of the mushroom can easily help to cultivate mushrooms on different substrates with varying nutrient compositions and also help to improve the nutritional composition of the mushrooms. Due to the good absorptive power of mushrooms, the concentration of the desired nutrient that is lacking in most of the food material can be made available by growing mushrooms in that nutrient-rich substrate. For example, *Sargassum* seaweed is known to have a rich content of iodine, and growing mushrooms on this substrate can produce iodine-rich mushrooms.

Currently, the mushroom industry is focusing on dual benefits of producing mushrooms as source of nutrient rich food and various compounds that are having medicinal importance like supplement for anticancer and having antiviral properties. Polysaccharides, the bioactive compound produced in higher basidiomycetes is known to have anticancer properties and the production of this compound can be increased by using the modern biotechnology. The other compound like nutriceuticals, can be extracted from mycelium or fruiting body of the mushroom. The promising method of production of pharmaceutical substances is by submerged mushroom cultivation and medicinal mushroom produced by submerged fermentation (SMF) on the different substrate (grain byproducts, winery waste) results in fast mushroom growth with high biomass production (Philippoussis, 2009).

12.3   RESTORATION OF DAMAGED ENVIRONMENT

Improper disposal of organic waste can lead to various problems (nutrient loss, eutrophication), but cultivating mushrooms on the same waste can reduce the burden on the environment. Enzymes are produced by mushroom mycelia which convert the complex substances into simple one thus results in the degradation of the waste and making this technique as an ecofriendly one. Mushroom mycelia (Saprotrophic, endophytic, mycorrhizal, or even parasitic fungi/mushrooms) can also restore the damaged environment by using different processes like mycofiltration, mycoforesty, mycoremediation, and mycopesticide. These processes are cleaning the environment by converting waste into useful products and restoration of forest, controlling insect pest population, and removing toxic pollutants from the contaminated environment (Stamets, 2005). Mushroom exhibits extraordinary abilities to transform recalcitrant pollutants and also degrades a broad spectrum of structurally diverse toxic environmental pollutants. The two main major problems that the whole world is facing are the continuous generation of waste and scarcity of protein-rich food, but both problems can be easily solved by using waste as the substrate for mushroom cultivation. The process of removing nutrients and toxic pollutants from the waste is known as mycoremediation. Mycoremediation by biodegradation and biosorption can help in cleaning the environment, thus representing a cheap and ecofriendly approach for the restoration of the environment.

12.4   BIODEGRADATION

Biodegradation is the degradation of complex substances into a simple and easily useable form. Mushrooms produce extracellular substances (peroxidises, lignases) that are capable of degrading recalcitrant pollutants, dyes, nitrotoluenes, and also plastic.

Fungi, along with bacteria, are known to show good results in the soil contaminated with polycyclic aromatic hydrocarbon (PAH), which can remove maximum concentration by converting PAH into carbon dioxide and water.

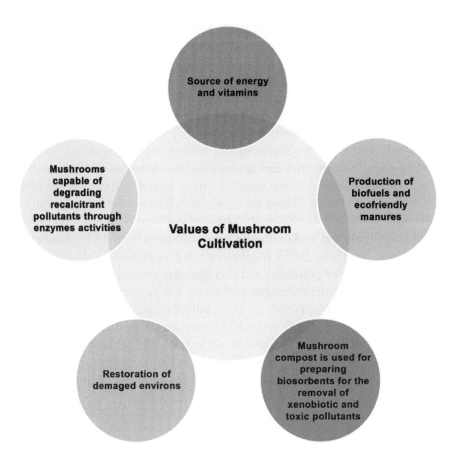

Source of energy
and vitamins

Mushrooms
capable of
degrading
recalcitrant
pollutants through
enzymes activities

**Values of Mushroom
Cultivation**

Production of
biofuels and
ecofriendly
manures

Restoration of
demaged environs

Mushroom
compost is used for
preparing
biosorbents for the
removal of
xenobiotic and
toxic pollutants

FIGURE 12.2   Various benefits of mushroom cultivation in disturbed environs.

12.5   BIOSORPTION

Biosorption is the removal of toxic pollutants from the contaminated environment. It is a second alternative process for the remediation of toxic pollutants from the effluent. Dead or alive mushroom biomass is tolerant to toxic pollutants and removes these pollutants from the contaminated sites by sorption. Mushroom mycelium or compost is used for preparing biosorbents for the removal of xenobiotic and toxic pollutants from the contaminated sites. Removal of toxic pollutants occurs by two processes, i.e., bioaccumulation, and biosorption. In bioaccumulation, energy is used to accumulate the toxic pollutants inside cell and intracellular components while in biosorption pollutants bind with the biomass without any energy usage. The use of biodegradation and biosorption is important as it helps to solve two major problems of restoration of the contaminated environment and production of the protein rich food. As biodegradation degrade the toxic substances and convert into the form that can be easily utilized through biosorption process for the biomass preparation. It is important to select the species which can degrade the toxic pollutants, and preference should also be given for the degradation of pollutants.

12.6   CONCLUSION

Growing organic waste is a big problem growing day-by-day throughout the world. The unscientific disposal of organic waste into the water bodies is troublesome and makes the water unfit for human use. Mushroom cultivation reduces not only the burden of organic waste generation but also the production of nutrient-rich food. Using waste as a resource for the cultivation of mushrooms can also become an extra source of income for the farmers. Thus, it is high time to use this technique widely in order to provide the necessary information and techniques necessary for the cultivation of mushrooms on organic waste.

KEYWORDS

- **bioconversion**
- **biodegradation**
- **environment management**
- **mushroom**
- **organic waste**
- **polycyclic aromatic hydrocarbon**

REFERENCES

Chang, S. T., & Miles, P. G., (2004). *Mushrooms: Cultivation, Nutritional Value, Medicinal Effect, and Environmental Impact* (IInd edn.). CRC Press, New York.
Chang, S. T., (2008). Overview of mushroom cultivation and utilization as functional foods. In: Cheung, P. C. K., (eds.), *Mushrooms as Functional Foods* (pp. 1–34). John Wiley & Sons, Inc.

Jebapriya, G. R., Daphne, V., Gnanasalomi, V., & Gnanadoss, J. J., (2013). Application of mushroom fungi in solid waste management. *International Journal of Computing Algorithm, 02*, 279–285.

Kulshreshtha, S., Mathur, N., & Bhatnagar, P., (2014). Mushroom as a product and their role in mycoremediation (Mini Review). *AMB Express, 4*, 29–36.

Lal, R., (2008). Crop residues as soil amendments and feedstock for bioethanol production. *Waste Manag., 28*(4), 747–758.

Lohani, H., (2012). *Training Manual on Mushroom Cultivation Technology.* United Nations-Nations Unies Economic and Social Commission for Asia and the Pacific Asian and Pacific Centre for Agricultural Engineering and Machinery (APCAEM), Beijing-100029., P.R. China.

Petre, M., & Petre, V., (2013). Environmental biotechnology for bioconversion of agricultural and forestry wastes into nutritive biomass. In: Petre, M., (ed.), *Environmental Biotechnology-New Approaches and Prospective Applications* (pp. 1–22). In Tech Publishers Janeza Trdine 9, 51000 Rijeka, Croatia.

Philippoussis, A. N., (2009). Production of mushrooms using agro-industrial residues as substrates In: Nigam, P. S., & Pandey, A., (eds.), *Biotechnology for Agro-Industrial Residues Utilization* (pp. 163–196).

Phulippoussis, A., & Diamantopoulou, P., (2011). Agro-food industry wastes and agricultural residues conversion into high value products by mushroom cultivation. In: *Proceedings of the 7ᵗʰ International Conference on Mushroom Biology and Mushroom Products (ICMBMP7)* (pp. 339–351). Section: Waste conversion, substrates and casing.

Stamets, P., (2005). *Mycelium Running: How Mushroom Can Help Save the World* (p. 574). Ten Speed Press, Berkeley and Toronto.

Suman, B. C., & Sharma, V. P., (2007). Uses of Mushrooms. In: Suman, B. C., & Sharma, V. P., (eds.), *Mushroom Cultivation and Uses* (pp. 18, 035, 110). Daya Publishing House (Delhi, India).

Tanyol, M., Tepe, Q., & Uslu, G., (2014). *Assessment of Agro-Industrial Waste: Biotechnological Potential.* Eur Asia Waste Management Symposium. YTU 2010 Congress Center, İstanbul Turkey.

Waseer, S. P., (2010). Medicinal mushroom science: History, current status, future trends, and unsolved problems. *Int. J. Med. Mush., 12*(1), 1–16.

Zhang, Y. H. P., (2008). Reviving the carbohydrate economy via multi-product lignocellulose biorefineries. *J. Ind. Microbiol. Biotechnol., 35*, 367–375.

CHAPTER 13

An Essay on Some Biotechnological Interventions in Agricultural Waste Management

RUKHSANA AKHTAR,[1] ADIL FAROOQ WALI,[2] SAIEMA RASOOL,[3] SABHIYA MAJID,[4] HILAL AHMAD WANI,[4] MUNEEB U. REHMAN,[4,5] SHOWKAT AHMAD BHAT,[4] SHABHAT RASOOL,[4] SHAFAT ALI,[6] and REHAN KHAN[7]

[1] *Department of Biochemistry, University of Kashmir, Hazratbal Srinagar – 190006, Jammu and Kashmir, India*

[2] *RAK College of Pharmaceutical Sciences, RAK Medical and Health Sciences University, Ras Al Khaimah, P.O. Box – 11172, United Arab Emirates*

[3] *Forest Biotech Laboratory, Department of Forest Management, Faculty of Forestry, University of Putra Malaysia, Serdang, Selangor – 43400, Malaysia*

[4] *Department of Biochemistry, Government Medical College (GMC), Karan Nagar, Srinagar – 190010, Jammu and Kashmir, India*

[5] *Department of Clinical Pharmacy, College of Pharmacy, King Saud University, Riyadh – 11451, Saudi Arabia, E-mail: muneebjh@gmail.com*

[6] *Centre of Research for Development, University of Kashmir, Srinagar, Jammu and Kashmir, India*

[7] *Department of Nano-Therapeutics, Institute of Nano-Science and Technology (DST-INST), Mohali, Punjab, India*

ABSTRACT

The human population is growing rapidly throughout the globe. To feed the growing population; there has been a rapid increase in intensive agricultural practices. As a result, the volume and types of agricultural waste biomass have gone up. The common traditional approach to manage agro-waste has been its release to the environment either with or without any treatment. Such practices increase the risk of environmental pollution and public health hazards. With the advancement of biotechnology, many new techniques for the management of agricultural wastes (AWs) have been developed, which had enabled one to reduce the toxic and hazardous effects of waste. These techniques convert the agro-waste into eco-friendly and value-added products having the potential of providing sustainable raw material for utilization of living organisms, including humans. Some of the potential applications of agricultural residues that had been exploited with the aid of biotechnology had been discussed in this review.

13.1 INTRODUCTION

Globally human population growth increases to 1.1% annually. The population has grown from 1 billion from 1800 to 7.616 billion in 2018. To feed the growing population; there has been a rapid increase in intensive agricultural practices. As a result, the volume and types of agricultural wastes (AWs) biomass have gone up. Globally, 140 billion metric tonnes of biomass is generated every year from agriculture (Singhania et al., 2017). The AWs exist in variety of forms like solid, liquid, and slurries (Table 13.1). The composition of the waste varies with the type of agricultural activity. Agricultural waste most commonly belongs to the following three categories: food processing waste, crop waste (field residue), and hazardous and toxic agricultural waste like pesticides, insecticides, and herbicides (Obi et al., 2016). Among the three, crop waste and processing residues contribute to a considerable proportion of total waste. Field residues or crop waste are remains of the crop that are left in an agricultural field during the practice of crop harvesting. These residues comprises of stalks and stubble or stems, leaves, and seedpods. Such residues either are plowed with the soil or are burnt to add to the fertility of the field. During the processing of crops in the valuable form, some residues such as husks, seeds, bagasse, molasses, and roots are taken out. This unusable crop processing waste is most commonly used to feed the livestock. In some alternative cases, such waste are dumped and kept stagnant for some months which are later on used as soil amendment to add the fertility of soil. This waste can be used as irrigation control which proves quite helpful in prevention of soil erosion (Sadh et al., 2018).

TABLE 13.1 Different Types of Agricultural Waste

	Agricultural Waste		
SL. No	Crop Waste (Field Residue)	Food Processing Residues	Toxic Agricultural Waste
1.	Leaves	Husks	Pesticides
2.	Stems	Seeds	Insecticides
3.	Stalks	Roots	Herbicides
4.	Seed pods	Bagasse	Manure
5.	Stubble	Molasses	Fertilizer waste

The composition of AWs varies from being starch or cellulose rich to the ones quite rich in nitrogen. The majority of the agricultural biomass is rich in cellulose (40% approx.), followed by hemicelluloses (almost 30%) and lignin (25%). The composition of the most commonly generated crop waste is shown in Table 13.2.

TABLE 13.2 Chemical Composition of Commonly Used Agricultural Waste

Type of Crop	Chemical Composition (% w/w)				
	Cellulose	Hemicellulose	Lignin	Ash	References
Sugarcane bagasse	30.2	56.7	13.4	1.9	Nigam et al. (2009)
Rice straw	39.2	23.5	36.1	12.4	El-Tayeb et al. (2012)
Corn stalks	61.2	19.3	6.9	10.8	El-Tayeb et al. (2012)
Cotton stalks	58.5	14.4	21.5	9.98	Nigam et al. (2009)
Wheat straw	32.9	24.0	8.9	6.7	Martin et al. (2012)
Barley stalks	33.8	21.9	13.8	11.0	Nigam et al. (2009)
Soya stalks	34.5	24.8	19.8	10.39	Motte et al. (2013)
Sunflower stalks	42.1	29.7	13.4	11.17	Motte et al. (2013)

This agricultural waste has attractive potential of providing sustainable raw material for utilization of living organisms including humans. Most of starch rich agricultural biomass is being utilized as animal fodder. Starches are polysaccharides with a large number of glucose molecules linked together by alpha 1–4 linkages as well as alpha 1–6 linkages at branching points. These energy storage compounds are easily digestible. Cellulose is most abundant organic biomass utilized as animal feed and an energy source. The field residue rich in cellulose are burnt directly and the remaining amount available can be used as raw material for conversion into value-added products. As a result, cellulosic AWs had gained much interest of environmental engineers.

13.2   AGRICULTURAL WASTE MANAGEMENT

The common traditional approach to manage agro-waste has been its release to the environment either with or without any treatment. Such practices increase the risk of environmental pollution and public health hazards (Wright et al., 1998). The dumping of organic waste causes volatilization of ammonia which affects the overall quality of air (due to the emission of gases like NH_3, NO, and CH_4), water (due to release of high nitrogen and phosphorous), and soil (because of loading of potassium and phosphorous). The stagnation of AWs provides a medium for the growth and multiplication of flies, which are the causative agents of various diseases (Fabian et al., 1993). In order to clear lands, agricultural waste is openly burnt in the fields, which emit CO_2, CH_4, and other pollutants in the air, thus contributing the climate change. The improper management of waste is an alarming problem of the century. There is a dire need to recognize waste as a potential resource rather than leaving it as discarded products. This will not only provide us a renewable energy source but also proves helpful in the prevention of contamination of air, water, and land resources. It has been found that some AWs have a potential of establishing markets. Most of them can be extensively used as soil nutrient recycling and soil upgrading purposes. These wastes possess the potential of replacing the synthetic fertilizers. Besides this, biomass wastes possess the high value with respect to material and energy outputs; as a result this can be used as a raw material for large-scale industries (Zafer et al., 2014). In the developing countries it has been found that, AWs have great potential to be used as raw material that has a capacity of generating 50 billion tons of oil and may offer renewable energy to millions of houses which still lack access to basic facility of electricity (Buyukgungor et al., 2009).

To circumvent these problems, significant efforts are being made by many governments and non-government agencies throughout the globe. Biotechnology has provided a wide platform to manage waste agricultural biomass and to convert it into a material resource.

13.3   APPLICATION OF BIOTECHNOLOGY IN THE MANAGEMENT OF WASTE

Biotechnology had played an important role in the management of agricultural waste in terms of economical and environmental benefits. Biotechnological applications and approaches had enabled one to reduce the toxic and hazardous effects of waste and by converting them into value added products that are environmental friendly. Some of the biotechnological applications viz. biofiltration, biosorption, bioreactor, bioleaching, composting, and phytoremediation, etc., had contributed immensely in mitigating agricultural waste and making it beneficial for mankind. Since agricultural residues are starch and cellulose rich, it therefore possesses an inexhaustible raw material for the generation of energy. Some of the potential applications of agricultural residues that had been exploited with the aid of biotechnology had been discussed in this review.

13.3.1 AGRICULTURAL WASTE AS A RAW MATERIAL FOR CONVERSION OF ENERGY

Agricultural biomass has a potential to be used as a resource for energy and other value added products. There are several routes for the conversion of agricultural biomass into energy. The two major routes are thermochemical conversion and biochemical conversion (Figure 13.1).

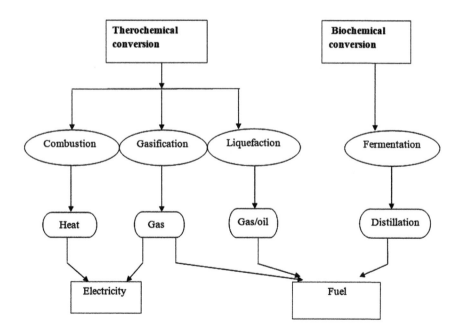

FIGURE 13.1 Schematic representations of different routes of conversion of waste into energy.

13.3.1.1   THERMOCHEMICAL CONVERSION

Thermochemical conversion is characterized by very high temperature and is best suited for low moisture containing biomass. The three principal methods of thermochemical conversion are combustion, gasification, and pyrolysis which are discussed one by one.

13.3.1.2   COMBUSTION

There are certain properties of agricultural residues which lend them as an important tool for biotechnologists for generation of electricity. The potential of power generation varies from few kilowatts up to several megawatts. According to the study conducted in the year 1980 it has been reported that capacity of power generation in sugarcane growing countries was about 70% more than the total electricity generated in these countries from all sources on that time. The biomass residues are characterized by high volatile matter which makes it quite easy to burn and ignite (Hallet al., 1991). The agricultural residues mainly consist of plant materials that had been periodically harvested. During the growth and development, these plants have lost CO_2 to atmosphere through the process of photosynthesis. This adds the extra benefit of using agricultural residue as partial substitution of fossil fuels as their combustion is CO_2 neutral which may have otherwise aroused the issue of global warming (Werther et al., 1995; Oganda et al., 1996).

   Besides the generation of electricity, biotechnologists have also used the agricultural biomass for production of low-pressure steam ranging from 0.1 to 1 MPa and high-pressure steam which ranges from 4 to 10 MPa. The low pressure steam finds its use in certain industrial processes like food preparations, lumber, and pulp manufacture and kiln operations. The high-pressure steam are being used to expand turbines either connected to electric generators or connected to direct machine drives. To achieve this end, biotechnology had provided a helping hand in the designing and operation of systemic machinery. With the advancement of biotechnology, different types of combustors for agricultural residues have been designed from time to time. Among these, grate-fired systems is one of the first combustors used in industrial settings, while fluidized bed systems are the versatile units prevailing from the last 10 years. Suspension burners being one of the recent combustors designed for the special application of dry fuel. Depending upon the purpose of the process and the size and property of available fuel, the selection of a suitable system is made (Tilman et al., 1997).

13.3.1.3   GASIFICATION

It is a process of choked or incomplete combustion that uses heat and pressure to convert any carbon rich material into synthetic gas known by several names like wood gas, syngas, producer gas, etc. This synthetic gas mainly composed of CO and H still retain the combustion potential and is passed through the pipes to other places where it is used for at large scale purposes. Gas produced by this method is known by several names as wood gas, syngas, producer gas, etc. In this way, gasification adds value to agricultural waste by converting them to marketable fuels. The generation of syngas depends upon the availability of types of waste. Usually, in farmlands, the feedstock's like cane stock are used for it, but in the urban areas, wastes like garbage and tires are also utilized for it. According to one of the studies conducted in the US (Department of Energy), gasification can be considered for the production of hydrogen for transportation fuel. Apart from being a building block for a broad range of chemical products, it is also be supplied to power-generating fuel cells and power-generating turbines (Singhania et al., 2017).

13.3.1.4   PYROLYSIS

Pyrolysis is defined as the process of the irreversible thermochemical decomposition of material at increased temperature (300 to 600°C) in the absence of oxygen to vaporize a portion of the material, leaving a char behind (Figure 13.2). In the process of pyrolysis low energy materials are converted into high-energy compounds, which include liquids like bio-oil and solids like biochar, and gases like syngas (the production of syngas from pyrolysis is actually the gasification is discussed above). The compounds thus produced are used for a wide range of purposes, starting from the burning of the cooking stove to the running of automobiles (Laird et al., 2009). The three forms of pyrolysis products, i.e., liquid (bio-oil), solid (biochar), and gas (syngas) are apportioned according to the condition, type of reaction, and availability of biomass. Broadly pyrolysis is categorized into three categories: fast, slow, and gasification. In the process of fast pyrolysis, bio-oil is produced; thus, this process is employed when the production of liquid fuel is concerned. In contrast to fast, slow pyrolysis maximizes the production of biochar

(Venderbosch and Prins, 2010; Butler et al., 2011; Jarboe et al., 2011). The bio-oil produced from pyrolysis is characterized by certain properties like non-oxidative reactivity and presence of oxygenates like organic acids. The oil is also high composed of high moisture and high heteroatom content. Therefore, in order to use the bio-oil as transportation fuel, prior upgrading operations like distillation and hydroprocessing are required (Elliot, 2007; Mortensen et al., 2011; Balat, 2011; Sorrell et al., 2010).

The other technique that is involved in the thermal conversion of biomass waste is catalytic pyrolysis. In the process of catalytic pyrolysis zeolite catalyst is added and mixed with the waste which is then subjected to pyrolysis. In some cases, catalyst is not directly mixed with biomass instead are placed under the pyrolysis reactor in a fixed form (Carlson et al., 2009). Vapors from the mixture get absorbed in the pores of the zeolite. Within the pores, these vapors undergo rearrangement reactions to produce olefins and aromatic compounds (Carlson et al., 2011). In this process proteins and oxygenated compounds get converted into hydrocarbons without adding any amount of hydrogen. The nitrogen from proteins gets converted to ammonia (Wang and Brown, 2013).

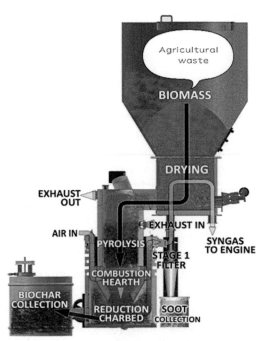

FIGURE 13.2 Thermochemical conversion of agricultural biomass through different processes.

13.4   BIOCHEMICAL CONVERSION OF AGRICULTURAL BIOMASS

This is an environment friendly and sustainable method for the conversion of agricultural biomass into fuel especially bioethanol. This technique is more suitable for wet wastes containing lignocellulosic biomass. Three principal methods of biochemical conversion route are pretreatment, hydrolysis, and fermentation (Figure 13.3).

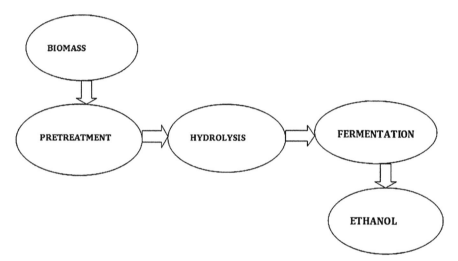

FIGURE 13.3   Schematic representation of biochemical conversion of biomass into ethanol.

13.4.1   PRETREATMENT

Pretreatment is the process which aims to decrease the crystalline nature of cellulose to make it more accessible to enzymes. During pretreatment lignin cellulose interaction is broken, hemicellulose is removed, and cellulose structure is opened up to increase its exposed surface area. This enables efficient and rapid bioconversion of cellulose into free sugar molecules (Panday et al., 2015). Several biowaste pretreatments are done which includes:

- Physical treatment;
- Dilute acid pretreatment;
- Dilute alkali treatment;
- Hot water treatment; and
- Steam explosion treatment.

Physical pretreatment aims to reduce the particle as small as possible. It includes techniques like milling and grinding. All these preliminary treatments are meant to increase hydrolysis efficiency. The pretreatment varies with the availability and type of biomass (Murthy et al., 2011). Dilute acid pretreatment is one of the extensively used treatments. In it, the biomass is subjected to acid treatment at high temperature whose concentration varies with the type and composition of waste. This method is more suitable for large-scale treatments owing to its characteristics of short residence which makes biomass ready for next treatment quickly. In the presence of dilute acid, hemicellulose present in the biomass gets solubilized by means of hydrolyzation, while cellulose and lignin components are left behind in solid fraction. This allows one to separate the hemicellulose from cellulose. At high temperatures or high acid concentrations the hemicellulose gets converted into furans which inhibit the subsequent treatments (Nguyen et al., 1999). Dilute alkali pretreatment is rare though it also denatures the biomass structure. In presence of alkali, at high temperature, lignin is degraded. The lignin-cellulose interaction is broken down due to the reaction of alkali with the esters of lignin. This makes lignin soluble and removable from the mixture and sugar-containing compounds remain in solution. The degradation of biomass varies with concentration of alkali and a good balance is required to achieve the good result (Murthy et al., 2011). Hot water pretreatment is an auto catalyzed process in which biomass is dissolved in high temperature water. This allows the degradation of branched structure of hemicellulose. In this process acetic acid is generated which lowers the pH thus facilitates structural degradation of biomass. This method has not been used on commercial scale since it requires high pressure for the water to be kept liquid (Kotiranta et al., 1999). The steam explosion treatment is more suitable for large-sized biomass and is energy saving since no energy is needed to reduce the size of biomass. In it, biomass is heated under high pressure with water vapor. The pressure is released at once which facilitates the degradation of biomass (Sun et al., 2002). All the aforementioned pretreatment methods have its merits and drawbacks. Once the agricultural biomass undergoes pretreatment, it is subsequently subjected to hydrolysis by enzymes.

13.4.2 HYDROLYSIS

The enzymatic hydrolysis or saccharification is the second step involved in the process of ethanol production from agricultural waste. This step is carried out with the aid of certain enzymes like cellulases and hemicel-lulases that are produced from a variety of fungal strains. These cellulases are enzyme complexes of three main enzymes as: endocellulase, exocellu-lase, beta-glucosidase. All three enzymes cooperate to hydrolyze cellulose into monomeric sugars (Singhania et al., 2009). Besides this, enzymes like xylanase and ligninase are also found to work in association. It has been found that all of these three enzymes are produced by several fungal microorganisms among *Trichoderma reesei* is one mostly studied cellu-lase producer strain of fungi. There are several other filamentous fungal strains that had been adopted for the commercial production of cellulases and all other components required for hydrolysis of waste (Gusakov et al., 2013). Depending upon the type of waste and fermentation process, different fungal species are used for cellulase production, as shown in Table 13.3.

TABLE 13.3 Different Fungal Species Used in Fermentation Process

Fungi	Waste Category	Fermentation Process	References
Aspergillus terreus	Rice straw	SSF	Narra et al. (2012)
Aspergillus fumigates ABK9	Wheat bran, rice straw	SSF	Das et al. (2013)
Aspergillus protuberus	Rice husk	SSF	Yadav et al. (2016)
T. asperellum	Wheat bran	SSF	Raghuwanshi et al. (2014)
Aspergillus	Wheat straw	SSF	Pensupa et al. (2013)
Aspergillus fumigates NITDGPKA3	Rice straw	SSF	Sarkar and Aikat (2012)
Aspergillus niger KK2	Rice straw	SSF	Kang et al. (2004)
Aspergillus niger NCIM548	Wheat bran, Corn bran, Kinnow peel	SMF	Kumar et al. (2011)
Penicillium echinulatum	Sugar cane bagasse	SMF	Camassola and Dillon (2014)
Trichoderma viride VKF3	Sugarcane bagasse	SSF	Nathan et al. (2014)
Rhizopus oryzae CCT7560	Rice husk and rice bran	SSF	Kupski et al. (2014)

In the whole process of ethanol production from biomass this step accounts maximum part of the cost of technology because of cost of enzyme (Mosier et al., 2005). In order to reduce the cost and make this technology economically feasible recycling of enzyme is an important strategy that needs to be adopted (Tu et al., 2007). Zhao et al. (2012) reviewed certain chemical and physical factors that had been found to affect enzyme hydrolysis. Some of the key factors are:

- Amount of lignin content in the pretreated waste;

- Activity of the cellulases and hemicellulases;

- Optimum temperature and pH for action of enzyme;

- Optimal concentration and amount of enzyme and substrate loading;

- Time duration of the enzymatic hydrolysis;

- Inhibitory compounds (Furfurals) generated from lignin degradation.

All the above-mentioned factors should be overcome to maximize the efficiency and yield of the process. In order to tackle the problem of enzyme loading, it has been seen that addition of surfactants during the hydrolysis had reduced the surface tension and changed the surface properties of soil (Taherzadeh and Karimi, 2007).

Currently separate hydrolysis and fermentation (SHF) and simultaneous saccharification and fermentation (SSF) are highly adopted technologies for bioethanol production from biomass. However, Onsite production of enzyme using low substrate could be the forward step to develop economically feasible technique (Singhania et al., 2014).

13.4.3 FERMENTATION

The hydrolyzed or saccharified AWs are subsequently subjected to incubation with yeast and bacteria and this process is known as fermentation. This is the third and final step of ethanol production from biomass. In this process, several microorganisms, especially yeasts and some bacteria, as shown in Table 13.4, are used to reduce hexoses and pentoses to ethanol. In most of the cases, this process is carried out at moderate temperature (25–37°C), but it has been found that high temperatures (45–55°C) yield good results (Antoni et al., 2007). Since the microorganisms lack the ability to ferment both pentoses and hexoses, this makes the process more costly and is one of the hurdles in lowering ethanol prices.

TABLE 13.4   Different Microorganisms Which Produce Ethanol by Fermentation

Microorganism	Species	References
Yeast	1. 24860-*S. cerevisiae*	Valet et al. (1996)
	2. 27774-*Kluyveromyces fragilis*	
	3. 30016-*Kluyveromyces marxianus*	
	4. 30091-*Candida utilis*	
Fungi	1. *Mucor* sp. M105	Ingram et al. (1998)
	2. *Fusarium* sp. F5	
Bacteria	1. *Clostridium sporogenes*	Miyamoto et al. (1997)
	2. *Clostridium indoli*	
	3. *Clostridium sphenoides*	
	4. *Clostridium sordelli*	
	5. *Spirochaeta stenostrepta*	
	6. *Spirochaeta litorali*	
	7. *Erwinia amylovora*	
	8. *Streptococcus lactis*	
	9. *Leuconostoc mesenteroides*	
	10. *Escherichia coli* KO1	

Fermentation can be carried out through two different ways as:

1. **Separate Hydrolysis and Fermentation (SHF):** In SHF, the two processes of hydrolysis and fermentation are carried out separately in two different units, each at its own optimal conditions of temperature, pH, and enzyme substrate concentration. This accounts for efficiency of both of the steps. However, main demerit of this method is that glucose formed as a result of hydrolysis inhibits further activity of enzymes. Moreover, this process is time consuming which leads to the risk of contamination (Taherzadeh and Karimi, 2007).

2. **Simultaneous Saccharification and Fermentation (SSF):** In this method both the time and labor are reduced. In SHF the two steps of hydrolysis and fermentation are merged and carried out in one-step. As a result pentoses or hexoses formed from hydrolysis gets quickly converted into ethanol which does not inhibit any enzyme activity. Over SHF this process is more advantageous owing to its low cost, low labor, less time, minimum equipment requirement and less risk of contamination (Brethauer et al., 2010).

13.4.4 BIOCHEMICAL CONVERSION OF AGRICULTURAL WASTE VIA ANAEROBIC DIGESTION (AD)

Anaerobic digestion (AD) is a sequence of biological processes that occur in oxygen-free conditions in which microorganisms break down biodegradable material. The anaerobic process is economical that can also produce energy at low cost treatments. The process is accomplished in four steps as hydrolysis, acidogenesis, acetogenesis, and methanogenesis (Singhania etal., 2014).

13.5 AGRICULTURAL BIOMASS AS AN OPTIMIZED CARBON SOURCE

Agricultural biomass is exploited in various bioprocesses by using it as a carbon source. Besides the important enzymes starchy residues have been found to be rich in various components that make them the excellent carbon source in various bioprocesses. Agricultural biomass like Oil cakes, rice bran, soya bean hulls, barley husk, maize are being widely used as optimized carbon sources in many bioprocesses like solid state fermentation (SSF) and submerged fermentation (SMF).

13.5.1 SOLID STATE FERMENTATION (SSF)

This fermentation is carried out in absence of free water. In SSF solid agricultural biomass containing enough moisture like sugarcane bagasse, coffee pulp, oilseed residue which are economical and easily available are used as source of nutrients to support the microbial activities. This solid matrix of agricultural biomass acts as a source of carbon and other nutrients for the metabolism of microbes. The efficiency of SSF is dependent on various factors viz. temperature, aeration, microorganisms, and type of fermenter used (Pandey et al., 2003). The different cultures of microorganisms used in SSF are as:

- Single pure culture;
- Mixed identifiable culture;
- Consortium of mixed indigenous microorganisms.

SSF is a multistep process which is carried out in following steps:

1. Substrate selection;
2. Pretreatment of substrate either mechanical or chemical;
3. Hydrolysis of substrates;
4. Fermentation process;
5. Downstream processing.

13.5.2   SUBMERGED FERMENTATION (SMF)

This process involves the growth of the microorganisms (mostly fungi) as a suspension in liquid medium in which various nutrients are either dissolved or suspended as a particulate solid. Besides SSF, this process also utilizes agriculture wastes as carbon source. The various substrates that serve as cellulosic substrate are used as carbon source rather than pure cellulose production thus shows the potential of bringing down the cost of cellulase remarkably.

13.6   PRODUCTION OF XYLOOLIGOSACCHRIDES (XOS) FROM AGRICULTURAL WASTE

Biotechnology had played a major role in value addition of agroresidues. One of the major developments in this aspect is transformation of this waste into taste. For this lignocellulosic biomass has gained much attention in the production of nutraceutical compounds known as xylooligosacchrides (XOS) (Collard and Blin, 2014). XOS are the polymers of xylose sugars that are produced from lignocellulosic biomass. This acts as prebiotic and selectively feeds the beneficial microflora like *Bifidobacteria* and *Lactobacilli* within the gastrointestinal tract (Samanta et al., 2015). The consumption of XOS have been found to induce certain health benefits like immune system stimulation, reduction of blood glucose and cholesterol, increased mineral absorption, and laxation. The sweet taste of XOS has allowed it to be used as artificial sweeteners (Singhania et al., 2017).

With the advancement and development of biotechnological applications, certain approaches are available for the production of XOS. These include:

1. **Direct Hydrolysis of Substrate:** This process is carried out with the help of certain enzymes like hemicellulases and commercial enzymes like Celtic Htec2 to convert lignocellulosic biomass into xylose sugars under mild conditions. The use of commercial

enzymes reduces the cost of process and also allows one to simultaneously evaluate the effect of factors on the process through the use of Design expert software (Golshani et al., 2013).

2. **Autohydrolysis:** This process is carried out under high temperature and pressure which is accommodated by special equipment. During this process, the corrosive chemicals are eliminated (Aachary and Prapulla, 2011).

13.7 CONCLUSION AND FUTURE DIRECTIONS

Agricultural biomass is an abundant natural resource that has an immense potential to be used for the welfare of mankind from the generation of electricity to the production of food. There is a need to create awareness among the masses about the toxic effects of direct burning of agricultural biomass as well as the potential of these wastes to be employed for conversion into energy. Although biotechnology had played an important role in mitigating the waste into taste but still more ecofriendly techniques need to get flourished. To accomplish this aspiration, collective global efforts are required.

ACKNOWLEDGMENT

The special thanks are due to the Research Centre, College of Pharmacy, King Saud University, Riyadh, and Deanship of Scientific Research, King Saud University, Kingdom of Saudi Arabia.

KEYWORDS

- agricultural waste
- biomass
- biotechnology
- separate hydrolysis and fermentation
- submerged fermentation
- xylooligosaccharides

REFERENCES

Aachary, A. A., & Prapulla, S. G., (2011). Xylooligosaccharides (XOS) as an emerging prebiotic: Microbial synthesis, utilization, structural characterization, bioactive properties, and applications. *Compr. Rev. Food Sci. Food Saf., 10*, 2–16.

Antoni, D., Zverlov, V. V., & Schwarz, W. H., (2007). Biofuels from microbes. *Appl. Microbiol. Biotechnol., 1*, 23–35.

Balat, M., (2011). Production of bioethanol from lignocellulosic materials via the biochemical pathway: A review. *Energy Convers. Manag., 52*, 858–875.

Brethauer, S., & Wyman, C. E., (2010). Continuous hydrolysis and fermentation for cellulosic ethanol production. *Biores Technol., 101*, 4862–4874.

Brown, T. R., & Brown, R. C., (2013). A review of cellulosic biofuel commercial-scale projects in the United States. *Biofuels, Bioproducts, and Biorefining, 7*, 235–245.

Butler, E., Devlin, G., Meier, D., & McDonnell, K., (2011). A review of recent laboratory research and commercial developments in fast pyrolysis and upgrading. *Renew Sust. Energ. Rev., 15*, 4171–4186.

Buyukgungor, H., & Gurel, L., (2009). The role of biotechnology on the treatment of wastes. *African J. Biotechnol., 8*, 7253–7262.

Camassola, M., & Dillon, A. J. P., (2014). Effect of different pretreatment of sugar cane bagasse on cellulase and xylanases production by the mutant *Penicillium echinulatum* 9A02S1 grown in submerged culture. *Biomed. Res. Int.* Article ID: 720740.

Carlson, T. R., Cheng, Y. T., Jae, J., & Huber, G. W., (2011). Production of green aromatics and olefins by catalytic fast pyrolysis of wood sawdust. *Energy Environ. Sci., 4*, 145–161.

Carlson, T. R., Tompsett, G. A., Conner, W. C., & Huber, G. W., (2009). Aromatic production from catalytic fast pyrolysis of biomass-derived feedstocks. *Top Catal., 52*, 241–252.

Collard, F. X., & Blin, J., (2014). A review on pyrolysis of biomass constituents: Mechanisms and composition of the products obtained from the conversion of cellulose, hemicelluloses and lignin. *Renew Sustain. Energy Rev., 38*, 594–608.

Das, A., Paul, T., Halder, S. K., Jana, A., Maity, C., Mohapatra, P. K. D., Pati, B. R., & Mondal, K. C., (2013). Production of cellulolytic enzymes by *Aspergillus fumigatus* ABK9 in wheat bran-rice straw mixed substrate and use of cocktail enzymes for deinking of waste office paper pulp. *Biores, Technol., 128*, 290–296.

Elliot, D. C., (2007). Historical developments in hydro processing bio-oils. *Energy Fuels, 21*, 1792–1815.

El-Tayeb, T. S., Abdelhafez, A. A., Ali, S. H., & Ramadan, E. M., (2012). Effect of acid hydrolysis and fungal biotreatment on agro-industrial wastes for obtainment of free sugars for bioethanol production. *Braz. J. Microbiol., 43*, 1523–1535.

Fabian, E. E., Richard, T. K. D., & Allee, D., (1993). *RegensteinAgricultural Composting: A Feasibility Study for New York Farms.* Available at: http://compost.css.cornell.edu/feas.study.html (accessed on 30 October 2020).

Golshani, T., Jorjani, E., Chelgani, S. C., Shafaei, S. Z., & Nafechi, Y. H., (2013). Modeling and process optimization for microbial desulfurization of coal by using a two-level full factorial design. *Int. J. Min Sci. Technol., 23*, 261–265.

Gusakov, A. V., (2013). Cellulases and hemicellulases in the 21st-century race for cellulosic ethanol. *Biofuels, 4*, 567–569.

Hall, D. O., Rosillo-Calle, F., & Woods, J., (1991). Biomass, its importance in balancing CO_2 budgets. In: Grassi, G., Collina, A., & Zibetta, H., (eds.), *Biomass for Energy, Industry and Environment, 6th E.C. Conference* (pp. 89–96). London: Elsevier Science.

Ingram, L. O., Gomez, P. F., Lai, X., Moniruzzaman, M., Wood, B. E., Yomano, L. P., & York, S. W., (1998). Metabolic engineering of bacteria for ethanol production. *Biotechnol. Bioeng., 58*, 204–214.

Jarboe, L. R., Wen, Z., Choi, D. W., & Brown, R. C., (2011). Hybrid thermochemical processing: Fermentation of pyrolysis-derived bio-oil. *Appl. Microbiol. Biotechnol., 91*, 1519–1523.

Kotiranta, P., Karlsson, J., Siika-Aho, M., Medve, J., Viikari, L., Tjerneld, F., & Tenkanen, M., (1999). Adsorption and activity of *Trichoderma reesei* cellobiohydrolase I, endoglucanase II, and the corresponding core proteins on steam pretreated willow. *Appl. Biochem. Biotechnol., 81*, 81–90.

Kumar, D., & Murthy, G. S., (2011). Impact of pretreatment and downstream processing technologies on economics and energy in cellulosic ethanol production. *Biotechnology for Biofuels, 4*, 27.

Kumar, R., Tabatabaei, M., Karimi, K., & Horváth, I. S., (2016). Recent updates on lignocellulosic biomass derived ethanol: A review. *Biofuel Res. J., 9*, 347–356.

Kupski, L., Pagnussatt, F. A., Buffon, J. G., & Furlong, E. B., (2014). Endoglucanase and total cellulase from newly isolated *Rhizopus oryzae* and *Trichoderma reesei*: Production, characterization, and thermal stability. *Appl. Biochem. Biotechnol., 172*, 458–468.

Laird, D. A., Brown, R. C., Amonette, J. E., & Lehmann, J., (2009). Review of the pyrolysis platform for coproducing bio-oil and biochar. *Biofuels Bioprod. Bioref., 3*, 547–562.

Martin, J. G. P., Porto, E., Correa, C. B., Alencar, S. M., Gloria, E. M., Cabral, I. S. R., & Aquino, L. M., (2012). Antimicrobial potential and chemical composition of agro-industrial wastes. *J. Nat. Prod., 5*, 27–36.

Miyamoto, K., (1997). *Renewable Biological Systems for Alternative Sustainable Energy Production.* Food & Agriculture Organization.

Mortensen, P. M., Grunwaldt, J. D., Jensen, P. A., Knudsen, K. G., & Jensen, A. D., (2011). A review of catalytic upgrading of bio-oil to engine fuels. *Appl. Catal. A Gen., 407*, 1–19.

Mosier, N., Wyman, C., Dale, B., Elander, R., Lee, Y. Y., Holtzapple, M., & Ladisch, M., (2005). Features of promising technologies for pretreatment of lignocellulosic biomass. *Biores. Technol., 96*, 673–686.

Motte, J. C., Trably, E., Escudié, R., Hamelin, J., Steyer, J. P., Bernet, N., Delgenes, J. P., & Dumas, C., (2013). Total solids content: A key parameter of metabolic pathways in dry anaerobic digestion. *Biotechnol. Biofuels, 6*, 164.

Narra, M., Dixit, G., Divecha, J., Madamwar, D., & Shah, A. R., (2012). Production of cellulases by solid state fermentation with *Aspergillus terreus* and enzymatic hydrolysis of mild alkali-treated rice straw. *Biores. Technol., 121*, 55–361.

Nathan, V. K., Rani, M. E., Rathinasamy, G., Dhiraviam, K. N., & Jayavel, S., (2014). Process optimization and production kinetics for cellulase production by *Trichoderma viride* VKF3. *Springer Plus, 3*, 92.

Nguyen, Q. A., Keller, F. A., Tucker, M. P., Lombard, C. K., Jenkins, B. M., Yomogida, D. E., & Tiangco, V. M., (1999). Bioconversion of mixed solids waste to ethanol. *Appl. Biochem. Biotechnol., 7*, 455–472.

Nigam, P. S., Gupta, N., & Anthwal, A., (2009). Pre-treatment of agro-industrial residues. In: Nigam P. S., & Pandey, A., (eds.), *Biotechnology for Agro-Industrial Residues Utilization* (pp. 13–33). Springer, Heidelberg.

Obi, F. O., Ugwuishiwu, B. O., & Nwakaire, J. N., (2016). Agricultural waste concept, generation, utilization and management. *Niger J. Technol., 35*, 957–964.

Ogada, T., & Werther, J., (1996). Combustion characteristics of wet sludge in a fluidized bed: Release and combustion of the volatiles. *Fuel, 75*, 617–626.

Pandey, A., (2003). Solid state fermentation. *Biochem. Eng. J., 13*, 81–84.

Pandey, A., Negi, S., Binod, P., & Larroche, C., (2015). *Pretreatment of Biomass: Processes and Technologies* (p. 264). Elsevier, UK. Chapter 10 Biotechnology for Agricultural Waste Recycling.

Pensupa, N., Jin, M., Kokolski, M., Archer, D. B., & Du, C., (2013). A solid state fungal fermentation based strategy for the hydrolysis of wheat straw. *Bioresour. Technol., 149*, 261–267.

Raghuwanshi, S., Deswal, D., Karp, M., & Kuhad, R. C., (2014). Bioprocessing of enhanced cellulase production from a mutant of *Trichoderma asperellum* RCK2011 and its application in hydrolysis of cellulose. *Fuel, 124*, 183–189.

Sadh, P. K., Duhan, S., & Duhan, J. S., (2018). Agro-industrial wastes and their utilization using solid state fermentation: A review. *Biores. and Bioproc., 5*, 1.

Samanta, A. K., Jayapal, N., Jayaram, C., Roy, S., Kolte, A. P., Senani, S., & Sridhar, M., (2015). Xylooligosaccharides as prebiotics from agricultural by-products: Production and applications. *Bioactive Carbohydrates and Dietary Fiber, 1*, 62–71.

Sarkar, N., & Aikat, K., (2012). Cellulase and xylanase production from rice straw by a locally isolated fungus *Aspergillusfumigatus* NITDGPKA3 under solid state fermentation-statistical optimization by response surface methodology. *J. Technol. Innov. Renew Energy, 1*, 54–62.

Singhania, R. R., Patel, A. K., Saini, R., & Pandey, A., (2017). Industrial enzymes: β-glucosidases. In: *Current Developments in Biotechnology and Bioengineering* (pp. 103–125). Elsevier.

Singhania, R. R., Patel, A. K., Soccol, C. R., & Pandey, A., (2017). Recent advances in solid-state fermentation. *Biochem. Engin. J., 44*, 13–18.

Singhania, R. R., Saini, J. K., Saini, R., Adsul, M., Mathur, A., Gupta, R., & Tuli, D. K., (2014). Bioethanol production from wheat straw via enzymatic route employing *Penicilliumjanthinellum* cellulases. *Biores. Technol., 169*, 490–495.

Sorrell, S., Speirs, J., Bentley, R., Brandt, A., & Miller, R., (2010). Global oil depletion: A review of the evidence. *Energy Policy, 38*, 5290–5295.

Sun, Y., & Cheng, J., (2002). Hydrolysis of lignocellulosic materials for ethanol production: A review *Bioresource Technology, 83*, 1–11.

Taherzadeh, M. J., & Karimi, K., (2008). Pretreatment of lignocellulosic wastes to improve ethanol and biogas production: A review. *Int. J. Mol. Sci., 9*, 1621–1651.

Tillman, D. A., (1987). Biomass combustion. In: Hall, D. O., & Overend, R. P., (eds.), *Biomass* (pp. 203–219). New York: Wiley.

Tu, M., Chandara, R. P., & Saddler, J. N., (2007). Evaluating the distribution of cellulases and the recycling of free cellulases during the hydrolysis of lignocellulosic substrates. *Biotechnol. Prog., 23*, 398–406.

Vallet, C., Said, R., Rabiller, C., & Martin, M. L., (1996). Natural abundance isotopic fractionation in the fermentation reaction: Influence of the nature of the yeast. *Bioorg. Chem., 24*, 319–330.

Venderbosch, R. H., & Prins, W., (2010). Fast pyrolysis technology development. *Biofuels Bioprod. Biorefin., 4*, 178–208.

Wang, K., & Brown, R. C., (2013). Catalytic pyrolysis of microalgae for production of aromatics and ammonia. *Green Chem., 15*, 675–681.

Werther, J., Ogada, T., Borodulya, V. A., & Dikalenko, V. I., (1995). Devolatilization and combustion kinetic parameters of wet sewage sludge in a bubbling fluidized bed furnace. In: *Proceedings of the Institute of Energy's 2nd International Conference on Combustion and Emission Control* (pp. 149–158). London, UK.

Wright, R. J., (1998). *Executive Summary*. Available at: www.ars.usda.gov/is/np/agbyproducts/agbyexecsummary.pdf (accessed on 30 October 2020).

Yadav, P. S., Shruthi, K., Prasad, B. V. S., & Chandra, M. S., (2006). Enhanced production of β-glucosidase by new strain *Aspergillus protuberus* on solid state fermentation in rice husk. *Int. J. Curr. Microbiol. App. Sci., 5*, 551–564.

Zafar, S., (2014). Overview of biomass pyrolysis. *Bioenergy Consult*. Obtained from: http://www.bioenergyconsult.com/tag/pyrolysis (accessed on 21 November 2020).

CHAPTER 14

Bioremediation Technologies for the Management of Agricultural Waste

MONICA BUTNARIU,[1] IOAN SARAC,[1] and ALINA BUTU[2]

[1] Banat's University of Agricultural Sciences and Veterinary Medicine, "King Michael I of Romania" from Timisoara – 300645, Calea Aradului 119, Timis, Romania, E-mail: monicabutnariu@yahoo.com (M. Butnariu)

[2] National Institute of Research and Development for Biological Sciences, Splaiul Independentei, 296, Bucharest – 060031, Romania

ABSTRACT

By assimilating the concepts, methods, and technologies of bioremediation of natural or anthropic ecosystems, cognitive support is provided for the design of integrated environmental management, including the assessment of environmental expenditure, investments for depollution or ecological reconstruction of biotechnical and ecological systems. The objectives of this review are to list a number of basic theoretical and practical information on the bioremediation of pollutants from agricultural waste (biological depollution, biodiversity reconstruction, or restoration of the biocenotic balance affected by destructive factors in natural and anthropic ecosystems). The aim is to present the methods and techniques of restoration and reconstruction by means of bioremediation, to acquire the necessary skills to analyze and evaluate the main causes of natural and anthropic imbalance and dysfunctionality, to analyze new relationships between environmental components and to establish specific technologies for depollution and bioremediation.

14.1 INTRODUCTION

Depollution, decontamination, remediation, and reintroduction in the normal circuit of contaminants/pollutants from agricultural waste, namely the transformation of so-called "brownfields" into "greenfields," are one of the major tasks envisaged on multiple levels: legislative, technical, social, etc. Due to the virtually unlimited number of contaminants from agricultural waste and distinct soil structures, there is no generally applicable method for soil remediation. Choosing a remediation technique is a complex activity, which involves taking into account many factors: the type of contaminants, the amount of contaminants, the dynamics of the contaminants, the hydrogeological characteristics of the soil, the climatic factors. Last but not least, the economic aspects, respectively the expenditure of the remediation (O'Connor et al., 2019).

While remediation ensures the destruction or reduction of the quantity and quality of contaminants from agricultural waste, securing is aimed at lifting barriers to prevent the spread of contaminants over wider areas. As the source of pollution remains and the obstructions are subject to degradation and aging, securing is only a temporary measure, and the remediation remains the procedure to be applied. Remediation techniques can be categorized according to their place of implementation and depending on the type of methods involved. Thus, in the first case we distinguish methods *ex-situ* and *in situ*, and in the second case, we distinguish thermal, physicochemical, and biological methods. *Ex-situ* methods require the excavation of the polluted soil followed by treatment either on land (on-land remediation) or in an external soil treatment plant (off-land remediation). *In situ* treatment is carried out directly into the polluted land without the need to excavate the soil. Thermal remediation methods are substantiated on the transfer of contaminants from agricultural waste from the soil matrix to the gaseous phase by heat input. The contaminants are propagated from the soil by vaporization and then are incinerated, the residual gases being subsequently purified. Physico-chemical methods are generally wet extraction and/or wet ranking. The principle of *ex-situ* washing of soils includes in concentrating the contaminants from agricultural waste into a residual fraction as low as possible, the water being the most frequently used extraction agent (Jin et al., 2019). For the transfer of soil contaminants/pollutants to the extractant, two mechanisms are important:

- Causing powerful shear forces induced by mixing, pumping, vibration, or by using high-pressure water jets that break down the agglomerations of infected and uninfected particles and disperse the contaminants/pollutants in the extraction phase; and

- Dissolution of contaminants/pollutants by the components of the extraction phase. *In situ* extraction includes of the percolation of an aqueous extractant agent through polluted soil.

Percolation can be achieved by surface ditches, horizontal drains, or vertical depth wells. Dissoluble contaminants/pollutants from soil are dissolved in the percolating liquid which is pumped and subsequently treated on land. Biological methods are substantiated on the action of microorganisms (MO), i.e., a microscopic organism, especially a bacterium, virus, or fungus that have the capacity to transform organic contaminants primarily into CO_2, water, and biomass, or to immobilize contaminants by binding to the humic fraction of the soil.

Degradation is usually performed under aerobic conditions or, more rarely, in anaerobic conditions. In order to increase technique performance, it is essential to optimize the conditions for the development of MO (such as oxygen intake, pH, water content, etc.).

Stimulation of biological activity can be achieved by active aeration, humidification soil homogenization, or addition of nutrients or substrates, drying, heating, inoculation with MO. Biological methods require a much lower energy supply than thermal or physico-chemical methods, but require longer treatment periods (Masi et al., 2019).

14.2 BIOREMEDIATION

Bioremediation is the usage of living organisms (MO, plants, etc.), to improve and restore the ecological status of an infected or degraded substrate (aquifer, soil, etc.), towards a better-quality parameter, favorable to life, harmless, non-polluting, or to return it to the state before pollution. It is a modern technology for treating contaminants from agricultural waste that uses biological factors (MO) to convert certain chemicals into less harmful/dangerous endpoints, ideally CO_2 and H_2O, which are nontoxic and propagated without substantially altering the balance of ecosystems. Bioremediation is substantiated on the capacity of chemical compounds to be biodegraded (Orellana et al., 2018). The concept of biodegradation is accepted as a sum of the methods of decomposition of natural or synthetic constituents by activating strains of specialized MO resulting in end-products useful or environmentally acceptable (Fan et al., 2017). Example of bioremediation: Biological treatment of wastewater and sewage by using MO to restore it to the original situation. In the last decades, the term bioremediation is used in a more specific way, which is reflected by the two specific definitions:

- The usage of MO to degrade environmental contaminants, to prevent pollution or waste treatment; and

- Implementation of biological methods for the cleaning, decontamination, and degradation of dangerous substances.

Bioremediation can be applied *in situ* (on the land, on the infected substrate, where the contamination occurred) or *ex situ* (in specially designed systems/installations where the infected substrate is transported and treated by biological methods) (Karig, 2017).

14.2.1 ADVANTAGES OF BIOREMEDIATION

Addressing pollution complications by biotechnological techniques through *in situ* bioremediation techniques has the advantage of: requiring a moderate level of investment capital, environmental safety, reduced waste, and self-sustainability. Biotechnological techniques for toxic effluent treatment compete with existing techniques in terms of efficiency and economic efficiency. It techniques for toxic waste treatment are designed to replace current techniques of depositing and detoxifying of new xenobiotic compounds (resulted from human interventions). It is important, however, to limit the generation of dangerous and non-dangerous waste, as well as to usage recycling techniques when the waste is produced (Yeo et al., 2018).

Other advantages of bioremediation compared to other technologies:

- Permanent disposal of waste (limitation of compliance complications);
- Eliminates transportation and compliance expenditure;
- Minimal disturbance of the land;
- Can be combined with other treatment technologies;
- Can be done on land;
- Positive from the point of view of public acceptance.

Bioremediation offers, in many cases, a permanent solution to the problem, and is cost-effective.

14.2.2 ON THE SPOT TREATMENT OF THE CONTAMINATION

Traditional cleaning technologies involve the removal and storage of polluted soils, which make up most of the treatment expenditure. Since bioremediation can be done locally (*in situ*) by adding nutrients to polluted soils, this does not entail removal/storage expenditure.

14.2.3   EXPLOITATION OF NATURAL METHODS

On specific lands, natural microbial methods can remove contaminants/ pollutants, even without human intervention. In such cases where bioremediation (natural pollution mitigation) is appropriate, substantial cost reductions can be achieved.

14.2.4   REDUCING ENVIRONMENTAL DISTURBANCE

Bioremediation minimizes land disruption compared to conventional technologies, with post-treatment expenditure substantially diminished. Bioremediation is generally more cost-effective than the two widely used techniques, namely burial or incineration (Ushani. et al., 2018).

14.3   TECHNOLOGIES INVOLVED IN BIOREMEDIATION

A number of specific terms and methods are used to describe the activity of MO and pathways through which they are used in bioremediation such as:

- **Biodegradation:** It is the breakage or fragmentation of a compound or substance, produced by MO, bacteria or fungi that may be indigenous to that area or may be introduced.

- **Biostimulation:** It is the method by which MO (natural or introduced) activity is enhanced by nutrient, biotechnological engineering, or other techniques, to increase the speed of natural remediation methods.

- **Bioaugmentation:** It is the method by which specific MO are added to a land or material to achieve a desired bioremediation effect.

- **Biorestoration:** It is the restoration of the original state or of a state close to the original state by the usage of living MO. Bioremediation technologies can be applied locally, *in situ* or *ex situ* (by transporting the infected substrate to special treatment facilities). The technology used to treat an infected land is land specific and pollutant dependent.

The various *in situ* bioremediation methods are linked to a number of technologies (Sharma et al., 2018), such as: *in situ* bioremediation (ISB), accelerated ISB, enhanced bioremediation (EB), naturally monitored attenuation.

14.3.1   IN SITU BIOREMEDIATION (ISB)

In situ **bioremediation**(ISB) is the usage of MO to degrade the occurring contaminants/pollutants on the land of pollution (*in situ*) in order to produce non-dangerous final compounds. ISB is applied for the degradation of contaminants/pollutants in saturated soils and underground waters, and in unsaturated areas, also.

The technique has been developed to be less costly, more efficient than standard pumping and treatment techniques used to degrade aquifers and soils polluted with chlorinated solvents, petroleum hydrocarbons, explosives, nitrates, and toxic metals. ISB has advantages such as: complete destruction of contaminants/pollutants, lower health risks for land workers, lower expenditure for installation and operation. ISB can be classified according to metabolism or depending on the degree of human intervention.

The two types of metabolism are aerobic metabolism and anaerobic metabolism. The type of metabolism for a particular ISB system will be set as target according to the type of contaminants/pollutants to be destroyed. Some contaminants/pollutants are degraded by aerobic treatment (such as petroleum) others by anaerobic treatment (such as carbon tetrachloride), while other contaminants/pollutants can be biodegradable both by aerobic and anaerobic routes (such as trichloroethene). Since, the soil treatment doesn't require excavation and transport; the treatment expenditure is significantly reduced, thus posing a great advantage. However, this remediation requires longer periods, and the uniformity of treatment is less reliable given the variability of soil and aquifer characteristics. In addition, it is more difficult to control the effectiveness of the technique.

Bioremediation techniques are destructive techniques aimed at stimulating the multiplication of MO by using contaminants/pollutants as sources of food and energy.

Causing favorable conditions for the development of MO usually involves providing certain combinations of oxygen, nutrients, and humidity, as well as controlling the temperature and pH.

Sometimes, to improve the technique, MO adapted to degrade certain contaminants/pollutants is added. The usage of biological remediation methods is usually done with low expenditure. Contaminants/pollutants are destroyed and rarely additional residue treatment is required (Majoneet al., 2015).

Some disadvantages arise with specific contaminants/pollutants. Such as, biodegradation of polycyclic aromatic hydrocarbon (PAH) leads to ground degradation to PAH with large, recalcitrant, and potentially carcinogenic masses. Polyhalogenated compounds are highly biodegradable, and some of them are transformed by biodegradation into secondary and more toxic products (such as, conversion of trichloroethene to vinyl chloride). If adequate control techniques will not be used, these by-products may be mobilized by groundwater.

ISB requires a thorough characterization of soil, aquifer, and contaminants/pollutants. Sometimes it may be necessary to extract and treat the groundwater, and then the low polluted groundwater may be recirculated through the treated area to provide the necessary humidity (Němeček et al., 2018).

14.3.1.1 ACCELERATED "IN-SITU" BIOREMEDIATION

Accelerated "in-situ" bioremediation stimulates the growth of a bacterial consortium through by adding substrate or nutrients to the aquifer. Despite the usual target bacteria being indigenous to the land, a good alternative is represented by external enriched cultures from other lands. These cultures should be especially effective in degrading a certain contaminant (bioamplification). Accelerated "in-situ" bioremediation is used for specific areas, where an increase in the rate of biotransformation of the contaminant is desired, the speed being limited by the lack of nutrients, the lack of donors or electron acceptors. The type of modification required depends on the target metabolism and the target contaminant.

In-situ aerobic bioremediation may require only an addition of oxygen (electron acceptor), whereas the anaerobic can require both the addition of electron donors (such as lactate, benzoate) and electron acceptors (ex. nitrate, sulfate).

Chlorinated solvents often require the addition of carbon substrate to stimulate reductase degradation. The goal pursued in accelerated "in-situ" bioremediation technique is to increase the amount of biomass inside the polluted aquifer and thereby to achieve effective biodegradation of the dissolved and absorbed contaminant.

Although it usually is a faster solution compared to others, accelerated "in-situ" bioremediation does normally require a bigger investment in equipment, labor work and materials (Tong et al., 2015).

14.3.1.2 ADVANTAGES OF ACCELERATED IN SITU BIOREMEDIATION (ISB)

Contaminants can be fully turned into harmless substances such as water, ethane, and carbon dioxide. The volumetric treatment in-situ bioremediation provides is able to treat both saponified and dissolved contaminants. The length of ISB treatment of underwater soil pollution can often be faster than if pumping-treatment technologies are used. ISB tends to expenditure less, whereas other remediation options are often more expensive due to the technologies used. Since bioremediation treatment follows the aquifer movement, otherwise inaccessible areas are reached; therefore the area of treatment is greater than when using different remediation technologies (Abel and Akkanen, 2019).

14.3.1.3   RESTRICTIONS OF ACCELERATED IN-SITU BIOREMEDIATION

Some contaminants could be only partially turned into harmless compounds, depending on the specificity of the land. Intermediate products which formed in the biotransformation technique can potentially have a higher mobility or toxicity than the parent compound.

Moreover, certain contaminants are recalcitrant to biodegradation, meaning they are not biodegradable. If it is inappropriately applied, the addition of nutrients, electron acceptors, and/or donors could lead to intense microbial growth, which in turn could lead to clogging of the injection wells.

Since nutrient transport is limited in aquifers with low permeability, it is more difficult to apply accelerated in-situ bioremediation. The activity of indigenous MO can be inhibited by toxic concentrations of organic compounds and heavy metals. Populations of acclimated MO, which are usually required for ISB, cannot be developed for recalcitrant compounds or for recent waste MO (Carrara et al., 2011).

14.3.2   ENHANCED BIOREMEDIATION (EB)

It is a technique in which indigenous or inoculated MO (bacteria, fungi, etc.), metabolize organic contaminants from soil or groundwater, with the formation of stable, non-polluting products. To improve the technique, or for desorption of contaminants from agricultural waste from underground materials, nutrients, oxygen, other amendments can be added.

EB may involve the usage of specially cultivated microbial cultures for the degradation of certain contaminants or groups of contaminants, or to withstand particularly severe environmental conditions. Sometimes the MO in the land undergoing remediation are collected, cultivated separately and then reintroduced into the land as a means of rapidly increasing the microbial population in that land. Another method involves that other types of MO can be added at distinct stages of the remediation technique as a result of changing the composition of contaminants from agricultural waste as the bioremediation technique evolves.

If the degradation of contaminants from agricultural wastes (AWs) is an aerobic technique, BI can be achieved by percolation or injection into soil of groundwater or unpolluted water containing nutrients and saturated with dissolved oxygen. Instead of the dissolved oxygen another oxygen source can be used, such as H_2O_2. In case of soils polluted in the superficial layer, the injection wells are replaced with infiltration galleries or surface irrigation systems (Xu et al., 2018).

Because low temperatures slow down bioremediation, the soil can be covered with various heating or temperature maintenance devices to speed up the technique. If anaerobic degradation results in more dangerous intermediates or products than the original contaminants (such as anaerobic degradation of trichloroethene to vinyl chloride), it is recommended to create aerobic conditions to neutralize them. EB has been successfully applied for the remediation of soils, sludge, and groundwater polluted with petroleum hydrocarbons, solvents, pesticides, preservatives for wood, and other organic substances.

Pilot studies have shown the efficiency of the technique of TNT anaerobic degradation in soils polluted with ammonia residues, especially after the source has been removed and the pollutant concentration in the soil is low. Frequently removed contaminants by this technique are HAP, non-halogenated COSV and benzene-toluene-ethylbenzene-xylenes (BTEX) fractions from infected lands with wood preservatives (creuzots) or refinery locations (Suet al., 2012).

However, EB also has a number of restrictions, such as:

- Inefficient if the soil matrix does not allow contact between contaminants and micro-organisms;

- The circulation of aqueous solutions through the soil can lead to increased mobility of contaminants from agricultural waste;

- Preferential colonization of micro-organisms may cause clogging of water/nutrient injection wells;

- Preferential flow can greatly reduce the contact of fluids injected with contaminants-the technique is not recommended for loamy, highly layered or heterogeneous soils;

- High concentrations of heavy metals, highly chlorinated compounds, long chain alkanes, inorganic salts are toxic to MO;

- Lowering the technique speed at lower temperature;

- The need extract the groundwater to surface and treat it (air stripping or active carbon treatment) before re-injection into soil or storage.

EB can be considered a long-term technique, cleaning a land that can last between 6 months and 5 years, depending on local specifics. Expenditure of technique varies between 30–100 Euro/m^3 of treated soil (McHugh et al., 2014).

14.3.3   MONITORED NATURAL ATTENUATION (INTRINSIC BIOREMEDIATION)

ISB can be applied through multiple methods, one of them being intrinsic bioremediation. Using indigenous MO in order to degrade the involved contaminants without nutrient additions, therefore without human intervention, represents a component of natural attenuation.

Implementation of natural mitigation requires good land knowledge and long-term monitoring. Knowledge of the land involves a characterization of the extent of pollution and the characteristics of the aquifer. With this information, a predictive transport model can be developed, which will predict the pollution danger and when the target receptors will be affected by the contaminants. The transport and hazard of the contaminants will then be compared to the predicted ones through long-term monitoring.

Then, by refining the transport model, there will be better predictions given. Natural pollution mitigation methods normally take place on all lands but the degree of their effectiveness is distinct and depends on the types and concentrations of contaminants/pollutants present and on the physical, chemical, and biological characteristics of soil and groundwater.

Natural attenuation is substantiated on natural methods of decontamination or attenuation of pollution in soil and groundwater. Naturally, several methods can occur in the soil leading to a decrease of the concentration of contaminants from agricultural waste, below the permissible limit. Such methods may be: dilution, volatilization, adsorption, chemical transformation, and biodegradation.

Although natural mitigation results in most of the infected lands, there is a need for appropriate depollution conditions, otherwise it will be incomplete or insufficiently fast. It is necessary to test or monitor these conditions to verify the feasibility of natural attenuation. This method is best suited for usage in areas where the source of pollution has been removed (Naidu et al., 2012).

14.3.4   MONITORED NATURAL ATTENUATION

It is not synonymous with "not taking any measures," although this is the most common perception. Compared to other remediation technologies, this method has a number of advantages such as:

- Generating a smaller volume of waste, reducing the potential for environmental transfer of the associated contaminants/pollutants by *ex situ* methods and a reduced risk for the exposure of human beings to the polluted environment. It can be used in combination with other remediation measures or in addition to the remediation targets.

- Low impact on lands (no built-in structures):

- Total or partial opportunity in a particular land, depending on the actual conditions and the purpose of the remediation;

- The possibility of using it together or after other active remediation measure;

- Lower overall expenditure than in the case of an active remediation.

Contaminants susceptible to *monitored natural attenuation* are volatile and semi-volatile organic compounds (VOC, SVOC) as well as hydrocarbons present in fuels, certain categories of pesticides, and several heavy metals (such as Cr) if there are conditions to immobilize them by modifying oxidation state.

Among the disadvantages of monitored natural attenuation can be mentioned:

- Longer duration compared to active remediation measures;

- Long-term monitoring and its associated expenditure;

- The need to collect precise data used as input parameters in technique modeling;

- The possibility of immobilizing some pollutants/contaminants (such as Hg) without being able to degrade them;

- The possibility of migrating contaminants/pollutants before their degradation;

- The possibility of modified hydrological and geochemical conditions over the time, which could lead to the restoration of the mobility of pollutants/contaminants from agricultural waste previously immobilized;

- The possibility that intermediate degradation products may be more mobile or more toxic than the original contaminant;

- The reluctance of public opinion to such "passive" measures of depollution (Adetutu et al., 2015).

Natural mitigation methods can reduce the potential risk posed by land contaminants; this risk mitigation being achieved through the following pathways:

- The contaminant can be converted into a non-toxic form by destructive methods: biodegradation or abiotic transformation.

- Potential exposure levels can be reduced by decreasing concentrations (as a result of destructive methods or by dilution and dispersion).

The mobility and bioavailability of the contaminant can be reduced by sorption on soil or rock particles. The usage of accelerated *in situ* bioremediation or natural attenuation for a particular land will depend on the properties of the aquifer, the concentrations of chemical contaminants/pollutants, the purpose of the remediation project, and the economic aspects of each option. The rate of degradation of the contaminant is lower for natural attenuation than for active bioremediation because of the higher bacterial concentration in active methods and the rate of biodegradation is proportional to the amount of biomass. Thus, natural mitigation usually requires a long period of time to become complete (Takahata et al., 2006).

14.4 BIOAERATION

Bioaeration is a technique by which aerobic biodegradation *in situ* is stimulated by additional oxygen intake to soil bacteria. Unlike the vapor extraction technique in the soil, bioaeration uses low airflows, to support microbiological activity. Usually oxygen is added to the soil by direct injection of air into the polluted land. Injection of air can be done in vertical wells or in horizontal channels. In addition to accelerating degradation, bioaeration also has a side effect, namely to move volatile contaminants through activated soil. The technique usually applies to the unsaturated soil area and is suitable for all compounds that can be aerobically biodegradable (Preston et al., 2011).

14.5   PHYTOREMEDIATION

Under the generic name of phytoremediation are those methods that usage plants to remove, transfer, stabilize, and destroy contaminants/pollutants from soil, water, sediments. Phytoremediation techniques offer significant potential for certain implementations and allow lands that are much larger than would be possible if traditional remediation technologies were used. A large number of plant species (over 400), starting with pteridophyte ferns and ending with angiosperms such as sunflower or poplar, can be used to remove contaminants from agricultural waste through several mechanisms. Phytoremediation mechanisms include intensified biodegradation in the rhizosphere (rhizodegradation), phytoextraction (phytoaccumulation), phytodegradation, and phytostabilization (Singh and Singh, 2017).

14.5.1   RHIZODEGRADATION

Rhizodegradation takes place in the soil part that surrounds the roots of the plants. Natural substances propagated from plant roots serve as a substrate for MO present in the rhizosphere, thus accelerating the degradation of contaminants/pollutants. Plant roots loosen the soil, leaving room for water transport and aeration. This technique tends to push the water to the surface area and dehydrate the lower saturated areas (Wang et al., 2017).

14.5.2   PHYTOEXTRACTION

Phytoextraction is the technique by which plant roots absorb water and nutrients together with soil contaminants/pollutants (such as especially metals). Contaminants/pollutants are not destroyed, but accumulate in the roots, stems, and leaves of plants that can be harvested to remove and destroy contaminants/pollutants. The extraction technique depends on the capacity of plants to grow in soils with high concentrations of metals and their capacity to extract the soil contaminants under specific climatic conditions.

For phytoextraction may be used either plants with exceptional natural capacity to accumulate metals, so-called hyperaccumulators, or plants that produce high amounts of biomass (corn, barley, peas, oats, rice, Indian mustard) chemically assisted with additives that improve the capacity to extract contaminants (Tauqeer et al., 2019).

Additives of citric acid, oxalic acid, gallic acid, vanilic acid, classical chelating agents such as ethylenediaminetetraacetate (EDTA) and diethy-lenetriaminepentaacetate (DTPA) or biodegradable chelating agents such as ethylenediamine disuccinate (EDDS), methylglycine diacetate (MGDA) substantially improve soil extraction of Zn, Cd, Cu, and Ni.

However, care must be taken, as these additions have the risk of mobi-lizing metals in groundwater. The number of hyperaccumulators in the plant kingdom is reduced: about 400 species of vascular plants, the vast majority having a particular affinity for Ni. By definition, hyperaccumulators should accumulate at least 100 mg/g of Cd or As, 1000 mg/g of Co, Cu, Cr, Ni or Pb, 10,000 mg/g Mn or Ni. Certain species of ferns have a special accumulation capacity for As-up to 23,000 mg/kg in the *Pteris vitata* species. Common buckwheat (*Fagopyrum esculentum*Moench) can accumulate up to 4200 mg/kg Pb, being the first Pb hyperaccumulator species that also has high productivity in biomass.

Other plants with the potential for phytoextraction are those of *Brassica* genus: *Brassica juncea* (Indian mustard) for Cr, Cd (VI), ^{137}Cs, Cu, Ni, Pb, *Brassica napus* for Pb, Se, Zn, *Brassica oleracea* (ornamental cabbage) for As, ^{137}Cs, Ni, Tl. Hg bioavailable from the soil can be extracted with barley, wheat, yellow lupine (*Lupinus luteus*), dog grass (*Cynodon dactylon*) (Ashraf et al., 2019).

14.5.3 PHYTODEGRADATION

Phytodegradation is the technique of metabolizing contaminants/pollutants in plant tissues. Plants produce enzymes (such as oxygenases and dehaloge-nases) that promote catalytic degradation of contaminants/pollutants in plant tissue. It is studied the possibility of simultaneous degradation of aromatic compounds and chlorinated aliphatic compounds by this method (Dolphen and Thiravetyan, 2015).

14.5.4 PHYTOSTABILIZATION

Phytostabilization is the technique substantiated on the capacity of certain plants to produce chemical compounds that can bind significant amounts of toxic compounds (especially heavy metals) to the root-soil interface in an inactive form, thus preventing their spreading in groundwater or other environments. Typically, phytostabilizing soil is plowed and treated with various modifications for rapid fixation of metals (lime, phosphate fertilizer, Fe or Mn oxyhydroxides, clay minerals, etc.), after which it is seeded with plants known as weak metal translocators, so they do not they reach the parts of the plant that can be eaten by animals.

Agrostis tenuis and *Festuca rubra* are used in commercial implementations for the phytostabilization of soils polluted with Cu, Pb, or Zn. *Risofiltration* is similar to phytostabilization, with the observation that it applies only to liquid effluents. Plants are grown without soil and transported in polluted areas. As the roots are saturated with contaminants/pollutants, they are harvested and stored (Wu et al., 2019).

14.5.5 PHYTOVOLATILIZATION

Phytovolatilization is the technique by which plants absorb water polluted with organic compounds, which then removes them into the atmosphere through the leaves. Some metals (Hg, As, Se) can be removed as gaseous compounds, but their toxicity casts doubt on the efficacy of this method. Genetically modified tobacco plants (*Nicotiana tabacum*) have been used for the sorption of mercury and methylmercury in the soil, followed by their release into the atmosphere as mercury oxide. Hydraulic influence is the technique by which trees especially facilitate remediation methods, influencing the movement of groundwater. Trees act as natural pumps when their roots reach under the groundwater mirror, establishing a dense network of roots that take up large amounts of water.

Such as, a mature poplar (*Populus deltoides*) is able can absorb up to 1.3 m^3 of water, daily. In conclusion, the phytoremediation can be defined as the in-situ usage of live plants for the treatment of soils, muds, and groundwater, by removing, degrading or immobilizing contaminants from existing agricultural waste. Phytoremediation techniques are suitable for areas where contamination is low to moderate, close enough to the surface, and in a shallow area. With these restrictions, phytoremediation can be applied to distinct categories of contaminants: various organic compounds, crude oil, explosives, metals, PAH, pesticides, solvents, leakage from household waste disposal (Limmer and Burken, 2016). The plant species commonly used in phytoremediation projects is poplar. This tree grows rapidly, can survive under varying climatic conditions, and in comparison, with other species it can extract large amounts of water from the aquifers or from the soil, thus extracting the solubilized contaminants from the polluted environment. Phytoremediation is a new technique, still in its development phase, its practical implementations being relatively recent. The first researches were made in the early 1990s, a number of techniques being applied with reasonable results in some infected lands.

Among the disadvantages of phytoremediation can be mentioned:

- Biodegradation products may be mobilized in groundwater or bioaccumulate in the animal kingdom via the trophic chain;

- Can transfer pollutants/contaminants between various media (from soil to air, e.g.); is not effective for highly adsorbed contaminants/pollutants, such as PCBs;

- High concentrations of hazardous/dangerous materials may be toxic to plants;

- Limiting the depth of the treated area depending on the plants used-in most cases the technique is applicable to near-surface pollutants/contaminants;

- May be seasonal, depending on the geographical location of the bioremediated area;

- Presents the same restrictions on mass transfer as other bioremediation technologies;

- The toxicity and bioavailability of biodegradation products is not always known;

- Restrictions being still in the demonstration phase, is relatively unfamiliar for the legislative bodies.

Phytoremediation expenditure are low: for the removal of contaminants from agricultural waste from a soil layer of 50 cm deep the expenditure variable between 30–50 Euro/m^3 (about 150000–250000 Euro/ha). One hectare of land polluted to 50 cm, *ex situ* treated by biodegradation in excavated layers expenditure between 0.99 and 4.2 million Euros (Arnold et al., 2007).

14.6   PRACTICAL IMPLEMENTATION OF BIOREMEDIATION

The rapid expansion and growing complexity of the chemical industry in the last century, and especially in the last 30 years, has resulted in an increasing amount and increasing complexity of toxic effluent waste. At the same time, fortunately, regulators have paid more attention to environmental pollution complications. Industrial companies have become more attentive to the pressure of political, social, and environmental regulations to prevent the discharge of effluents into the environment. Implications of major pollution incidents (such as Exxon Valdez oil spill, Union Carbide (Dow) Rhopal disaster, large-scale pollution of the Rhine River, progressive deterioration of aquatic habitats and coniferous forests in the north-eastern USA, Canada, and a part of Europe or radioactive material scrap in the chernobyl accident, etc.), and massive publicity on the environmental issues that followed resulted in public awareness of potential long-term disaster results. Environmental policies will make continued efforts to put pressure on industry to reduce the production of toxic waste and bioremediation presents opportunities for the detoxification of a whole range of industrial effluents. Bacteria can be adapted or modified to produce certain enzymes that metabolize the components of industrial waste that are toxic to other animals, and new ways of biodegradation of distinct wastes can be developed. If waste management itself is a well-defined industry, genetics, and enzymology can very well accomplish it through the existing engineering experience in these fields. Thus, it can be concluded that bioremediation can be applied to a wide range of chemical waste (Manheim et al., 2019).

14.6.1   BIOREMEDIATION OF CHLORINATED HYDROCARBONS

Chlorinated hydrocarbons are one of the most common groups of compounds that require bioremediation technologies. Chlorinated hydrocarbons may be biotransformed by three mechanisms: the usage of the chlorinated compound as an electron acceptor, the usage of the chlorinated compound as an electron donor, or cometabolism (reactions that do not produce the benefits of MO). On one given land, one or more of these mechanisms can be achieved (Jesus et al., 2016).

14.6.2   REACTIONS AS AN ELECTRON ACCEPTOR

The usage of chlorinated compounds as electron acceptors has been demonstrated in nitrate- and iron-reducing or sulfate-reducing and metanogenic conditions (with the highest biodegradation rates affecting the chlorinated aliphatic hydrocarbon domain.) This biotransformation technique produces reductive dehalogenation and requires an electron donor carbon originating from natural organic matter, from anthropogenic sources (such as petroleum hydrocarbon companion contamination), or from the deliberate introduction of organic carbon into the aquifer interior (such as, accelerated *in situ* bioremediation) (Vogt and Richnow, 2014).

14.6.3   REACTIONS AS AN ELECTRON DONOR

In this case, the chlorinated hydrocarbon is used as the primary substrate (electron donor) and the microorganism gets the energy and the organic carbon from the chlorinated hydrocarbon. This can be done under aerobic conditions and under certain anaerobic conditions. The less oxidized chlorinated compounds (such as, vinyl chloride-PVC, DCE, or 1,2-dichloroethane) are more favorable to undergo this biotransformation. It should be noted that petroleum hydrocarbons are biodegraded with this method because they can be used as an organic carbon source. Note: Electron donor/acceptor terminology gives a more explicit description than the terms "reductive dehalogenation," "direct degradation," and "primary substrate." Generally, reductive dechlorination is the technique by which a chlorine atom is removed from the chlorinated compound and is replaced by a hydrogen atom. Direct dechlorination is usually associated with the chlorinated hydrocarbon, which acts as an electron donor. The primary substrate also usually refers to the electron donor.

14.6.3.1   COMETABOLISM

When a hyper chlorinated aliphatic hydrocarbon is biodegraded by cometabolism, degradation is catalyzed by an enzyme or cofactor produced by organisms for other purposes. By cometabolism, the MO does not obtain any known benefit from the degradation of the chlorinated compound. The biotransformation of the chlorinated compound may be both inhibitory/harmful to MO. Cometabolism is best known for aerobic media, but it can also be achieved under anaerobic conditions (Jin et al., 2014).

14.7   BIOREMEDIATION AS A BUSINESS

In recent years, several companies have decided to develop and market biodegradation technologies. The existence of such companies has now become economically justifiable due to the explosive increase in the cost of traditional treatment technologies due to the increasing resistance of the public to some traditional technologies (from the Love Canal to the ENSCO incinerator plans in the past years) and due to the increasingly demanding regulations.

The interest of the commercial business environment in the usage of MO for the detoxification of effluents, soils, etc., is reflected in "bioremediation" that has become a common word in waste management. Companies that specialize in bioremediation (biodegradation technologies) need to develop a viable integration of microbiology with engineering systems. An example of a bioremediation company is Envirogen (NJ) that has developed MO that degrades PCBs (polychlorobiphenyls) and has good stability and survival of soil populations. This Company has developed bacteria that naturally exist and degrade trichloroethylene (TCE) in the presence of toluene, a toxic organic solvent that kills many other MO. A large number of such companies can be found using WEB search and a keyword (such as bioremediation) (Bjerketorp et al., 2018).

The MO has been successfully applied to remove oil spills from Exxon Waldez. Some species of MO can produce usage oil as a source of food, and many of them produce surface surfactants that can emulsify the oil in water and thus facilitate the removal of oil. Unlike chemical surfactants, the microbial emulsifier is nontoxic and biodegradable. "Fertilizers" have been used to increase the rate of growth of the indigenous bacteria population that can degrade the oil. The usage of MO for bioremediation is not limited to the detoxification of organic compounds. In many cases, selected MO may be used to reduce cation toxicity (such as selenium) by converting them to less toxic and less dissoluble forms. Thus, the bioremediation of surface water with significant contamination of heavy metals can be achieved (Bastida et al., 2016).

14.8   EXAMPLES OF SUCCESSFUL IMPLEMENTATION OF BIOREMEDIATION

In Hanahan, a quiet suburb of Charleston, South Carolina, in 1975, a massive drain from a military depot discharged approximately 5920 m^3 of kerosene.

Immediate measures to recover and isolate leakage could not prevent certain penetrations into the permeable sandy soil that reached the groundwater level. Soon soil water leached some toxic substances like benzene from the saturated soil with fuel and transported them to the residential area. By 1985 the contamination had reached the residential area raising serious environmental complications.

The removal of polluted soil was technically unproductive, and the removal of polluted groundwater did not solve the source of contamination. How could polluted underground water from the leakage to the residential area be stopped? A possible solution was bioremediation technique. Studies conducted by the US Geological Survey (USGS) have shown that MO that is naturally present in soil has actively consumed toxic compounds derived from fuel by converting them into unsafe carbon dioxide. Moreover, these studies have shown that the rate of biotransformation could be greatly increased by the addition of nutrients. By stimulating the natural microbial community as a result of the addition of nutrients, it was theoretically possible to increase the rate of biodegradation and hence protect the residential area from future contamination.

In 1992 this theory was put into practice by scientists. Nutrients were introduced into polluted soils through infiltration galleries, polluted underground water was removed through a series of extraction wells, and a rigorous monitoring activity of contamination levels began. At the end of 1993, contamination in the residential area was reduced by 75%. In the immediate vicinity of the infiltration galleries (nutrient sources), the results were even better. Groundwater that previously contained more than 5000 ppb of toluene now had an undetectable level. Bioremediation has worked! The success of the Hanahan bioremediation project has been the result of many years of intense efforts by a large number of scientists. In the early 1980s, little was known about how toxic wastes interaction with the hydrosphere.

This lack of knowledge required special efforts for remediation of the polluted environment under the new Comprehensive Environmental Response, Compensation, and Liability Act. For solving this problem, Congress mandated the USGS to lead a program to address this lack of knowledge. This program, known as the Toxic Substances Hydrology Program, has systematically investigated the most important categories of waste on lands in the United States. One of the main revelations of this program was that microorganisms in small-scale aquifers reduce the danger and transport for almost all kinds of toxic substances (Johnsen et al., 2014).

14.8.1   OTHER SUCCESSFUL EXAMPLES OF BIOREMEDIATION

The oil spill from Bemidji, Minnesota in 1979, a crude oil pipeline broke and infected the aquifer below. USGS studying the land found that toxic contaminants from crude oil were rapidly degraded by natural microbial populations. Significantly, it was shown that the groundwater contamination spot stopped expanding after only a few years as the microbial degradation rate equaled (counterbalanced) the leaching rate of the contaminant. This was the first and best-documented example of intrinsic bioremediation in which naturally occurring microbial methods recover polluted underground water without human intervention (Islam et al., 2016).

14.8.2   SEWAGE EFFLUENT FROM CAPE COD, MASSACHUSETTS

Storage of sewage sludge is a common practice in the USA. Systematic studies on a sewage effluent patch at the Massachusetts Military Reserve (Otis Air Force Base) led to the obtaining for the first time of laboratory and field data on the rapidity with which microbial natural populations degraded nitrate pollution (denitrification) in a small depth aquifer (Miller and Smith, 2009).

14.8.3   CHLORINATED SOLVENTS, NEW JERSEY

Chlorinated solvents are very common contaminants/pollutants in the highly industrialized northeast. Due to the fact that their metabolic methods are highly adaptable, MO can usage chlorinated compounds as oxidants when they do not have other oxidants. Such transformations, which can lead to natural remediation of groundwater solvents contamination, have been extensively documented by the USGS at Pacatinny Arsenal, New Jersey (Shapiro et al., 2017).

14.8.4   PESTICIDES, THE SAN FRANCISCO BAY ESTUARY

Pesticide contamination of rivers is common across the USA. Laboratory and field studies in the Sacramento River and San Francisco Bay have shown the effects of biological and non-biological methods in the degradation of pesticides commonly used such as carbofuran, methylparation, molinate, and thiobencarb (Hoenicke et al., 2007).

14.8.5   AGRICULTURAL CHEMICALS IN THE CENTRAL AREA

Agricultural chemicals affect the quality of groundwater in many Western Midwestern states. Studies in this area have highlighted the route of nitrogen fertilizer and pesticides in groundwater and surface waters. These studies have shown that many common contaminants/pollutants such as the atrazine herbicide are degraded by biological methods (microbial degradation) and nonbiological methods (photolytic degradation) (Smith et al., 2015).

14.8.6   CONTAMINATION WITH GASOLINE, GALLOWAY NEW JERSEY

Gasoline is probably the most common pollutant of underground water in the United States. Studies on this land have demonstrated rapid microbial degradation of gasoline contaminants/pollutants and have shown the importance of these methods in the unsaturated zone regarding the degradation of contaminants/pollutants (Belchior and Andrews, 2016).

14.8.7   CREOSOTE CONTAMINATION, PENSACOLA, FLORIDA

Creosote and chlorinated phenols have been widely used to treat wood in the USA. Contaminants/pollutants washed and circulated in the aquifer via several ponds were transported in the direction of Pensacola Bay. Studies on this land have shown that MO can adapt to extremely harmful chemical conditions and that microbial degradation has reduced migration of contamination stain. These studies have contributed to the technical foundation that made Bioremediation possible at Hanahan (Elder and Dresler, 1988).

14.9   TECHNOLOGIES FOR SOIL TREATMENT BY BIOREMEDIATION

Biological *in situ* methods are as follows:

- bioventilation;
- bioaugmentation; and
- phytoremediation.

14.9.1 BIOVENTILATION

Oxygen is introduced into unsaturated polluted soils through forced air circulation (extraction or air injection) to increase oxygen concentration and to stimulate biodegradation. Bioventilation is a technique that stimulates the in-situ biodegradation of any aerobic degradable compound by supplying oxygen to existing MO in the soil. As opposed to suction vapor extraction, bioventilation uses poor airflows to provide sufficient oxygen just to support microbial activity. Oxygen is most often applied by direct injection into the residual contaminants in the soil. In addition to the degradation of adsorbed fuel residues, volatile components are biodegraded as vapors circulate slowly through the biologically active soil (Li et al., 2017).

1. **Opportunity:** Bioventilation is a medium and long-term technique. Cleaning can take several months to a few years. Bioremediation by bioventilation techniques have been successfully used to remediate polluted soils with petroleum hydrocarbons, unchlorinated solvents, certain pesticides, wood preservatives and other organic chemical compounds. Bioremediation can be used to change the valence state of inorganic substances and to cause adsorption, assimilation, accumulation, and concentration of inorganic substances in micro or macro-organisms. These bioremediation techniques are promising in terms of stabilizing or removing inorganic substances in the soil, while biodegradation cannot degrade inorganic contaminants.

2. **Restrictions:** Among the factors that can limit the field of implementation and efficiency of the bioventilation technique are: The water mass of a few decimeters below the soil surface, saturated soil lens, or low permeability soil reduces the efficiency of bioventilation. Vapors may collect in pools within the range of air injection wells. This difficult situation can be eliminated by sucking the air near the structure. A low degree of soil moisture can limit biodegradation and bioavailability efficiency. It is necessary to monitor the waste gases at the surface of the soil. The aerobic biodegradation of certain chlorinated compounds may not be effective unless there is a co-metabolite or an aerobic cycle. Low temperatures can slow down the remediation, although successful bioremediation was achieved in extremely cold environments (Zhang and Bennett, 2005).

3. **Conditionings:** A successful bioventilation is substantiated on the fulfillment of two basic criteria. First, the air must penetrate into

the soil in sufficient quantity to maintain aerobic conditions; and second, MO that naturally degrades hydrocarbons must be present in concentrations sufficiently large to achieve the appropriate biodegradation rates. Initial trials aim at both determining soil permeability and breathing rates *in situ*. Soil granulation and its humidity have an important influence on soil gas permeability. The greatest restriction on air permeability is the excessive moisture of the soil. The combination of elevated level of groundwater, high humidity, and low-granulated soils has prevented successful bio-ventilation in certain locations. Among the soil properties that have an impact on microbial activity are pH, humidity, and nutrients (nitrogen and phosphorus) and temperature. It has been calculated that the optimal soil pH for microbial activity ranges between 6 and 8.

Optimal humidity is established for each soil type. Too much moisture can reduce air permeability and decrease oxygen transfer capacity. A too low humidity will prevent microbial activity. Several bioventilation tests have shown good biodegradation rates with moisture levels between 2% and 5% by weight.

However, in extremely cold climates, it is possible to increase the rate of biodegradation by irrigation or wetting with injected air (Saingam et al., 2018). Contaminants degrade faster by bioventilation in the summer, but the remediation can also occur at temperatures of 0°C. Biodegradation rates of hydrocarbons are always estimated substantiated on the percentage of oxygen usage, assuming that oxygen loss is due to the mineralization of hydrocarbons by means of microbes. Oxygen, as the final acceptor of electrons, uses not only the degradation of organic matter, but also the oxidation of inorganic compounds reduced by MO that produce energy through chemical oxidation.

Measuring the usage of oxygen in an unpolluted neighboring area, as control area, is used to establish inorganic oxidation reactions. When used along with other microbial activity or biodegradation indicators, respiration tests can provide at least one of the converging directions in the separate evidence of qualitatively documented biodegradation.

4. **Efficiency Data:** Bioventilation has become a common technique, with most technical components already available. Bioventilation

gained a lot of attention from the scientific research community, especially with regard to the usage of this technique along with soil vapor aspiration (SVE). As with all biological technologies, the time required to fix a bioventilation land depends largely on the soil characteristics and chemical properties of the polluted environment.

5. **Expenditure:** The main cost elements are as follows:

   i.   **Surface Area; Number of Injection/Extraction wells Installed:** the number of wells and related expenditure increases proportionally with the surface.

   ii. **Type of Soil:** Soil types with sand and gravel content reduce expenditure due to the lower number of injection/extraction wells to be installed.

Indicative bioventilation prices are between 25–200 Euros per cubic meter of soil. Expenditure can be influenced by the type of soil and its chemical properties, the type and amount of modifications used the type and size of the contamination (Cheng et al., 2008).

14.9.2 BIOAUGMENTATION

Bioaugmentation is a technique in which local or inoculated MO (such as fungi, bacteria, and other microbes) degrade (metabolize) organic contaminants in soil and/or groundwater and neutralize their harmful effect. The activity of naturally occurring microbes is stimulated by aqueous solutions that circulate through polluted soils and which increase the degree of biological degradation of contaminants from organic agricultural waste *in situ* or the immobilization of inorganic ones. Nutrients, oxygen, and other amendments can be used to increase bioremediation and desorption of contaminants from agricultural waste from underground material (Tao et al., 2019).

1. **Bioaugmentation in Aerobic Technique:** In the presence of sufficient oxygen (aerobic) and other nutrients, MO will turn many organic contaminants into carbon dioxide, water, and microbial cell masses. Biogementation of a soil normally involves the percolation or injection of groundwater or unpolluted water mixed with nutrients and saturated with dissolved oxygen. Sometimes acclimated MO (bio-augmentation) and/or other oxygen sources such as hydrogen peroxide can be added. Irrigation by spill infiltration is regularly

used in low-polluted soils and injection wells for deep polluted soil. Although ISB proved to be successful in cold climates, low temperatures slow down the remediation technique. Low-temperature polluted soils can be used to cover the surface of the soil to increase its temperature and degradation rate (Saez et al., 2018).

2. **Bioaugmentation in the Anaerobic Technique:** In the absence of oxygen (anaerobic conditions), organic contaminants will turn into methane, small amounts of carbon dioxide and low amounts of hydrogen. Under sulfate-reducing conditions, sulfate is converted to elemental sulfide or sulfur, and in the nitrate-reducing conditions it finally produces hydrogen sulfide. Contaminants can sometimes degrade in intermediate or final products more or less dangerous than the original pollutant. Such as, TCE (triclorethylene) can be anaerobic biodegraded to vinyl chloride that is more persistent and more toxic. To avoid such complications, most bioremediation projects are done *in situ*. Vinyl chloride can be further degraded under aerobic conditions. Biogementation is a long-term technique that can take years to clean a pollutant (Cheng et al., 2015).

3. **Opportunity:** Bioremediation techniques have been successfully used in the remediation of soils and sludge, remediation of groundwater's infected with petroleum hydrocarbons, solvents, wood preservatives, and other organic chemical products. Bioremediation is especially effective in tackling low-level residual contamination related to the removal of the source of pollution. The most infected groups of contaminants are PAH, non-halogenated SVOC (without PAH), and BTEX, i.e., benzene, ethylbenzene, toluene, xylene (volatile organic compounds). Types of polluted lands treated most often have been infected by methods or by waste resulting from wood preservation and refining of crude oil. Preservation of wood involves the usage of creosote containing a high concentration of PAH and other non-halogenated SVOCs. Similarly, crude oil refining and reuse methods are frequently substantiated on BTEX. Given that SVOC (PAH and other non-halogenated SVCs) are the most commonly treated pollutant groups by bioremediation, their treatment with ASV-type volatility technologies (aspiration of soil vapor) may prove difficult. Bioremediation treatment frequently does not require thermal treatment, involves few cost-effective items such as nutrients, and does not normally generate residues

that require other methods or disposal. In addition, when done *in situ*, it does not require excavation of the polluted environment. As opposed to other technologies, bioremediation is advantageous in treating non-halogenated SVOCs. Although bioremediation cannot degrade inorganic contaminants, it can instead be used to change the valency of inorganic substances and cause adsorption, immobilization into soil particles, precipitation, assimilation, accumulation, and concentration of inorganic substances in micro- and macro-organisms. These techniques are promising for stabilizing or removing inorganic substances from the soil (Ikeda-Ohtsubo et al., 2013).

4. **Restrictions:** Among the factors that impede the opportunity and efficiency of the technique are:

   i.   Circulation of aqueous solutions through soil can increase the mobility of contaminants from agricultural waste and may require groundwater treatment.

   ii. Cleaning targets cannot be achieved if soil base soil prevents the contact of pollutant-microorganism.

   iii. Preferential colonization with microbes may cause nutrients and water injection probes to clog.

Preferred circular pathways can greatly reduce contact between injected fluids and contaminants throughout the polluted area. The system should not be used for heavily stratified clay or for heterogeneous underground environments because of oxygen transfer limits (or other electron acceptors). High concentrations of heavy metals, strong chlorinated organic substances, long hydrocarbon chains or inorganic salts may be toxic to MO. Bioremediation slows down at low temperatures. Concentrations of hydrogen peroxide greater than 100–200 ppm in groundwater prevent the activity of MO. A surface treatment system such as air stripping or carbon adsorption may require scrubbing of underground water before re-injection or removal. Many of the above factors can be controlled with particular attention to good technical practice. The duration of treatment can be 6 months to 5 years and depends on land-specific factors (Inoue et al., 2012).

5. **Conditionings:** The main traits of contaminants from agricultural waste that need to be identified in feasibility studies for bioaugmentation are the potential for infiltration (such as water solubility and soil sorption); their chemical reactivity (such as, tendency towards non-biological reactions such as hydrolysis, oxidation, and polymerization); and, most importantly, their degree of biodegradation. The soil characteristics include the depth and area of pollution, the concentration of contaminants from agricultural waste, the type of soil and its properties (such as organic substance content, texture, pH, permeability, water retention capacity, moisture, nutrient level); oxygen competition (such as redox potential); presence or absence of toxic substances for MO; concentration of other electron acceptors, nutrients; and the capacity of MO in soil to degrade contaminants. Feasibility tests are performed to determine whether bioaugmentation is achievable under certain given conditions and to determine the time frame of the remedy and its parameters. Field-testing can be done to determine the influence radius and well spacing as well as to obtain anticipated preliminary expenditure (Park et al., 2008).

6. **Efficiency Data:** The main advantage of an in-situ technique is that it allows the soil to be treated without excavation and transportation, which does not disturb the activities carried out on the land. If Bioaugmentation can achieve its cleaning purpose over a compatible timeframe, it can significantly reduce expenditure without excavation and transportation. In addition, both polluted underground waters and soil can be treated simultaneously, which is another cost advantage. *In situ* procedures generally require longer periods of time, and there is no certainty about the uniformity of methods due to soil inherent variability, aquifer characteristics, and monitoring process difficulties. Remediation methods can sometimes take years, mostly due to pollutant degradation rates derived from specific agricultural waste, land features, and climate. It may take less than a year to clean certain contaminants, while higher molecular weight compounds need more time to degrade. There is a risk of increasing the mobility of contaminants from agricultural waste and their infiltration into groundwater. Bioaugmentation was selected for corrective and emergency actions on a large number of polluted lands. Generally, petroleum hydrocarbons can be bioremediated immediately at a relatively low cost by stimulating local nutrient MO.

7. **Expenditure:** Indicative prices for bioaugmentation fall between 25–220 Euros per cubic meter of soil. Expenditure can be influenced by the type of soil and its chemical properties, the type and amount of modifications used the type and size of the contamination (Gentry et al., 2001).

14.9.3   SOIL PHYTOREMEDIATION

Phytoremediation is a process of using plants to remove, transfer, stabilize, and destroy contaminants from agricultural waste from soil and sediments. Contaminants may be organic or inorganic. Phytoremediation mechanisms include advanced biosynthesis of the rhizosphere, phytoaccumulation, phytodegradation, and phytostabilization.

The advanced biodegradation of the *rhizosphere* occurs in the soil surrounding the roots of the plants. The natural substances propagated from the roots provide nutrient MO, which increases their biological activity. The roots of the plant loosen the soil and afterwards they dry leaving room for the water circulation and venting. This technique tends to push water to the surface area and to dry the lower saturated areas. The most commonly used flora in phytoremediation projects are poplars because these trees grow fast and can survive in many climate types. In addition, poplars can draw large amounts of water (compared to other plant species) when passing through the soil or directly from the aquifer (Scotti et al., 2019).

This means absorbing a large amount of dissolved contaminants from polluted environments and reducing the amount of contaminants scattered through or outside the soil or the aquifer. Phytoaccumulation is the assimilation of contaminants from agricultural waste by the plant roots and the mobilization/accumulation (phytoextraction) of them into trunks and leaves. Phytodegradation is the metabolism of contaminants from agricultural waste into plant tissues. Plants produce enzymes such as dehalogenase and oxygenase that help catalyze degradation. Investigations will determine whether aromatic and chlorinated compounds respond to phytodegradation.

Phyto stabilization is a phenomenon of plants production of chemical compounds that serve for the immobilization of contaminants from agricultural waste to roots contact with the soil (Elshamyet al., 2019).

1. **Opportunity:** Phytoremediation can be applied for the remediation of metals, pesticides; solvents, crude oil, PAH, and landfill infiltration. Some plant species have the capacity to store metals in their

roots. These plants can be transplanted to polluted lands to filter out the metals in the wastewater. When the roots are loaded with metal contaminants, these plants can be removed. Plants that accumulate large quantities can remove or store significant amounts of metallic contaminants. At present, trees are tested to determine their capacity to remove contaminants from organic farming waste from underground water, translocation, and perspiration, and their possible metabolism into CO_2 or vegetal tissues.

2. **Restrictions:** Soil phytoremediation may be limited by:
   - The depth of the treatment area is determined by the plants used in phytoremediation. In most cases, this technique can be used on low-soil soils.
   - High concentrations of dangerous substances can be toxic to plants.
   - It can transfer contaminants between environments such as soil to air.
   - It is not effective for highly absorbed contaminants (such as polychlorinated biphenyls (PCBs)) and those poorly absorbed.
   - Toxicity and bioavailcapacity of degradation products are not always known.
   - Products may be mobilized in underground or bioaccumulated waters in animals.

3. **Conditionings:** Detailed information is needed to determine the soil types used in phytoremediation projects. Flow rate, oxygen-lowering concentrations, root growth, and structure affect plant growth and should be considered when implementing phytoremediation.

4. **Efficiency Data:** Several phytoremediation demonstrations are currently being carried out (such as, oak species were planted in the middle of a TCE infected area to assess TCE sweat and TCE transformation rates). Evaporation-sweating rates were measured equally. This is a long-term test on the capacity of trees to control the circulation rate of groundwaters.

5. **Expenditure:** The most important elements of this bioremediation method expenditure are the surface of polluted area. Phytoremediation is a long-term remedial technique. Expenditure are mainly

caused by plant purchases, tree-related investments with test expenditure and subsequent sampling. Expenditure can range from 10 Euros for a slightly polluted land up to 150 Euros for a difficult land per cubic meter treated (Vocciante et al., 2019).

14.10   SOIL BIOTECHNOLOGIES FOR BIOREMEDIATION

Biological *ex situ* methods are as follows:

- Biopiles;
- Composting;
- Cultivation of land (including excavation).

14.10.1   BIOPILES

Excavated soils are mixed with amendments and placed on surface. It is a composting technique with airy static mounds in which the compost is raised in piles and ventilated with blowers or vacuum pumps. Biopiles treatment is a widespread technique by which excavated soils are mixed with amendments and placed in treatment areas that include infiltration water collection systems and some forms of ventilation. It is used for reduction of concentrations of petroleum components in soils excavated. Humidity, heat, nutrients, oxygen, and pH can be controlled to stimulate biodegradation. The treatment area will generally be covered or kept with an impermeable liner to minimize the risk of leakage of contaminants from agricultural waste to unpolluted soil. The system itself can be treated in a bioreactor prior to recycling. Suppliers have developed patented nutrients and additive formulas, as well as techniques for incorporating formulas into soil to stimulate biodegradation. The formulas are usually modified from according to soil characteristics. Piles of soil usually have an air distribution system buried underground to allow air to pass through the ground either through vacuum or through positive pressure. The soil piles can have a maximum height of 2–3 meters. These piles can be covered with plastic to control scattering, evaporation, and volatilization, and to stimulate solar heating. If there is VOC in the soil that will evaporate in the air, the air that is emitted from the soil can be treated to remove or destroy VOCs before entering the atmosphere. Biopiles is a short-term technique that can last from a few weeks to a few months. Treatment options include static methods such as the preparation of treatment layers, biotratment cells, soil piles, and composting (Coulon et al., 2010).

1. **Opportunity:** Biopiles treatment has been applied to non-halogenated VOCs, hydrocarbons from fuels, halogenated VOC, and SVOC. Pesticides can also be treated, but the efficiency of the technique will vary and may be applicable to only a few compounds within these pollutant groups.

2. **Restrictions:** Among the factors that can limit the opportunity and efficiency of the technique are: Excavating soils is required; Testing of the degree of treatment should be done to determine the biodegradability of contaminants from agricultural waste, adequate oxygenation, and nutrient loading rates. Solid-phase methods have a probable efficacy on halogenated components and may prove ineffective in degradation of explosive transformation products. Similar Biopiles sizes require more time to be cleaned than mud phase biopiles. Static treatment methods may result in less uniform treatment than methods involving periodic mixing.

3. **Conditionings:** The first steps in preparing a well-grounded project for biotreatment of polluted soil include: Soil characteristics, contaminants from agricultural waste characteristics, the existence of Laboratory and/or field trials, pilot tests and/or demonstrations on the ground. The characteristics of the land, soil, and contaminants from agricultural waste serve to: identify and quantify contaminants; determine the need for organic and inorganic amendments; identify possible security issues; determine the need for excavation and transportation of polluted soil; determine the availability and location of utilities (electricity and water). Laboratory studies or field feasibility studies are needed to identify: mixtures of the amendment that best promote microbial activity; the possible toxic products derived from degradation; the percentage reduction and the low limit of the pollutant concentration that can be achieved; the possible degradation rate.

4. **Efficiency Data:** Biopiles treatment has already been demonstrated on several fuel-infected lands, with good results.

5. **Expenditure:** It depends on the contaminants, the procedure to be followed, the need for additional treatment or post-treatment, and the need for sophisticated equipment. Biopiles are quite simple and require little staff for maintenance and handling. Typical indicative expenditure with one layer and one linearly prepared fall between 35 Euro and 100 Euro per cubic meter (Whelan et al., 2015).

14.10.2 COMPOSTING

Polluted soil is excavated and mixed with fillers and organic amendments such as hay, natural fertilizers, plant waste (such as potatoes) and wood remnants. Selecting the right amendments ensures sufficient porosity and provides a balance between carbon and nitrogen to promote thermophilic microbial activity. Composting is a biologically controlled technique by which organic contaminants (such as PAHs) are transformed by MO (under aerobic and anaerobic conditions) into harmless products, stabilized derivatives. Normally, the thermoelectric conditions (54–65°C) must be maintained to properly treat the soil polluted with dangerous organic contaminants. High temperatures result from the heat produced by MO during the degradation of organic matter from waste. In most cases, this is done by using local MO. The soils are excavated and mixed with fillers and organic amendments such as chips, animal, and vegetable waste, in order to increase the porosity of the mixture to be decomposed. Maximum degradation efficiency is achieved by maintaining oxygenation (such as daily return of the biopiles), irrigation if necessary, and careful monitoring of soil moisture and temperature. There are three types of methods used in composting:

- Composting in mechanically agitated vessels (the compost is placed in the reactor vessel where it is mixed and ventilated);

- Composting on the furrows (the compost is placed in as furrows which are mixed periodically with mobile equipment); and

- The stacking of static aerated biopiles (composting in piles and blowing with blowers or suction pumps).

Composting on furrows is commonly considered as the most cost-effective composting option. However, if VOC or SVOC are present in the soil, it may be necessary to control residual gases (Liao et al., 2019).

1. **Opportunity:** The composting technique can be applied to soils infected with biodegradable organic compounds. Aerobic, thermophilic compost can be used in soil polluted with PAH. The consumable materials and equipment used in composting is already available on the market.

2. **Restrictions:** Among the factors that can limit the opportunity and efficiency of the technique are as follows: a large space is required; it is necessary to excavate polluted soils that can cause uncontrolled VOC emissions; composting leads to an increase in volume of material due to the addition of amendments, although metal levels can be reduced by dilution, heavy metals cannot be treated by this method; large concentrations of heavy metals can be toxic to MO.

3. **Conditionings:** Among the specific data needed to evaluate the composting technique are the concentration of contaminants from agricultural waste, the need for excavation, the availability, and cost of the necessary amendments to the compost mixture, the space required for treatment, the soil type, and the pollutant response to composting.

4. **Efficiency Data:** Composting has been proven to be an efficient technique for treating soils polluted with explosives. The need for simple equipment combined with these performance results make this technique an attractive economic and technical solution.

5. **Expenditure:** Expenditure may vary depending on the amount of soil to be treated, the soil compost, the availability of the amendments, the type of pollutant, and the type of technique project used. The cost of providing a treatment base with the collection of infected infiltration water is included. The main cost elements are:

    • The pollutant type is the key element in determining composting expenditure.

    • Soil type/total organic content (TOC); soils with higher density (generally small gravel sands and gravel) cost less on composting, while higher TOC soils would require more expenditure. Density influences the soil mass to be treated, and the TOC percentage indicates the level of contamination. Typical indicative expenditure with one-layer fall between 35 Euro and 100 Euro per cubic meter (Lu et al., 2018).

14.10.3   CULTIVATION OF THE LAND

Polluted soil, sediment, or shoreline is excavated, applied in aligned layers that are periodically reversed or are plowed for soil ventilation. Land cultivation is a widely used bioremediation technique that typically involves the installation of pipelines and other techniques for controlling pollutant leakage from agricultural waste, which requires the excavation of polluted soils. Soil conditions are often controlled to optimize the rate of degradation of contaminants from agricultural waste. Normally controlled conditions include: moisture content (usually by irrigation or spraying), ventilation (by sprays at specified intervals, soil being mixed and aerated), pH (limited around neutral pH by the addition of cracked limestone or agrocalcar), other amendments (such as soil fillers, nutrients, etc.). Infected soil is usually treated in elevations up to 0.50 m thick. Upon reaching the desired treatment level, the elevation is removed and another biopile is being built. It is recommended the removal of the top of the pile and adding the new elevation by adding more polluted material to the remaining material and mixing it. This leads to the immunization of the new material added with a microbial culture that actively degrades and can shorten treatment time.

Land treatment is a widely used bioremediation technique by which polluted soils are reversed (such as, by plowing) and allowed to interact with the soil and the climate at that specific land. Soil, climate, and biological activity interact dynamically as a unitary system, degrading, transforming, and immobilizing pollution constituents. A land for the treatment of land should be administered competent to prevent complications both on and outside the land related to groundwater, surface water, or air pollution. Additional monitoring and environmental precautions are required. Cultivation of the land is a medium and long-term technique (Azabet al., 2016).

1. **Opportunity:** *Ex situ* cultivation of the land has been successfully tested in the treatment of petroleum hydrocarbons. Since lighter and more volatile hydrocarbons such as gasoline are successfully treated by methods substantiated on volatilization (such as SVE), the usage of remediation technique in question is usually limited to heavier hydrocarbons. As an approximate formula, the higher the molecular weight (PAH), the lower the rate of degradation and the more chlorinated and nitrated the compound is, the more difficult it is to degrade. Contaminants successfully treated by cultivation are diesel fuel, heavy fuel oil, oily gravel, wood preservatives (PCP and creosote), and some pesticides (Wang et al., 2016).

2. **Restrictions:** Among the factors that can limit the opportunity and efficiency of the technique are: broad space is needed; conditions that influence the biodegradation of contaminants from agricultural waste (temperature, rainfall) cannot be controlled, which prolongs the completion of the remediation; inorganic contaminants will not be degraded. Volatile contaminants, such as solvents, must be treated in advance because they could evaporate into the atmosphere and produce pollution. Dust control must be considered, especially during drilling and other material handling operations. Leakage monitoring installations should be lifted and controlled. Topography, erosion, climate, soil stratigraphy, and land permeability should be evaluated to determine the optimal design of the installation (Ameen et al., 2015).

3. **Conditionings:** The following aspects of the contaminants should be analyzed: their types and concentrations, the depth and distribution profile, the presence of contaminants from toxic agricultural waste, the presence of VOC and that of contaminants from inorganic AWs such as metals, the geological and hydrogeological characteristics of the underground, the temperature, the precipitation, the speed and the direction of the winds, the water availability, the type and the texture of the soil, its humidity, the geological and hydrogeological characteristics of the land and the soil, soil organic matter content, cation exchange capacity, water retention capacity, nutrient content, pH, atmospheric temperature, permeability, and MO (degradation populations present on the land).

4. **Performance Data:** Numerous large-scale operations have been used, especially for sludge from the oil industry. As with other biological methods, cultivating the land under proper conditions can turn contaminants into harmless substances.

   The efficiency of their removal depends in any case on the type of pollutant and its concentrations, soil type, temperature, humidity, residue rates, implementation frequency, ventilation, volatilization, and other factors.

5. **Expenditure:** Distinct types of expenditure can occur: Indicative cost for preparing the base layer (treatment and *ex situ* placement of the soil on a prepared line): 30 Euro per cubic meter for a large land and 65 Euro for a small one (Roberts et al., 2015).

14.11   CONCLUSIONS AND RECOMMENDATIONS

The question that arises is which species are suitable for land rehabilitation, how to choose specific specie, and what are the best techniques of sowing? From the point of view of rehabilitation of a degraded area, the most important characteristic to be considered must be the adaptation of plants to the conditions of the land (soil). On the other hand, some of these plants are easy to obtain through the seeds produced by specialized manufacturers, while others are wild plants. Wild plants should not be ignored, as they can have a number of extremely favorable attributes and can provide very useful services in planting and permanent planting activity. Within each species there are usually many varieties and varieties with distinct attributes of resistance, wintering, requirements for soil and climate conditions, etc. Searching for the most appropriate varieties is a useful expense because a correct choice must take into account all the differences regarding the persistence and easy maintenance of the final vegetal carpet.

Legume species are a crucial component in almost all herb mixtures because they contribute to maintaining a proper nitrogen influx and ensure the provision of an adequate amount of organic nitrogen in the new soil. They remove the need for nitrogen treatment for maintenance by increasing the amount of mineralizable nitrogen. Thus, clover is a better fertilizer than chemical fertilizer methods as it provides nitrogen gradually and continuously. In the case of legumes, the choice of species depends on the soil and climate conditions. Usually the most suitable legumes are those used in agriculture as they have a high rate of nitrogen fixation. It is easy to see that there is an enormous range of legumes, and in the last years a large number of varieties of plant species were biotechnologically set up. The best way to choose the most suitable legume species and varieties in the degraded land rehabilitation activity is to collaborate with the local agrotechnical authorities. Legumes are appropriate due to the symbiosis with the *Rhizobium* bacterial species that fixes itself in the roots and forms nitrogen-binding nodosities. In order to achieve this symbiosis, there is a need for compatibility between the species and the variety of the leguminous plants and the strain of bacteria present in the soil. It is possible that in a new technogenic soil, there is no *Rhizobium* strain compatible with the legume seedlings. That is why the usage of seeds already infected with compatible *Rhizobium* is most often used. In addition, the soil in which the legumes have already been grown can be used to bring an appropriate supply of suitable *Rhizobium* bacteria. Sometimes there are used the so-called nurse cultures. The species used for this purpose must be chosen according to the target.

Even if it brings with it a nutrient competition, its utility proves to protect the culture against factors such as erosion, instability, etc. It must leave enough light at the ground to ensure the growth of newly sown grasses.

The alternative to the stabilization method by sowing grass is the cultivation of trees or shrubs. Trees can come up with a commercial production if the species is well chosen. They can be planted on fields where farming cannot be done, such as on mine tailings.

Their outline can be arranged to contribute positively to the landscape. However, in the short term, they are less effective than grass in soil stabilization against erosion. As with herbaceous plants, the choice is immense. For planting on degraded land, it is necessary to choose species known to be well adapted to the local environment. Generally, the most valuable species are the natural pioneers. It is possible to identify species that are particularly adapted to the difficult conditions of nutrient deficiency, the deficiency that is present in the early natural growth of vegetation on degraded lands. As with herbs, species should also be chosen substantiated on particular land attributes such as pH and climate.

There are a number of trees in the temperate area such as *Robinia pseudacacia* (acacia), *Cercis siliquastrum, Gleditschia triacanthos, Gymnocladus dioica, Sophora japonica*, and in the warm climate the Acacia species, as well as other tree species that have the capacity to fix nitrogen. Several more examples are Alnus (anin), *Hippophae rhamnoides*.

These species grow relatively quickly without the addition of fertilizers (nitrogen) and are valuable tools in the rehabilitation of degraded or technogenic land, although sometimes it requires the addition of other nutrients. The choice of shrubs is also immense.

There is no question of commercial usage. Specifically, will be chosen those that have the most appropriate attributes to the field conditions and which perform well the fixing role. Among the very valuable shrubs with the capacity to fix the nitrogen are the legumes *Ulex, Sarothamnus, and Eleagnaceae Eleagnus* (salvia) and *Hippofaër hamnoides*. Other shrubs often used to fix degraded lands and heaps are *Cornussan guinea, Cornus mas, Lygustrum vulgare, Cytisus nigricans, Viburnum lantana*, and *Viburnum opulus*.

Colonization of vegetation on the tailings is achieved through particular stages of ecogenesis. This is done on the abiotic substrate resulting from the profound deterioration of the natural environment as a result of various economic activities such as the mining and construction industry, by the accumulation of tailings after the extraction of the ore, by the establishment of tailings ponds, highways, dams, reservoirs, etc. The tailing, of an initial abiotic origin, is populated with pioneer species, which have large amplitude of ecological tolerance to variations in environmental factors. When installing this vegetation on the tailings, interspecific competition is small.

Associations made up of a small number of species that fix the slope, such as *Rumex scutatum* with *Galeopsis angustifolia*, and on the northern slopes, with high humidity, *Rumex* with *Tussilago farfara*. On the slopes there is vegetation with a deep root system that can withstand the landslides, such as the *Clematis vitalba*.

On carbon substrates (which melt in the sun, having over 38 °C) there is a thermophilic and xerophilic vegetation. On metal-containing tailings, special metallophyte species are installed (for selenium-a series of leguminous, strontium-graminea, etc.).

Colonization and fixation of anthropic relief by phytocoenoses depends on the adaptive capacities of the plant communities, the ecotypic variability (genetic biodiversity) of the constituent species, the physico-chemical properties of the base material.

The vertical distribution of the species that colonize tailings dumps from coalmines is dependent on the temperature of the deposited material (23–57°C), the hemicriptophiles (Poa *pratensis* or *Trifolium arvense*), following the level with annual terophilicbioforms (*Vulpia, Daucus carota, Tunicicaprolifica*) and level with camefilebioforms (*Sedum acre*). The vertical distribution of the species that colonize tailings dumps from non-ferrous metal mines (Pb, Zn, and Cu) is dependent on the slope exposition.

To the north, from the base to the top are installed: *Acer campestre, Arrhenatherum elatius, Luzulaluzuloides, Urtica dioica, Agrostis tenuis, Deschampsiaflexuosa*. To the south, *Acer campestre, Arrhenatherum elatius, Deschampsia flexuosa, Galium verum, Dianthus carthusianorum,* and *Festuca ovina* are installed. The usage of legumes or other nitrogen-fixing species is an encouraging path for successful rehabilitation and relatively low expenditure. In the case of trees, legumes will be introduced after installation of tree species. If the crops are extracted annually, then the nutrients will be removed with them; as a result, it will be necessary to replace nitrogen and phosphorus at levels at least equal to those used on arable land.

Ecological reconstruction in agriculture is of major importance in the context of current climate changes, the system of agricultural holdings, and the applied crop technologies. Changes in the structure, abundance, and dominance of harmful species in agrobiocenoses occur. There are serious effects on the soil and the development of crop plants due to technological mistakes, pests, pollution, drought, and burning, storms, torrents, landslides, floods, etc., which can lead to the destruction of affected agroecosystems.

KEYWORDS

- **bioamplification**
- **biodegradation**
- **bioremediation**
- **biorestoration**
- **biostimulation**
- **ecological reconstruction**

REFERENCES

Abel, S., & Akkanen, J., (2019). Novel, activated carbon-based material for *in-situ* remediation of contaminated sediments. *Environ. Sci. Technol., 53*(6), 3217–3224.

Adetutu, E. M., Gundry, T. D., Patil, S. S., Golneshin, A., Adigun, J., Bhaskarla, V., Aleer, S., et al., (2015). Exploiting the intrinsic microbial degradative potential for field-based *in situ* dechlorination of trichloroethene contaminated groundwater. *J. Hazard. Mater., 300,* 48–57.

Ameen, F., Hadi, S., Moslem, M., Al-Sabri, A., & Yassin, M. A., (2015). Biodegradation of engine oil by fungi from mangrove habitat. *J. Gen. Appl. Microbiol., 61*(5), 185–192.

Arnold, C. W., Parfitt, D. G., & Kaltreider, M., (2007). Field note phytovolatilization of oxygenated gasoline-impacted groundwater at an underground storage tank site via conifers. *Int. J. Phytoremediation., 9*(1), 53–69.

Ashraf, S., Ali, Q., Zahir, Z. A., Ashraf, S., & Asghar, H. N., (2019). Phytoremediation: Environmentally sustainable way for reclamation of heavy metal polluted soils. *Ecotoxicol. Environ. Saf., 174,* 714–727.

Azab, E., Hegazy, A. K., El-Sharnouby, M. E., & Abd, E. H. E., (2016). Phytoremediation of the organic xenobiotic simazine by p450–1a2 transgenic Arabidopsis thaliana plants. *Int. J. Phytoremediation, 18*(7), 738–746.

Bastida, F., Jehmlich, N., Lima, K., Morris, B. E. L., Richnow, H. H., Hernández, T., Von, B. M., & García, C., (2016). The ecological and physiological responses of the microbial community from a semiarid soil to hydrocarbon contamination and its bioremediation using compost amendment. *J. Proteomics, 135,* 162–169.

Belchior, F., & Andrews, S. P., (2016). Evaluation of cross-contamination of nylon bags with heavy-loaded gasoline fire debris and with automotive paint thinner. *J Forensic Sci., 61*(6), 1622–1631.

Bjerketorp, J., Röling, W. F. M., Feng, X. M., Garcia, A. H., Heipieper, H. J., & Håkansson, S., (2018). Formulation and stabilization of an arthrobacter strain with good storage stability and 4-chlorophenol-degradation activity for bioremediation. *Appl. Microbiol. Biotechnol.*, *102*(4), 2031–2040.

Carrara, S. M., Morita, D. M., & Boscov, M. E., (2011). Biodegradation of di(2-ethylhexyl) phthalate in a typical tropical soil. *J. Hazard. Mater.*, *197*, 40–48.

Cheng, W. H., Hsu, S. K., & Chou, M. S., (2008). Volatile organic compound emissions from wastewater treatment plants in Taiwan: Legal regulations and costs of control. *J. Environ. Manage.*, *88*(4), 1485–1494.

Cheng, Z., Chen, M., Xie, L., Peng, L., Yang, M., & Li, M., (2015). Bioaugmentation of a sequencing batch biofilm reactor with *Comamonast estosteroni* and Bacillus cereus and their impact on reactor bacterial communities. *Biotechnol Lett.*, *37*(2), 367–373.

Coulon, F., Whelan, M. J., Paton, G. I., Semple, K. T., Villa, R., & Pollard, S. J., (2010). Multimedia fate of petroleum hydrocarbons in the soil: Oil matrix of constructed biopiles. *Chemosphere, 81*(11), 1454–1462.

Dolphen, R., & Thiravetyan, P., (2015). Phytodegradation of ethanolamines by *Cyperus alternifolius*: effect of molecular size. *Int. J. Phytoremediation.*, *17*(7), 686–692.

Elder, J. F., & Dresler, P. V., (1988). Accumulation and bioconcentration of polycyclic aromatic hydrocarbons in a nearshore estuarine environment near a Pensacola (Florida) creosote contamination site. *Environ. Pollut.*, *49*(2), 117–132.

Elshamy, M. M., Heikal, Y. M., & Bonanomi, G., (2019). Phytoremediation efficiency of *Portulaca oleracea* L. naturally growing in some industrial sites, Dakahlia District, Egypt. *Chemosphere*, *225*, 678–687.

Fan, X., Liang, W., Li, Y., Li, H., & Liu, X., (2017). Identification and immobilization of a novel cold-adapted esterase, and its potential for bioremediation of pyrethroid-contaminated vegetables. *Microb. Cell Fact.*, *16*(1), 149.

Gentry, T. J., Newby, D. T., Josephson, K. L., & Pepper, I. L., (2001). Soil microbial population dynamics following bioaugmentation with a 3-chlorobenzoate-degrading bacterial culture. Bioaugmentation effects on soil microorganisms. *Biodegradation*, *12*(5), 349–357.

Hoenicke, R., Oros, D. R., Oram, J. J., & Taberski, K. M., (2007). Adapting an ambient monitoring program to the challenge of managing emerging pollutants in the San Francisco Estuary. *Environ. Res.*, *105*(1), 132–144.

Ikeda-Ohtsubo, W., Miyahara, M., Yamada, T., Watanabe, A., Fushinobu, S., Wakagi, T., Shoun, H., Miyauchi, K., & Endo, G., (2013). Effectiveness of heat treatment to protect introduced denitrifying bacteria from eukaryotic predatory microorganisms in a pilot-scale bioreactor. *J. Biosci. Bioeng.*, *116*(6), 722–724.

Inoue, D., Yamazaki, Y., Tsutsui, H., Sei, K., Soda, S., Fujita, M., & Ike, M., (2012). Impacts of gene bioaugmentation with pJP4-harboring bacteria of 2,4-D-contaminated soil slurry on the indigenous microbial community. *Biodegradation*, *23*(2), 263–276.

Islam, A., Ahmed, A., Hur, M., Thorn, K., & Kim, S., (2016). Molecular-level evidence provided by ultrahigh resolution mass spectrometry for oil-derived doc in groundwater at Bemidji, Minnesota. *J. Hazard. Mater.*, *320*, 123–132.

Jesus, J., Frascari, D., Pozdniakova, T., & Danko, A. S., (2016). Kinetics of aerobic cometabolic biodegradation of chlorinated and brominated aliphatic hydrocarbons: A review. *J. Hazard. Mater.*, *309*, 37–52.

Jin, J., Wang, M., Lu, W., Zhang, L., Jiang, Q., Jin, Y., Lu, K., et al., (2019). Effect of plants and their root exudate on bacterial activities during rhizobacterium-plant remediation of phenol from water. *Environ. Int.*, *127*, 114–124.

Jin, L., Sun, X., Zhang, X., Guo, Y., & Shi, H., (2014). Co-metabolic biodegradation of DBP by *Paenibacillus* sp. S-3 and H-2. *Curr. Microbiol.*, *68*(6), 708–716.

Johnsen, A. R., Styrishave, B., & Aamand, J., (2014). Quantification of small-scale variation in the size and composition of phenanthrene-degrader populations and PAH contaminants in traffic-impacted topsoil. *FEMS Microbiol. Ecol.*, *88*(1), 84–93.

Karig, D. K., (2017). Cell-free synthetic biology for environmental sensing and remediation. *Curr. Opin. Biotechnol.*, *45*, 69–75.

Li, J., Li, R., & Li, J., (2017). Current research scenario for microcystins biodegradation: A review on fundamental knowledge, application prospects and challenges. *Sci. Total Environ.*, *595*, 615–632.

Liao, H., Friman, V. P., Geisen, S., Zhao, Q., Cui, P., Lu, X., Chen, Z., et al., (2019). Horizontal gene transfer and shifts in linked bacterial community composition are associated with maintenance of antibiotic resistance genes during food waste composting. *Sci. Total Environ.*, *660*, 841–850.

Limmer, M., & Burken, J., (2016). Phytovolatilization of organic contaminants. *Environ. Sci. Technol.*, *50*(13), 6632–6643.

Lu, Y., Gu, W., Xu, P., Xie, K., Li, X., Sun, L., Wu, H., et al., (2018). Effects of sulphur and *Thiobacillus thioparus* 1904 on nitrogen cycle genes during chicken manure aerobic composting. *Waste Manag.*, *80*, 10–16.

Majone, M., Verdini, R., Aulenta, F., Rossetti, S., Tandoi, V., Kalogerakis, N., Agathos, S., et al., (2015). In situ groundwater and sediment bioremediation: Barriers and perspectives at European contaminated sites. *N. Biotechnol.*, *32*(1), 133–146.

Manheim, D. C., Detwiler, R. L., & Jiang, S. C., (2019). Application of unstructured kinetic models to predict microcystin biodegradation: Towards a practical approach for drinking water treatment. *Water Res.*, *149*, 617–631.

Masi, M., Paz-Garcia, J. M., Gomez-Lahoz, C., Villen-Guzman, M., Ceccarini, A., & Iannelli, R., (2019). Modeling of electrokinetic remediation combining local chemical equilibrium and chemical reaction kinetics. *J. Hazard. Mater.*, *371*, 728–733.

McHugh, T. E., Kulkarni, P. R., Newell, C. J., Connor, J. A., & Garg, S., (2014). Progress in remediation of groundwater at petroleum sites in California. *Ground Water.*, *52*(6), 898–907.

Miller, D. N., & Smith, R. L., (2009). Microbial characterization of nitrification in a shallow, nitrogen-contaminated aquifer, Cape Cod, Massachusetts and detection of a novel cluster associated with nitrifying betaproteobacteria. *J. Contam. Hydrol.*, *103*(3, 4), 182–193.

Naidu, R., Nandy, S., Megharaj, M., Kumar, R. P., Chadalavada, S., Chen, Z., & Bowman, M., (2012). Monitored natural attenuation of a long-term petroleum hydrocarbon contaminated sites: A case study. *Biodegradation*, *23*(6), 881–895.

Němeček, J., Steinová, J., Špánek, R., Pluhař, T., Pokorný, P., Najmanová, P., Knytl, V., & Černík, M., (2018). Thermally enhanced in situ bioremediation of groundwater contaminated with chlorinated solvents: A field test. *Sci. Total Environ.*, *622–623*, 743–755.

O'Connor, D., Müller-Grabherr, D., & Hou, D., (2019). Strengthening social-environmental management at contaminated sites to bolster green and sustainable remediation via a survey. *Chemosphere*, *225*, 295–303.

Orellana, R., Macaya, C., Bravo, G., Dorochesi, F., Cumsille, A., Valencia, R., Rojas, C., & Seeger, M., (2018). Living at the frontiers of life: Extremophiles in chile and their potential for bioremediation. *Front Microbiol.*, *9* 2309.

Park, D., Lee, D. S., Kim, Y. M., & Park, J. M., (2008). Bioaugmentation of cyanide-degrading microorganisms in a full-scale cokes wastewater treatment facility. *Bioresour. Technol.*, *99*(6), 2092–2096.

Preston, L. J., Izawa, M. R., & Banerjee, N. R., (2011). Infrared spectroscopic characterization of organic matter associated with microbial bioalteration textures in basaltic glass. *Astrobiology*, *11*(7), 585–599.

Roberts, D. A., Paul, N. A., Cole, A. J., & De Nys, R., (2015). From wastewater treatment to land management: Conversion of aquatic biomass to biochar for soil amelioration and the fortification of crops with essential trace elements. *J. Environ. Manage*, *157*, 60–68.

Saez, J. M., Bigliardo, A. L., Raimondo, E. E., Briceño, G. E., Polti, M. A., & Benimeli, C. S., (2018). Lindane dissipation in a biomixture: Effect of soil properties and bioaugmentation. *Ecotoxicol. Environ. Saf.*, *156*, 97–105.

Saingam, P., Baig, Z., Xu, Y., & Xi, J., (2018). Effect of ozone injection on the long-term performance and microbial community structure of a VOCs biofilter. *J. Environ. Sci.* (China), *69*, 133–140.

Scotti, A., Silvani, V. A., Cerioni, J., Visciglia, M., Benavidez, M., & Godeas, A., (2019). Pilot testing of a bioremediation system for water and soils contaminated with heavy metals: Vegetable depuration module. *Int. J. Phytoremediation*, *25*, 1–9.

Shapiro, A. M., Evans, C. E., & Hayes, E. C., (2017). Porosity and pore size distribution in a sedimentary rock: Implications for the distribution of chlorinated solvents. *J. Contam. Hydrol.*, *203*, 70–84.

Sharma, B., Dangi, A. K., & Shukla, P., (2018). Contemporary enzyme-based technologies for bioremediation: A review. *J. Environ. Manage*, *210*, 10–22.

Singh, T., & Singh, D. K., (2017). Phytoremediation of organochlorine pesticides: Concept, method, and recent developments. *Int. J. Phytoremediation.*, *19*(9), 834–843.

Smith, D. R., King, K. W., Johnson, L., Francesconi, W., Richards, P., Baker, D., & Sharpley, A. N., (2015). Surface runoff and tile drainage transport of phosphorus in the Midwestern United States. *J. Environ. Qual.*, *44*(2), 495–502.

Su, Z. H., Xu, Z. S., Peng, R. H., Tian, Y. S., Zhao, W., Han, H. J., Yao, Q. H., & Wu, A. Z., (2012). Phytoremediation of trichlorophenol by Phase II metabolism in transgenic Arabidopsis overexpressing a *Populus glucosyltransferase*. *Environ. Sci. Technol.*, *46*(7), 4016–4024.

Takahata, Y., Kasai, Y., Hoaki, T., & Watanabe, K., (2006). Rapid intrinsic biodegradation of benzene, toluene, and xylenes at the boundary of a gasoline-contaminated plume under natural attenuation. *Appl. Microbiol. Biotechnol.*, *73*(3), 713–722.

Tao, K., Zhang, X., Chen, X., Liu, X., Hu, X., & Yuan, X., (2019). Response of soil bacterial community to bioaugmentation with a plant residue-immobilized bacterial consortium for crude oil removal. *Chemosphere*, *222*, 831–838.

Tauqeer, H. M., Ur-Rahman, M., Hussain, S., Abbas, F., & Iqbal, M., (2019). The potential of an energy crop "*Conocarpus erectus*" for lead phytoextraction and phytostabilization of chromium, nickel, and cadmium: An excellent option for the management of multi-metal contaminated soils. *Ecotoxicol. Environ. Saf.*, *173*, 273–284.

Tong, H., Yin, K., Giannis, A., Ge, L., & Wang, J. Y., (2015). Influence of temperature on carbon and nitrogen dynamics during in situ aeration of aged waste in simulated landfill bioreactors. *Bioresour. Technol.*, *192*, 149–156.

Ushani, U., Kavitha, S., Yukesh, K. R., Gunasekaran, M., Kumar, G., Nguyen, D. D., Chang, S. W., & Rajesh, B. J., (2018). Sodium thiosulphate induced immobilized bacterial disintegration of sludge: An energy efficient and cost-effective platform for sludge management and biomethanation. *Bioresour. Technol.*, *260*, 273–282.

Vocciante, M., Caretta, A., Bua, L., Bagatin, R., Franchi, E., Petruzzelli, G., & Ferro, S., (2019). Enhancements in phytoremediation technology: Environmental assessment including different options of biomass disposal and comparison with a consolidated approach. *J. Environ. Manage.*, *237*, 560–568.

Vogt, C., & Richnow, H. H., (2014). Bioremediation via in situ microbial degradation of organic pollutants. *Adv. Biochem. Eng. Biotechnol.*, *142*, 123–146.

Wang, B., Wu, C., Liu, W., Teng, Y., Luo, Y., Christie, P., & Guo, D., (2016). Levels and patterns of organochlorine pesticides in agricultural soils in an area of extensive historical cotton cultivation in Henan province, China. *Environ. Sci. Pollut. Res. Int.*, *23*(7), 6680–6689.

Wang, J., Song, X., Wang, Y., Bai, J., Li, M., Dong, G., Lin, F., et al., (2017). Bioenergy generation and rhizodegradation as affected by microbial community distribution in a coupled constructed wetland-microbial fuel cell system associated with three macrophytes. *Sci. Total Environ.*, *607–608*, 53–62.

Whelan, M. J., Coulon, F., Hince, G., Rayner, J., McWatters, R., Spedding, T., & Snape, I., (2015). Fate and transport of petroleum hydrocarbons in engineered biopiles in Polar Regions. *Chemosphere*, *131*, 232–240.

Wu, S., Liu, Y., Southam, G., Robertson, L., Chiu, T. H., Cross, A. T., Dixon, K. W., et al., (2019). Geochemical and mineralogical constraints in iron ore tailings limit soil formation for direct phytostabilization. *Sci. Total Environ.*, *651*(Pt 1), 192–202.

Xu, P., Lai, C., Zeng, G., Huang, D., Chen, M., Song, B., Peng, X., et al., (2018). Enhanced bioremediation of 4-nonylphenol and cadmium co-contaminated sediment by composting with *Phanerochaetechrysosporium*inocula. *Bioresour. Technol.*, *250*, 625–634.

Yeo, J. C. C., Muiruri, J. K., Thitsartarn, W., Li, Z., & He, C., (2018). Recent advances in the development of biodegradable PHB-based toughening materials: Approaches, advantages and applications. *Mater. Sci. Eng. C Mater. Biol. Appl.*, *92*, 1092–1116.

Zhang, C., & Bennett, G. N., (2005). Biodegradation of xenobiotics by anaerobic bacteria. *Appl. Microbiol. Biotechnol.*, *67*(5), 600–618.

Index

Printed and bound by CPI Group (UK) Ltd, Croydon, CR0 4YY

23/10/2024

01777701-0011